New Research on Railway Engineering and Transportation

Edited by Ali G. Hessami and Roderick Muttram

Published in London, United Kingdom

New Research on Railway Engineering and Transportation
http://dx.doi.org/10.5772/intechopen.105261
Edited by Ali G. Hessami and Roderick Muttram

Contributors
Hitoshi Tsunashima, Ryu Honda, Akira Matsumoto, Qiuyi Li, Luyao Li, Masoud Mohebbi, Rasul Mohebbi, Yuan Ma, Stefan Marschnig, Ursula Ehrhart, Muhammad Adeel, Aadil Raza, Muhammad Muaz, Ebu Bekir Aygar, Garazi Carranza Ruiz de Loizaga, Olatz Amorrortu, Oihane de la Rua Losada, Bérengère Nihoul, Nikolaos Matskanis, Jiri Hofman, Roman Cermak, Jiri Korinek, Peter Veit, Patrícia Silva, Joaquim Mendes, Eurico Seabra, Pedro Pratas, Raphael Pfaff, Manfred Enning, Bernd D. Schmidt, Ci Liang, Mohamed Ghazel, Rustam V. V. Rahimov, Bakhrom A. Abdullaev

First published in London, United Kingdom, 2024 by IntechOpen
IntechOpen is the global imprint of INTECHOPEN LIMITED, registered in England and Wales, registration number: 11086078, 5 Princes Gate Court, London, SW7 2QJ, United Kingdom
Printed in Croatia

British Library Cataloguing-in-Publication Data
A catalogue record for this book is available from the British Library

Additional hard and PDF copies can be obtained from orders@intechopen.com

New Research on Railway Engineering and Transportation
Edited by Ali G. Hessami and Roderick Muttram
p. cm.

This title is part of the Civil Engineering Book Series, Volume 3
Topic: Transportation Engineering
Series Editor: Assed Haddad
Topic Editor: Saúl Antonio Obregón Biosca

Print ISBN 978-1-83768-619-3
Online ISBN 978-1-83768-620-9
eBook (PDF) ISBN 978-1-83768-621-6
ISSN 3029-0287

For EU product safety concerns:
IN TECH d.o.o., Prolaz Marije Krucifikse Kozulić 3, 51000 Rijeka, Croatia,
info@intechopen.com or visit our website at intechopen.com.

IntechOpen Book Series
Civil Engineering
Volume 3

Aims and Scope of the Series

Civil engineering is a traditional field of engineering from which most other branches of engineering have evolved. It comprises traditional sub-areas like transportation, structures, construction, geotechnics, water resources, and building materials. It also encompasses sustainability, risk, environment, and other concepts at its core. Historically, developments in civil engineering included traditional aspects of architecture and urban planning as well as practical applications from the construction industry. Most recently, many elements evolved from other fields of knowledge and topics like simulation, optimization, and decision science have been researched and applied to increase and evolve concepts and applications in this field. Civil engineering has evolved in the last years due to the demands of society in terms of the quality of its products, modern applications, official requirements, and cost and schedule restrictions. This series addresses real-life problems and applications of civil engineering and presents recent, cutting-edge research as well as traditional knowledge along with real-world examples of developments in the field.

Meet the Series Editor

Professor Assed N. Haddad is a Civil Engineer with a degree from the Federal University of Rio de Janeiro (UFRJ) earned in 1986, as well as a Juris Doctor degree from the Fluminense University Center earned in 1993, and a Master's degree in Civil Engineering from the Fluminense Federal University (UFF) obtained in 1992. He completed his Ph.D. in Production Engineering from COPPE / Federal University of Rio de Janeiro in 1996. Professor Haddad's academic pursuits have taken him to postdoctoral stays at the University of Florida, USA in 2006; at the Universitat Politècnica de Catalunya, Spain in 2010; and at the University of New South Wales Sydney, Australia in 2019. Currently, he serves as a Full Professor at the Federal University of Rio de Janeiro. He has held visiting professorships at various institutions including the University of Florida, Universitat Politècnica de Catalunya, Universitat Rovira i Virgili, and Western Sydney University. His research expertise encompasses Civil, Environmental, and Production Engineering, with a primary focus on the following topics: Construction Engineering and Management, Risk Management, and Life Cycle Assessment. He has been the recipient of research grants from the State of Rio de Janeiro, Brazil: CNE FAPERJ from 2019 to 2022 and from 2023 to 2025. Additionally, his research grants obtained from the Brazilian Government CNP since 2012 last to this date. Professor Haddad has been involved in several academic endeavors, being the Guest Editor of the International Journal of Construction Management; MDPI's Sustainability, Energies, and Infrastructures; Associate Editor at Frontiers in Built Environment / Sustainable Design and Construction; Guest Editor at Frontiers in Built Environment / Construction Management; and Academic Editor of the Journal of Engineering, Civil Engineering Section of Hindawi. He is currently a Professor of the Environmental Engineering Program at UFRJ and the Civil Engineering Program at UFF.

Meet the Volume Editor

Ali Hessami is currently Director of R&D and Innovation at Vega Systems and a UK expert on the European Committee for Electrotechnical Standardization (CENELEC) and International Electrotechnical Commission (IEC) systems, hardware and software safety, and security standards. He chaired the Institute of Electronic and Electrical Engineers (IEE) 7000 standard "Addressing Ethical Concerns in System Design" and was later appointed as the VC and process architect of IEEE's Certification Program for Autonomous and Intelligent Systems (CertifAIEd). He also chairs BSI's GEL/9/6 subcommittee focused on machine learning (ML), artificial intelligence (AI), and cybersecurity standardization in railways. He is a visiting professor at London City University's Centre for Systems and Control and at Beijing Jiaotong University, China. He is also a fellow of the Royal Society of Arts (RSA) and the UK Institution of Engineering and Technology (IET), as well as a life senior member of the IEEE.

Roderick (Rod) Muttram is a system and safety engineer with more than 40 years of experience. He is a fellow of the Royal Academy of Engineering (REng) and the Institution of Engineering and Technology (IET), a fellow and vice president of the Institution of Railway Signal Engineers (IRSE), a senior member of the Institute of Electronic and Electrical Engineers (IEE), and an advisory professor at Beijing Jiaotong University, China. His rail career began in 1994 when he joined Railtrack as Director of Electrical Engineering and Control Systems. He was promoted to Director of Safety and Standards in 1997. In 2003, he set up the Rail Safety and Standards Board (RSSB) and joined Bombardier Transportation as a vice president as well as served as a lead auditor for critical projects worldwide. In May 2012, he left Bombardier and established Fourth Insight Ltd., an engineering consultancy.

Contents

Preface

Railways are unique among transport systems in that they use and maintain a wide range of technologies, some of which date back to the nineteenth century. In road and air travel enthusiasts maintain 'heritage assets' for people to enjoy recreationally, but in rail, particularly with respect to infrastructure, assets are very long lived (some are more than 100 years old) and continue to deliver public services. This applies not only to civil infrastructure such as bridges and track formations but also to other more technically complex assets. For example, in train control (signaling), every technology developed over the years, from mechanical interlockings based on rods and wires and electromechanical relays to computer-based systems and now even AI, remains in service somewhere. This requires the maintenance of a wide range of competences and supporting equipment, but it also offers many opportunities to apply new techniques and technologies to improve the efficiency of the maintenance and operation of such heritage systems. This book contains several examples of such innovations.

Railways are largely operated and delivered nationally, but railway equipment supply is increasingly global and dominated by a small number of big players. At the national level, requirements differ considerably, but all railways face challenges of performance, cost, and sustainability. In big cities like Hong Kong, Singapore, Beijing, Taipei, London, Paris, and others, the dominant concern is availability. Metros, in particular, have become an essential part of the way such big cities function, and the breakdown of even one line rapidly becomes a major problem leading to significant disruption. Such railways find it beneficial to invest in significant system redundancy and preventive maintenance measures to avoid or minimize failures. For rolling stock, onboard monitoring with continuous connectivity to a maintenance facility is now offered by all the major manufacturers, and increasingly, modern communications and sensors allow similar measures on infrastructure, particularly on Metros, given their small geographic scale and the needs described.

For rural passenger and freight railways, and for most railways in developing countries, the primary driver is cost. Rural railways in developed countries virtually never cover their infrastructure costs, and freight is always in tight competition with other modes in terms of cost per ton-mile/tonne-km. Improved monitoring and shifting maintenance from 'fix on fail' to 'predict and prevent' is a key strategy. The age of infrastructure is an important factor, with many of these railways more than 100 years old having structural issues due to corrosion or other aging effects. Climate change is also a consideration, with increased dry-wet cycles leading to issues such as embankments and bridge abutments moving, coupled with increased flood risk and degradation and failures driven by extremes of temperatures, humidity, and other climate conditions. As such, improved monitoring is necessary not only to control costs but also for safety reasons. Traditional monitoring using patrol inspections either on foot or using dedicated rail vehicles is expensive to implement and maintain.

Higher-density, long-distance passenger and commuter railways sit somewhere 'in the middle' and all these factors apply to some extent, as does a requirement for passenger comfort for long journeys.

Despite the maintenance issues mentioned, climate change also presents significant opportunities for rail transport. The key impact of transportation on the climate is driven by burning fossil fuels that largely produce CO2 as a greenhouse gas. On a comparative basis, rail occupies an advantageous position compared with other transportation modes in terms of carbon footprint. At the top of the scale is aviation; both domestic and long-haul flights contribute around 2% of the total anthropomorphic emissions of greenhouse gases. This is the current forecast, and with the improved greening of other sectors of the economy, the aviation industry's share is likely to increase.

In 1943, Thomas Watson the president of IBM, famously said that he thought that there was a world market for around five personal computers. Today, most of us carry a smartphone, which is a handheld computer with more power than Mr. Watson could have dreamed of. In 2021, around 15 billion smartphones were estimated to be in service.

The development and mass manufacture of such devices has made available compact, high-power microcomputers, high-quality voice and data communication, and very small sensors for (c.f.) acceleration, temperature, luminous intensity, sound (microphones and speakers), and, critically, Global Positioning Systems, all at very affordable costs.

These technologies have enabled distributed data acquisition at an affordable cost using either equipment on service trains or equipment distributed across the infrastructure. Low-cost and high-power signal processing allows for converting data gathered into useful information, including predicting maintenance needs in a way that allows prevention without wasting effort and money on 'planned maintenance' that is not needed.

This book includes many examples of advanced innovations and emerging techniques.

We think it is clear that 'predict and prevent' technologies will develop further. The increasing availability and power of autonomous and intelligent systems will further enhance the ability to extract useful information from data that at a simple level looks incoherent.

We extend our sincere appreciation to Publishing Process Managers Dajana Jusic and Elena Vracaric Managers at IntechOpen for their guidance and support in the preparation of this volume.

Ali G. Hessami
Director of R&D and Innovation,
Vega Systems,
UK

Roderick Muttram
Fourth Insight,
UK

Section 1

Infrastructure

Chapter 1

Track Condition Monitoring Based on In-Service Train Vibration Data Using Smartphones

Hitoshi Tsunashima, Ryu Honda and Akira Matsumoto

Abstract

Although track maintenance is important, many operators of regional railway with limited financial resources are unable to conduct sufficient track inspections. In response to this problem, a track condition diagnosis system using car-body vibration sensors has been developed. In this study, a track condition monitoring system using a smartphone for general use has been developed. A technique for identifying train location using global navigation satellite system (GNSS) speed is proposed. The results of field testing shows that track condition diagnosis is possible using a smartphone-based monitoring system.

Keywords: railway, track, condition monitoring, wavelet transform, Hilbert-Huang transform, smartphone

1. Introduction

Railway track management helps support and guide wheels and is very important in terms of comfort and safety. Track management is usually carried out by track maintenance workers and usually uses track geometry cars. However, these methods are labor-intensive and costly, and many operators cannot carry out sufficient track inspections of regional railways due to limited financial resources. To address this problem, a track diagnosis system using an exclusive onboard sensing device has been developed [1–5].

Currently, track maintenance and management on railways is based on measured data of track displacement. However, track displacement measurement requires expensive equipment such as track inspection vehicles and measuring devices. A more economical management method is required, especially for regional railways. Other methods of track management exist in addition to displacement measurement, e.g., train vibration inspection using vehicle vibration measurement, but none of them constitutes a fundamental inspection method such as track displacement inspection owing to the low reproducibility of measurement data.

Meanwhile, recent advances in the practical use of IoT devices such as smartphones incorporating accelerometers based on microelectro mechanical systems (MEMS) have been reported. These devices can be used as simple and inexpensive vehicle vibration measurement devices. Many studies on track monitoring systems using vehicle vibration measurement with such IoT devices were conducted.

In this study, a track condition monitoring system for use on a smartphone was developed to reduce the cost of such a system. Using two types of IoT devices for business use and a commercially available smartphone, we took measurements of an actual car, compared the performance of both IoT devices, and diagnosed the condition of the track.

2. Literature review on track condition monitoring based on in-serve train vibration

2.1 Track condition monitoring from an in-service vehicle

Track maintenance and management on the railway is based on measured data of track displacement. However, track displacement measurement requires expensive equipment such as track inspection vehicles and measuring devices. A more economical management method is required, especially for regional railways.

Other methods of track management exist in addition to displacement measurement, e.g., train vibration inspection using in-service vehicle vibration measurement [6, 7]. Many studies on track monitoring systems using in-service vehicle vibration measurement with on-board sensing devices were conducted both in Japan and abroad.

2.2 Axle-box-mounted sensors

Chen *et al.*, Karis *et al.*, and Tsai *et al.* analyzed the relation between axle-box accelerations and railway defects or irregularities [8–10].

Sun *et al.* proposed an on-board detection technique for longitudinal track irregularity that can be applied to commercial high-speed trains. The acceleration of the axle-box of the high-speed train was evaluated [11].

Chudzikiewicz *et al.* demonstrated the possibilities of estimating the track condition using axle-boxes and car-bodies motions described by acceleration signals. They presented the preliminary investigation on the test track and supervised runs on Polish Railway Lines of an Electric Multiple Unit (EMU-ED74), [12].

2.3 Bogie-mounted sensors

Some types of track faults were detected by measuring the acceleration of bogies. Weston *et al.* demonstrated track irregularity monitoring by using bogie-mounted sensors [13, 14].

Malekjafarian *et al.* investigated the use of drive-by train measurements for railway track monitoring on the Dublin-Belfast line with an in-service Irish Rail train. The measurements were taken with accelerometers and a global positioning system. They used the train bogie accelerations [15].

2.4 Car-body-mounted sensors

Tsunashima *et al.* developed a system to identify track faults by using accelerometers and GNSS placed on the car-body of in-service vehicles [1–5].

Bai *et al.* used low-cost accelerometers that were placed on or attached to the floors of operating trains for analyzing track quality [16].

A track condition monitoring based on the bogie and car-body acceleration measurements was presented and verified in Shang Hai metro Line 1 [17].

Balouchi presented a cab-based track monitoring system developed in the UK. They presented through comparison of vibration response from sites with known defects and outputs from Network Rail's New Measurement Train (NMT). Good agreement was reported for track faults in relation to vertical and lateral alignment and dip faults [18].

2.5 Signal processing

To extract a signal on faulty tracks from measured vehicle vibration, several techniques using nonmodel-based and model-based method were proposed.

Tsunashima *et al.* proposed a nonmodel-based technique using time-frequency analysis [4]. In this paper, detection performance using continuous wavelet transform (CWT) and Hilbert-Huang transform (HHT) were compared for identifying track faults from car-body vibration. They showed that track fault features can be identified in the time-frequency plane based on the analysis of simulation studies and field tests.

A Kalman filter-based method to estimate the track geometry of Shinkansen tracks from car-body motions was proposed [19]. The proposed Kalman filter-based estimation technique was modified and applied for conventional railways [20].

Tsunashima proposed a classifier based on machine learning techniques for identifying track faults automatically from measured car-body vibration [3]. It is shown that the degradation of track can be classified in the feature space consists of car-body vibration RMS.

A new method for automatically classifying the type and degradation level of track faults using a convolutional neural network (CNN) by imaging car-body acceleration on a time-frequency plane by continuous wavelet transform [5].

2.6 Smartphones-based system

Chellaswamy *et al.* proposed a method for monitoring the irregularities in railway tracks by updating the status of the tracks in the cloud. The IoT based Railway Track Monitoring System (IoT-RMS) is proposed for monitoring the health of the railway track [21].

Rodríguez *et al.* presents the use of mobile applications to assess the quality and comfort of a railway section track (narrow gauge) in northern Spain [22].

Cong *et al.* proposed an approach for using the smartphone as a sensing platform to obtain real-time data on vehicle acceleration, velocity, and location for monitoring the track condition during subway rail transit in China [23].

Paixão *et al.* presented an approach to use smartphones to perform constant acceleration measurements inside in-service trains to complement the assessment of the structural performance and geometrical degradation of the tracks. To demonstrate the applicability of smartphone's sensing capabilities for on-board railway track monitoring, they evaluated the accelerations inside the car-body of the Portuguese Alfa Pendular passenger train [24].

3. Constructing the track diagnosis system

3.1 Track irregularities and track faults

Major private railway companies and Japan Railways (JR) use track inspection vehicles to measure track displacement, and track management is based on such measurements. Track irregularities such as longitudinal level, alignment, gauge, cross level, and twist (depicted in **Figure 1**) should be controlled properly.

However, it is difficult for regional railway companies to introduce track inspection vehicles because of the cost. Moreover, manual inspection by track maintenance staff is inefficient and expensive.

3.2 Overview of track management with proposed system

Figure 2 depicts the track management method used in this study. A 3-axis accelerometer mounted on a smartphone was used to measure the vibration of the carbody, a 3-axis Gyro sensor was used to measure the angular acceleration, and a GNSS sensor was used to collect information about the position and traveling speed; all data are then transmitted to the server. By analyzing the transmitted data, the condition of the track can be diagnosed, and, based on the result, railway operators can prioritize track maintenance and work.

3.3 Measurement devices

A BL-02 IoT device for business use (hereafter referred to as Device B) and a commercial smartphone Galaxy S7-edge (hereafter referred to as Device G) were used for measurements. **Figure 3** shows a photograph of these devices, and **Table 1** details their specifications.

Both devices were equipped with a 3-axis accelerometer, a 3-axis Gyro sensor, a GNSS sensor that can determine the location and traveling speed, and 4G internet, which is required for data transmission and reception.

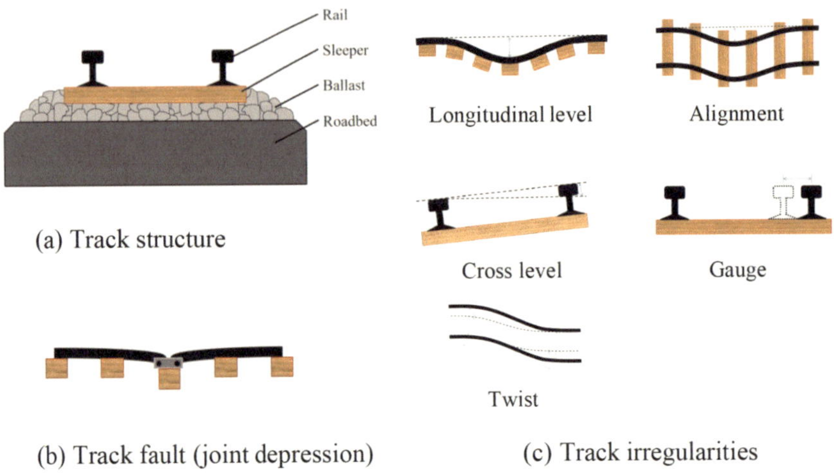

Figure 1.
Track structure and irregularities [5].

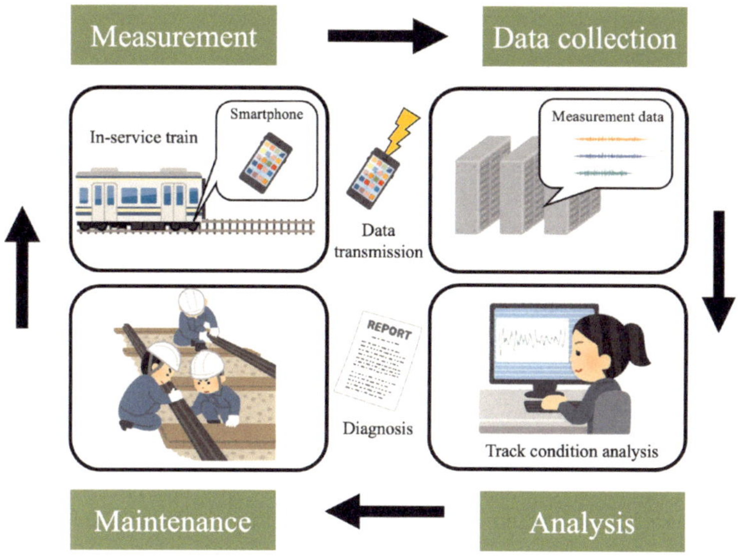

Figure 2.
Track condition management using car-body vibration.

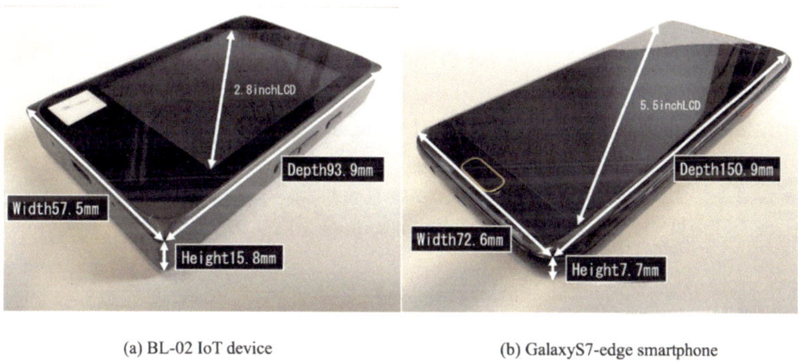

Figure 3.
Appearance of the BL-02 IoT device for business use and the GalaxyS7-edge smartphone for general use.

IoT Devices	Device B: BL-02	Device G: GalaxyS7-edge
CPU	Cortex-A7	Snapdragon820
OS	Android6.0	Andorid6.0
Display	2.8inch	5.5inch
Sensor	3-axis accelerometer, 3-axis Gyro sensor, GNSS sensor	3-axis accelerometer, 3-axis Gyro sensor, GNSS sensor
Sampling frequency	232 Hz	417 Hz
Size	94 × 58 × 16 mm	151 × 73 × 8.3 mm
Weight	102 grams	158 grams

Table 1.
Specifications of the device B and the device G.

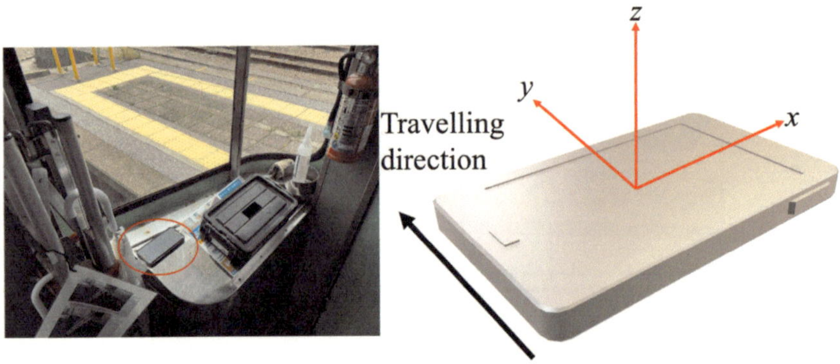

Figure 4.
Location of measurement devices.

At the time of measurement, data from Device B were measured at 232 Hz and data from Device G were measured at 417 Hz; both were down-sampled to 80 Hz at the time of acquisition from the server to reduce the amount of processing required for analysis.

Using these devices, we can measure and diagnose the vibration of the car-body. Considering convenience and GNSS reception environment, we installed the smartphone near the driver's cab, as shown in **Figure 4**.

3.4 Identifying the areas of interest from the vibration measurements

Smartphones are able to acquire latitude and longitude information; however, location detection errors increase when methods such as map matching are not employed. Therefore, we adopted a method to calculate the milage using the GNSS speed, which was, in turn, calculated using the Doppler effect of the GNSS carrier wave.

4. Verification of measurement data

4.1 Verification of vibration and angular velocity data

Figure 5 shows the measurements from Devices B and G installed on an actual car on Regional Railway A (line length: 30.5 km, Stations:17, Max. speed: 85 km/h) in December 2021. The data from both devices are almost identical in phase and amplitude.

Figure 6 shows the power spectral density (PSD) of the vertical acceleration of the vehicle. The frequency characteristics of both devices were consistent, and we can conclude that they yield sufficient accuracy as onboard sensing devices.

4.2 Identification of train location

4.2.1 Comparison of GNSS speeds

Identifying the location of a train is important for track management. In this system, the location of a train, D, is identified by integrating measured GNSS speed using a following equation.where $v_{GNSS}(t)$ is the measured GNSS speed.

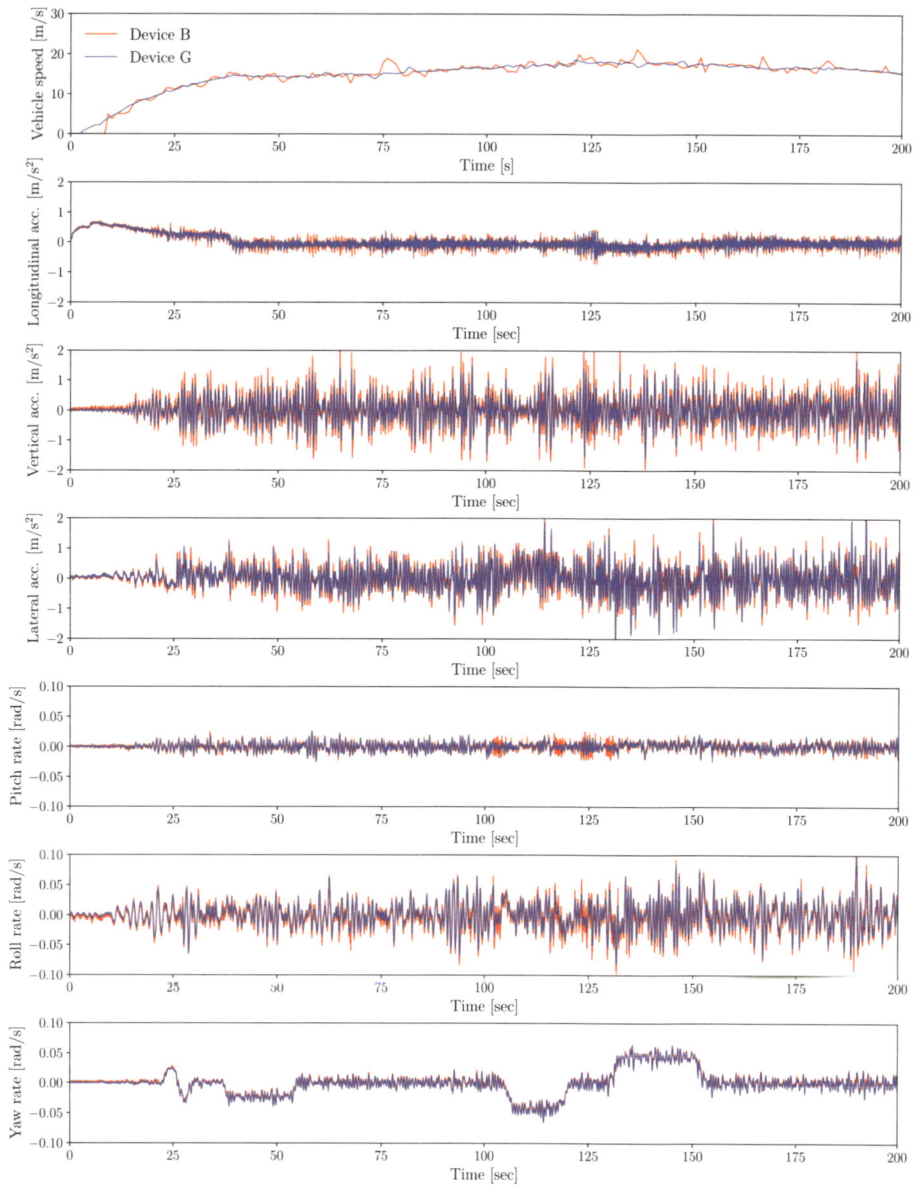

Figure 5.
Measured car-body vibration with Device B and Device G.

$$D = \int v_{GNSS}(t)dt, \quad (1)$$

The measured GNSS speed was shown in **Figure 7**. It should be noted that the measured GNSS speed was affected by multipath errors. Multipath is a major error source for GNSS receivers [25].

The location of a train can be estimated as shown in **Figure 8** using the measured raw GNSS speed. The location data are affected by the number of satellite navigation systems supported by the device. Device B supports few satellite positioning systems

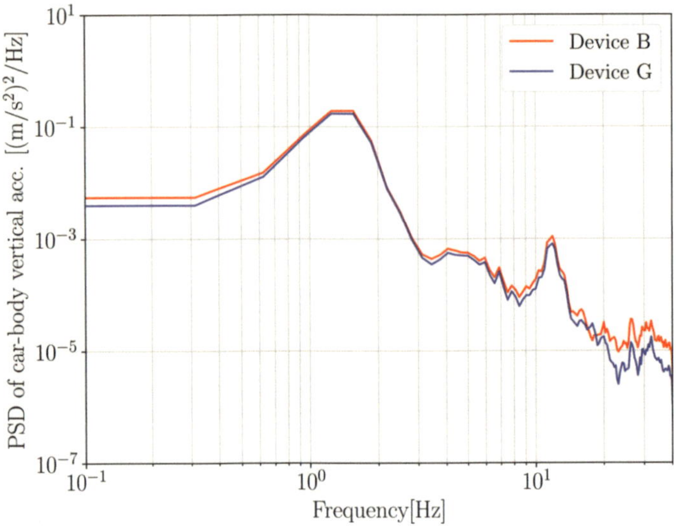

Figure 6.
Power spectral density (PSD) of measured car-body vertical acceleration by Device B and Device G.

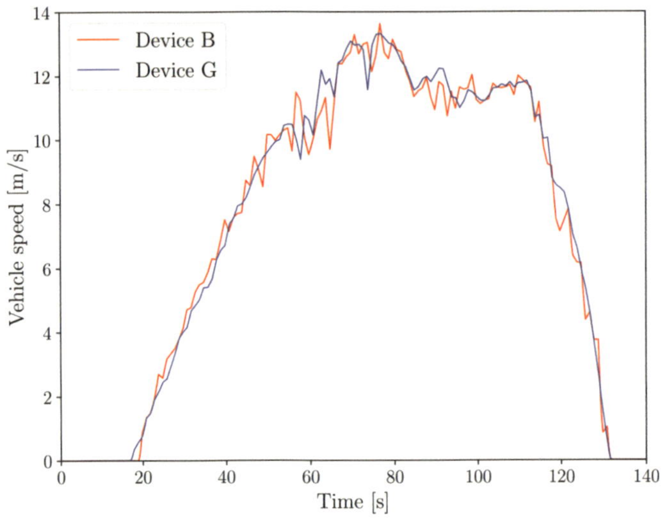

Figure 7.
Measured GNSS speed.

and is not compatible with A-GPS; therefore, the number of satellites it receives information from differs to that of Device G, and it is considered to be more susceptible to multipath errors.

We used a correction process that used a median filter for the GNSS speeds affected by multipath errors. By performing median filter processing with a window size of 800 data for approximately 5 seconds, and taking into account the magnitude of the effect of the multipath errors, we were able to improve the rapid decrease in speed due to multipath errors of the GNSS speeds measured by Device B, as shown in **Figure 9**.

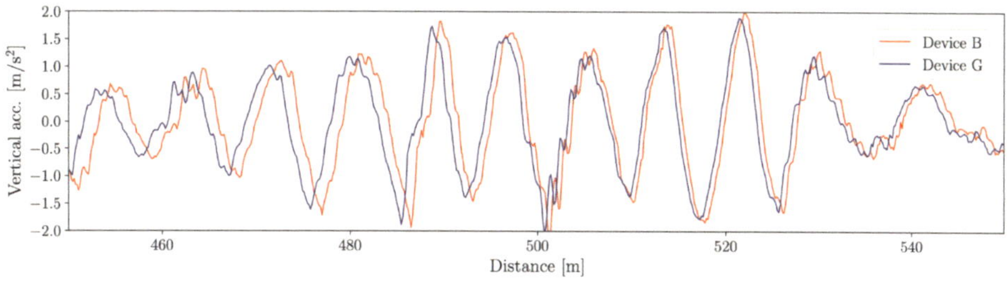

Figure 8.
Measured car-body vertical acceleration and vehicle location without GNSS speed filtering.

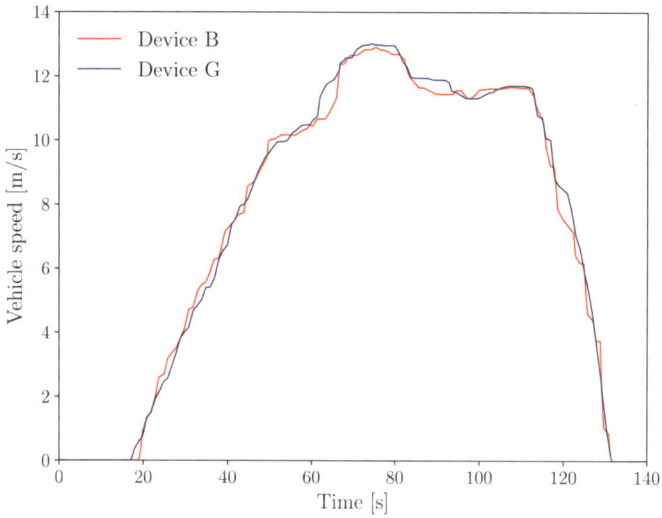

Figure 9.
Median filtered GNSS speed.

In addition, to evaluate the effect of the median filter processing on the accuracy of the location identification, we investigated the relationship between the vehicle location obtained by integrating the GNSS speeds and that obtained using the car-body vertical acceleration, as shown in **Figure 10**.

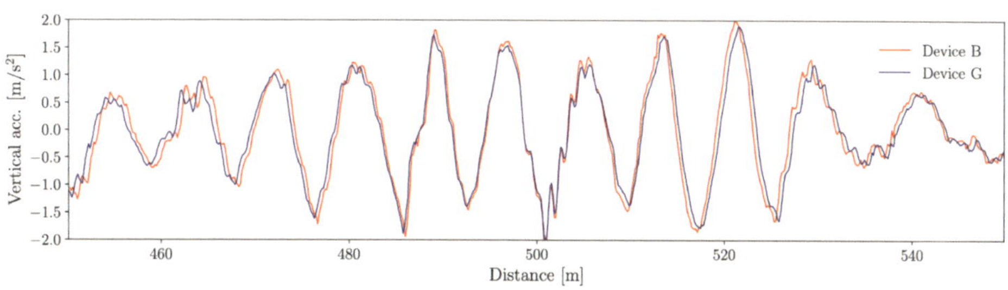

Figure 10.
Measured car-body vertical acceleration and vehicle location with GNSS speed filtering.

This figure shows that the position error is greatly improved between 450 and 550 m, which is a section that is particularly affected by multipath errors.

5. Track condition diagnosis using time-frequency analysis

5.1 Effect of track faults on time-frequency plane

5.1.1 Continuous wavelet transform (CWT)

The wavelet transform is well-known technique for analyzing nonstationary signals [26, 27]. A CWT gives simultaneous detection of the frequency and time characteristics for a nonstationary signals using a wavelet ψ, which is a function of zero average:

$$\int_{-\infty}^{\infty} \psi(t)dt = 0. \tag{2}$$

The CWT is calculated using the mother wavelet $\psi(t)$ as

$$W_{\psi}(a,b) = \int_{-\infty}^{\infty} \frac{1}{\sqrt{a}} \psi^{*}\left(\frac{t-b}{a}\right) x(t)dt, \tag{3}$$

where a and b correspond to the dilatation and location parameters, respectively.

Eq. (3) translates a source signal $x(t)$ using the mother wavelet transformed by a time shift b in time, and by $1/a$ in frequency. ψ^{*} indicates the complex conjugate of ψ.

In this study, the *Morlet* wavelet, which has a good performance between localization of time and frequency, was used [5, 28].

The CWT is subject to the uncertainty principle on time-frequency domain. In case of fault detection using CWT, if we are focusing on frequency related on the fault, the time when the fault occurred will be vague. If we are focusing on the time when the fault occurred, the frequency will be spread widely on the time-frequency plane.

5.1.2 Hilbert-Huang transform (HHT)

The Hilbert-Huang transform (HHT) has been proposed for analyzing nonlinear and nonstationary data by Huang *et al.* [29]. This method is not subject to the uncertainty principle on time-frequency domain mentioned above. Thus, more localized fault detection is possible.

The HHT consists of two operations. The first operation is the empirical mode decomposition (EMD) and the second operation is Hilbert transform.

The EMD operation breaks time domain data into intrinsic oscillatory modes called intrinsic mode functions (IMFs). The second operation is the Hilbert transform. Instantaneous amplitude, instantaneous phase, and instantaneous frequency of the IMFs are obtained by the Hilbert transform.

An IMF must satisfy the following requirements: (1) the number of local extrema and the number of zero crossings must either equal or differ by at most one. (2) the mean value of the envelopes of local maxima and local minima is zero at any point.

For extracting IMFs from the original signal, the iterative sifting process is applied. Once the first IMF is calculated, it is subtracted from the original signal to obtain a residual value. The EMD operation is applied again to the residual. This process repeats until the residual no longer contains any oscillation modes.

The original signal, $s(t)$, can be expressed by EMD operation as:

$$s(t) = \sum_{i=1}^{m} x_i(t) + R(t), \tag{4}$$

where $x_i(t)$ is the ith IMF and $R(t)$ is a residual.

Followed by the EMD operation, the analytical signal $z_i(t)$ is constructed on each IMFs component by:

$$z_i(t) = x_i(t) + jy_i(t) = a_i(t)e^{j\theta_i(t)}, \tag{5}$$

where $y_i(t)$ is a Hilbert transform of $x_i(t)$ calculated by:

$$y_i(t) = \frac{1}{\pi}\mathrm{PV}\int_{-\infty}^{\infty} \frac{x_i(t)}{t-\tau}d\tau, \tag{6}$$

where PV shows Cauchy principal value.

Instantaneous amplitude, $a_i(t)$, and instantaneous frequency, $\omega_i(t)$, can be obtained from the analytical signal $z_i(t)$ as:

$$a_i(t) = \sqrt{x_i(t)^2 + y_i(t)^2}, \tag{7}$$

$$\omega_i(t) = \frac{d\theta_i(t)}{dt}, \tag{8}$$

where

$$\theta_i(t) = \tan^{-1}\left(\frac{y_i(t)}{x_i(t)}\right). \tag{9}$$

This data-driven method is highly adaptive. However, intrinsic mode functions (IMFs) obtained by EMD strongly depend on the data itself. Thus, a small change in the data will appear on different decomposition level.

5.2 Track condition diagnosis for regional railway lines

5.2.1 Regional railway A

Time-frequency analysis was performed on the measured data to identify and evaluate the detailed location and type of track fault. When a train runs on a track where a fault exists, characteristic vibration corresponding to the type of track fault occurs. Therefore, one could identify the type of track fault and location of its occurrence by analyzing the time-frequency plane of measured car-body vertical acceleration.

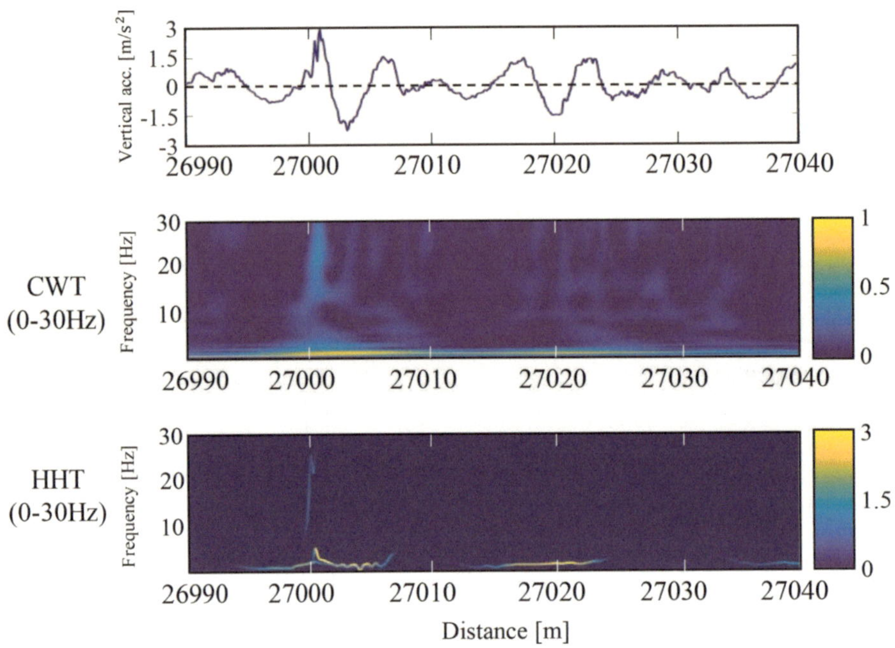

Figure 11.
Time-frequency analysis of date measured in December 23, 2021 in railway A.

Figure 11 shows the time-frequency analyses, CWT and HHT of data measured in December 23, 2021 in Railway A (line length: 30.5 km, Stations:17, Max. speed: 85 km/h). The data used for this analysis are data measured using Device G.

It can be seen from **Figure 11** that a high-frequency vibration appeared at 27000 m, which was caused by the joint depression [4]. Whereas a large vibration can be seen in low frequency in a 27,015–27,025 m section. This is caused by longitudinal-level track irregularities.

5.2.2 Regional railway B

Figures 12 and **13** show the time-frequency analyses, CWT and HHT of data measured in June and October 2022, respectively. In June 2022, vibrations due to longitudinal-level irregularity were detected at 1–2 Hz between 600 and 700 m but were no longer detected in October 2022 due to track irregularity correction. The data used for this analysis are data measured using Device G on Regional Railway B (line length: 6.4 km, Stations: 8, Max. speed: 40 km/h) in June 2022 and October 2022.

Figure 14 displays a photograph of the track section between 600 and 700 m in October 2022; the ballast was newly replenished, line maintenance work was carried out, and the longitudinal-level irregularity was eliminated. Thus, by performing time-frequency analysis using data measured by a smartphone, the type and location of track fault can be identified, and the effects of track irregularity correction can be confirmed.

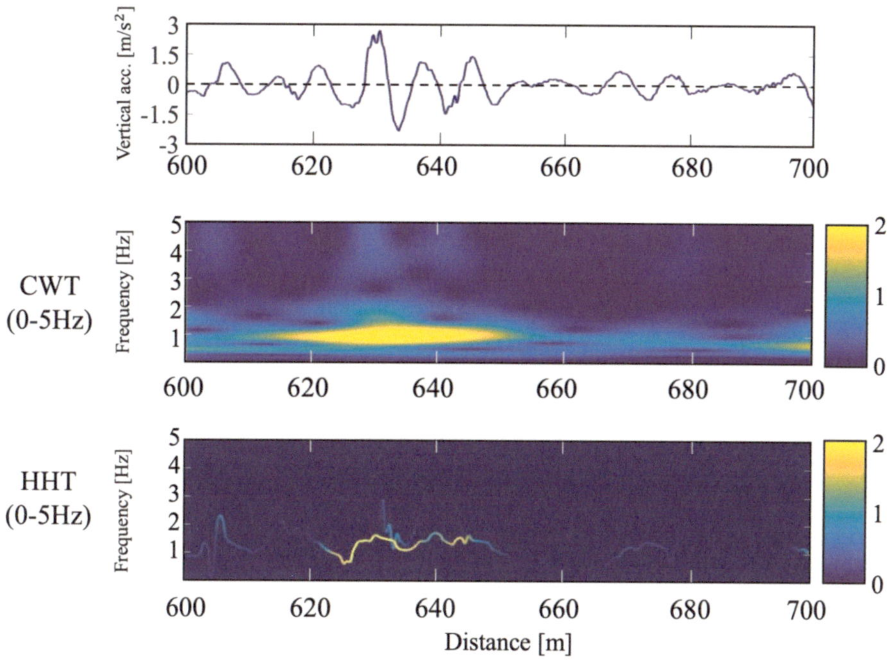

Figure 12.
Time-frequency analysis of date measured in June 2022 in railway B.

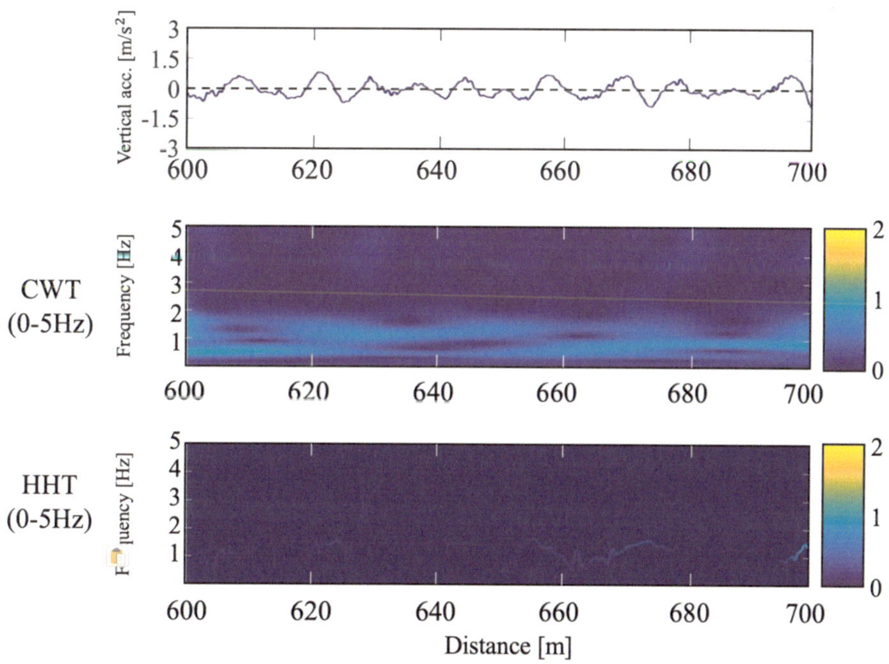

Figure 13.
Time-frequency analysis of date measured in October 2022 in railway B.

Figure 14.
Section where track maintenance work done.

6. Conclusion

In this study, we measured the car-body vibration of an in-service train using a smartphone and verified whether the track condition could be diagnosed. We were able to monitor and diagnose the track condition using both the IoT device for business use and the commercially available smartphone.

The accuracy of the GNSS speed, which is necessary to identify the location of the train, was reduced by the number of satellites received by the smartphone, that is, by the number of compatible satellite positioning systems. Therefore, when selecting a smartphone, the number of supported satellite positioning systems must be considered. In addition, we determined that the performance of devices susceptible to multipath errors can be improved by performing median filtering on the GNSS speed.

Time-frequency analysis of measured car-body acceleration obtained by a smartphone shows that proper diagnosis of track condition is possible using smartphone-based track condition monitoring system.

In the future, we plan to acquire data on a continuous basis and conduct track condition diagnosis.

Acknowledgements

This study was funded by JSPS KAKENHI Grant Number 20 K04368.

Conflict of interest

The authors declare no conflict of interest.

Abbreviations

CWT	continuous wavelet transform
HHT	Hilbert–Huang transform
GNSS	global navigation satellite system
MEMS	microelectro mechanical systems
IoT	internet of things
RMS	root mean square
EMD	empirical mode decomposition

Author details

Hitoshi Tsunashima*†, Ryu Honda† and Akira Matsumoto†
College of Industrial Technology, Nihon University, Chiba, Japan

*Address all correspondence to: tsunashima.hitoshi@nihon-u.ac.jp

† These authors contributed equally.

References

[1] Mori H, Ohno H, Tsunashima H, Saito Y. Development of compact size Onboard device for condition monitoring of railway tracks. Journal of Mechanical Systems for Transportation and Logistics. 2013;**6**(2):142-149. DOI: 10.1299/jmtl.6.142

[2] Tsunashima H, Mori H, Ogino M, Asano A. In: Zboinski K, editor. Development of Track Condition Monitoring System Using Onboard Sensing Device. London, UK: Railway Research; IntechOpen; 2015. DOI: 10.5772/61077

[3] Tsunashima H. Condition monitoring of railway tracks from car-body vibration using a machine learning technique. Applied Sciences. 2019;**9**(13): 2734. DOI: 10.3390/app9132734

[4] Tsunashima H, Hirose R. Condition monitoring of railway track from car-body vibration using time–frequency analysis. Vehicle System Dynamics. 2020;**60**(4):1170-1187. DOI: 10.1080/00423114.2020.1850808

[5] Tsunashima H, Takikawa M. Monitoring the condition of railway tracks using a convolutional neural network. In: Bulnes R, editor. Recent Advances in Wavelet Transforms and their Applications. London, UK: IntechOpen; 2022. DOI: 10.5772/intechopen.102672

[6] Weston P, Roberts C, Yeo G, Stewar E. Perspectives on railway track geometry condition monitoring from in-service railway vehicles. Vehicle System Dynamics. 2015;**53**(7):1063-1091. DOI: 10.1080/00423114.2015.1034730

[7] Kraft S, Causse J, Coudert F. Vehicle response based track geometry assessment using multi-body simulation. Vehicle System Dynamics. 2018;**56**(2): 190-220. DOI: 10.1080/00423114.2017.1359418

[8] Chen X, Chai X, Cao X. The time-frequency analysis of the train axle box acceleration signals using empirical mode decomposition. Computer Modelling and New Technologies. 2014;**18**(7):356-360

[9] Karis T, Berg M, Stichel S, Li M, Thomas D, Dirks B. Correlation of track irregularities and vehicle responses based on measured data. Vehicle System Dynamics. 2018;**56**(6):967-981. DOI: 10.1080/00423114.2017.1403634

[10] Tsai HC, Wang CY, Huang NE, Kuo TW, Chieng WH. Railway track inspection based on the vibration response to a scheduled train and the Hilbert-Huang transform. Proceedings of the Institution of Mechanical Engineers, Part F: Journal of Rail and Rapid Transit. 2014;**229**(1):1-15

[11] Sun X, Fei Y, Shi J, Zaitian K, Yunlai Z. On-board detection of longitudinal track irregularity. IEEE Access. 2021;**9**:14025-11437. DOI: 10.1109/ACCESS.2021.3052099

[12] Chudzikiewicz A, Bogacz R, Kostrzewski M, Konowrocki R. Condition monitoring of railway track systems by using acceleration signals on wheelset axle-boxes. Transport. 2019; **33**(2):555-566. DOI: 10.3846/16484142.2017.134210

[13] Weston P, Ling C, Goodman C, Roberts C, Li P, Goodall R. Monitoring vertical track irregularity from in-service railway vehicles. Proceedings of the Institution of Mechanical Engineers, Part F: Journal of Rail and Rapid Transit. 2007;**221**:75-88. DOI: 10.1243/0954409JRRT65

[14] Weston P, Ling C, Goodman C, Roberts C, Li P, Goodall R. Monitoring lateral track irregularity from in-service railway vehicles. Proceedings of the Institution of Mechanical Engineers, Part F: Journal of Rail and Rapid Transit. 2007;**221**:89-100. DOI: 10.1243/0954409JRRT64

[15] Malekjafarian A, OBrien E, Quirke P, Bowe C. Railway track monitoring using train measurements: An experimental case study. Applied Sciences. 2019;**9**:4859. DOI: 10.3390/app9224859

[16] Bai L, Liu R, Li Q. Data-driven bias correction and defect diagnosis model for In-service vehicle acceleration measurements. Sensors. 2020;**20**:872. DOI: 10.3390/s20030872, 2020

[17] Wei X, Liu F, Jia L. Urban rail track condition monitoring based on in-service vehicle acceleration measurements. Measurement. 2016;**80**:217-228. DOI: 10.1016/j.measurement.2015.11.033

[18] Balouchi F, Bevan A, Formston R. Development of railway track condition monitoring from multi-train in-service vehicles. Vehicle System Dynamics. 2021;**59**(9):1397-1417. DOI: 10.1080/00423114.2020.1755045

[19] Tsunashima H, Naganuma Y, Kobayashi T. Track geometry estimation from car-body vibration. Vehicle System Dynamics. 2014;**52**(sup1):207-219. DOI: 10.1080/00423114.2014.889836

[20] Odashima M, Azami S, Naganuma Y, Mori H, Tsunashima H. Track geometry estimation of a conventional railway from car-body acceleration measurement. Mechanical Engineering Journal. 2017;**4**(1) JSME, Paper No.16-00498:1-12. DOI: 10.1299/mej.16-00498

[21] Chellaswamy C, Geetha TS, Vanathi A, Venkatachalam K. An IoT based rail track condition monitoring and derailment prevention system. International Journal of RF Technologies. 2020;**11**:81-107. DOI: 10.3233/RFT-190210

[22] Rodríguez A, Sanudo S, Miranda M, Gomez A, Benavente J. Smartphones and tablets applications in railways, ride comfort and track quality. Transition Zones Analysis. Measurement. 2021;**182**: 1-12. DOI: 10.1016/j.measurement.2021.109644

[23] Cong J, Gao M, Miranda M, Wang Y, Chen R, Wang P. Subway rail transit monitoring by built-in sensor platform of smartphone. Frontiers of Information Technology & Electronic Engineering. 2020;**21**(8):1226-1238. ISSN 2095–9184 (print); ISSN 2095-9230 (online)

[24] Paixão A, Fortunato E, Calçada R. Smartphone's sensing capabilities for on-board railway track monitoring: Structural performance and geometrical degradation assessment. Advances in Civil Engineering. 2019;**2019**:1-13. Article ID 1729153. DOI: 10.1155/2019/1729153

[25] Hsu L. GNSS multipath detection using a machine learning approach. In: 2017 IEEE 20th International Conference on Intelligent Transportation Systems (ITSC). Nagoya, Japan: IEEE ITS Society; 2017

[26] Daubechies I. Ten lectures on wavelets. Society for Industrial and Applied Mathematics. Philadelphia. 1992. pp. 1-357. DOI: 10.1137/1.9781611970104

[27] Mallat S. A Wavelet Tour of Signal Processing. New York: Academic Press; 1998

[28] Le T. Use of the Morlet mother wavelet in the frequency-scale domain decomposition technique for the modal identification of ambient vibration responses. Mechanical Systems and Signal Processing. 2017;**95**:488-505. DOI: 10.1016/j.ymssp.2017.03.045

[29] Huang NE, Attoh-Okine NO. Hilbert-Huang Transforms in Engineering. Boca Raton, Florida, USA: CRC Press; 2005. ISBN 9780849334221

Chapter 2

Perspective Chapter: Intelligent Construction Technology of Ballastless Track for High-Speed Railway

Qiuyi Li and Luyao Li

Abstract

Intelligent construction technology is the only way to realize the construction and upgrading of high-speed railways, and strongly supports the construction of smart railways. In terms of intelligent track design technology, the digital, dynamic and visual design of track is realized, which changes the way the traditional design unit provides only construction drawings. The design unit then provides digital track design results in addition to construction drawings, facilitating intelligent manufacturing, construction, operation and maintenance, etc. In terms of the intelligent construction of track slabs, the traditional track construction technology is upgraded by using the results of track intelligent design, and data from the total station, rail inspection trolley, note books and related tooling are automatically collected, controlled and adjusted. Data collection, analysis, statistics and prediction mechanism of the plate ballastless track are established to form a track intelligent construction technology of "data collection, data penetration, data sharing and openness, data tracking and tracing". In terms of complete sets of software for track intelligent construction, a series of intelligent design, manufacturing, construction software and complete sets of intelligent construction equipment are proposed, which greatly improves the level of intelligent construction.

Keywords: high-speed railways, digital, governance platform, intelligent measurement and control terminal, artificial intelligence

1. Introduction

Ballastless track is one of the core technologies of high-speed railways. The track structure directly acts on the substructure, the deformation of the substructure will directly affect the track smoothness and stability, and the construction quality and accuracy will directly affect the safety and comfort of high-speed train operation. Track engineering has the following characteristics (**Figure 1**) [1, 2].

1. High precision requirement: The required accuracy of rail laying is ±1 mm, the manufacturing accuracy of track slab is ±0.5 to −1 mm; the accuracy of the

Figure 1.
Construction site laying of ballastless track for high-speed railway.

molds is 0.3 to 0.5 mm; The thickness accuracy of cast-in-situ concrete is 0 to 10 mm.

2. The short period of construction, and the time of completion is quite close to the opening of operation: After the completion of the track engineering construction, it will enter the joint trial and test period. The track quality needs to be strictly controled, and there is no time for mending and repair [3, 4].

3. There are abundant sources of construction data for track engineering. The measurement mainly adopts total station and rail inspection trolley; with the development of new technologies such as the Internet of Things, the integration of data interfaces has become possible [5–8]. Research on the conversion of data to information and intelligence for track construction and laying can be carried out to realize real-time data uploading and intelligent control, which can greatly improve construction quality and accuracy.

A new round of scientific and technological revolution and industrial transformation is emerging; artificial intelligence (AI), big data, cloud computing, Internet of Things (IoT), building information modeling (BIM), Beidou Navigation satellite system (BDS) and other new technologies are accelerating breakthrough applications, and human society will quickly enter the era of intelligence [9, 10].

With the rapid development of high-speed railways and artificial intelligence and other high-tech higher requirements have been put forward for the supervision ability, intelligence and recording of information for high-speed railway construction on site. At the same time, the development of high-speed railways is turning to pay more attention to delicacy management, the development quality, the service level, and solving the deep-seated and overall problems, further improving the modernization of construction and the management level of the high-speed railway.

The design realization of the traditional construction management mode of high-speed railway track engineering shall be delivered in the form of construction drawings, and the construction organization constructs according to the drawings. The construction level of information gathering and intelligence is too low. Research on intelligent construction technology is helpful to improve the quality of track construction, enhance the efficiency of operation, ensure project progress and control construction cost.

This chapter introduces the intelligent technology of ballastless track construction for high-speed railways, and intellisense, mobile internet intelligent transmission technology, intelligent analysis and control technology introduced to the track construction process. This mainly forms the dynamic design technology of the ballastless track construction process, the measurement and control technology of high-speed railway ballastless track construction based on data driven and the intelligent control technology which realize the high precision control and automatic measurement and control of ballastless track construction processes, lays the foundation for intelligent railways, and promotes the development of railway construction to an intelligent approach (**Figure 2**).

2. Process dynamic design technology for ballastless track construction in high-speed railway

Ballastless track is greatly affected by changes in the substructure, such as subgrade, bridges and tunnels. Realizing the intelligent update of design results and complete intelligent operation with dynamic data driven construction equipment is

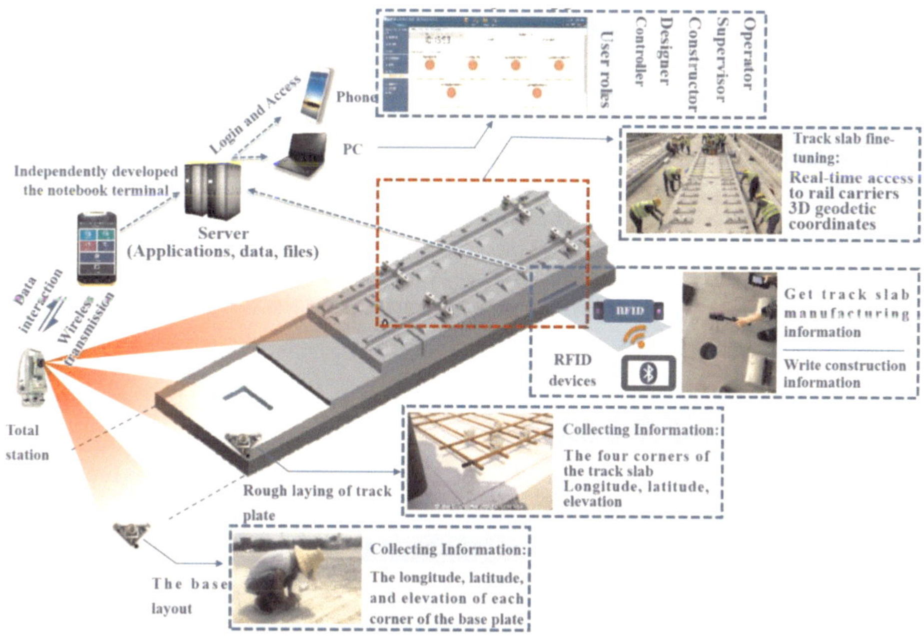

Figure 2.
The overall technical scheme of ballastless track intelligent construction.

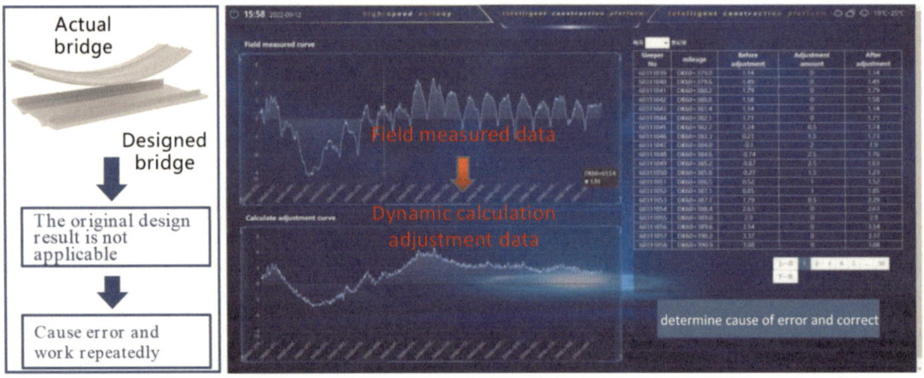

Figure 3.
Process dynamic design technology of ballastless track construction process.

the key to promoting the efficiency of ballastless track construction and controling quality. Based on the collected data of the whole construction process, taking into account the changes of the substructure such as subgrade, bridge and tunnel as the input conditions, the plane and elevation dynamic design of the concrete base and track slab is realized. Problems such as the fastener at beam end exceeding the standard requirement, the control of the concrete base plate hanging out, and the precise control of the geometric size of the ballastless track are solved. The dynamic updating of the design results in the construction process is realized, and dynamic design problems such as alignment changes at long-span Bridges, deformation joints and transition section position updating are all overcome.

Take the extremely long span bridge such as Ganjiang bridge of Chang-Gan Railway (main span 300 m), Yuxihe bridge of Shang-He-Hang Railway (main span 324 m) as an example, which were affected by bridge creep, temperature deformation, load on the bridge and wind speed and other factors, as a result of which the ballastless track line is difficult to control. The dynamic design technology was adopted to obtain the exact correspondence between bridge loads and cable force and design model (main beam line type) after the main bridge design was finally closed and preloaded. The main bridge line type was systematically adjusted and the digitized track results have been corrected in real time (**Figure 3**).

This mainly includes the layout algorithm of ballastless track, the coordinate cluster calculation method of ballastless track under the condition of complex line and the virtual preassembly technology of ballastless track.

2.1 The layout algorithm of ballastless track

During the construction period, due to the great difference between the actual working conditions of the substructure such as the subgrade, bridge and tunnel and the assumed design working conditions, the ballastless track needs to be dynamically adjusted according to the actual working conditions. The layout of the track is as follows:

1. By configuring different lengths of track slab and adjusting the gap of the track slab to meet the layout requirements, the layout of track slab is disconnected at the bridge joint, settlement joint, deformation joint of subgrade and bridge, subgrade and tunnel, bridge and tunnel boundary, etc.

2. The track slab lays on such as a 24, 32 m beam, and commonly used continuous beam and other areas of standardized layout, the track layout should consider the sleeper spacing uniformity and the maximum spacing of fastener at the beam end and other restrictions.

According to the above track slab layout principles, the track slab layout in the track slab design software is implemented as follows:

1. The left line and right line are arranged separately.

   Due to the different lengths of the left line and right line, the different beam joints at the center line of the track in curved sections, and the different beam lengths at the center line of the track when the bridge is designed into a curved beam in a small radius curve section, the layout of the track slab should be arranged separately on the left line and right line. According to the way that the left line and right line is laid separately, the more accurate data file of the track slab can be obtained.

2. The track layout mode of "the whole is zero and set zero for the whole" will adopted.

   The line is composed of straight line, front easing curve, round curve and back easing curve, etc. which is composed of subgrade, bridge and tunnel. The line of track layout design can be decomposed into different sections, each section is composed of many small sections with the same length, that is one line can be divided into several blocks. Each block is composed of a lot of simply supported beam and continuous beam bridge, for example: there are a lot of same span 32.6 m beam, the track layout on the 32.6 m beam takes the same parameters. Therefore, the detailed idea of the track layout design is: the parameters of the same length that compose the line are first laid out, for example, the 32.6 m simply supported beams are arranged first, and all the later simply supported beams of the same type are arranged in the same way.

Therefore, the layout design of track slab is designed to find the optimal layout scheme under the constraints of slab length, slab gap, sleeper spacing, etc., to achieve uniform layout of fastener spacing, the fastener spacing at beam end that meets the design requirements, and the optimal control of curve vector distance deviation on small radius curve (**Figure 4**).

2.2 Calculation method for ballastless track coordinate cluster under complex line conditions

Accurate measurement is the most important and difficult process to control in the construction of ballastless track. The whole measurement technology requires high precision and is influenced by many factors. The azimuth of the track slab centerline and the three-dimensional coordinates of the rail top center at the rail support platform used in the measurement process need to be accurately calculated, and digital results are automatically generated to guide the construction and laying of the base, track slab, etc.

According to the plane data of the route, the tangent offset method (i.e. rectangular coordinate method) is used to calculate the coordinate values of each point on the

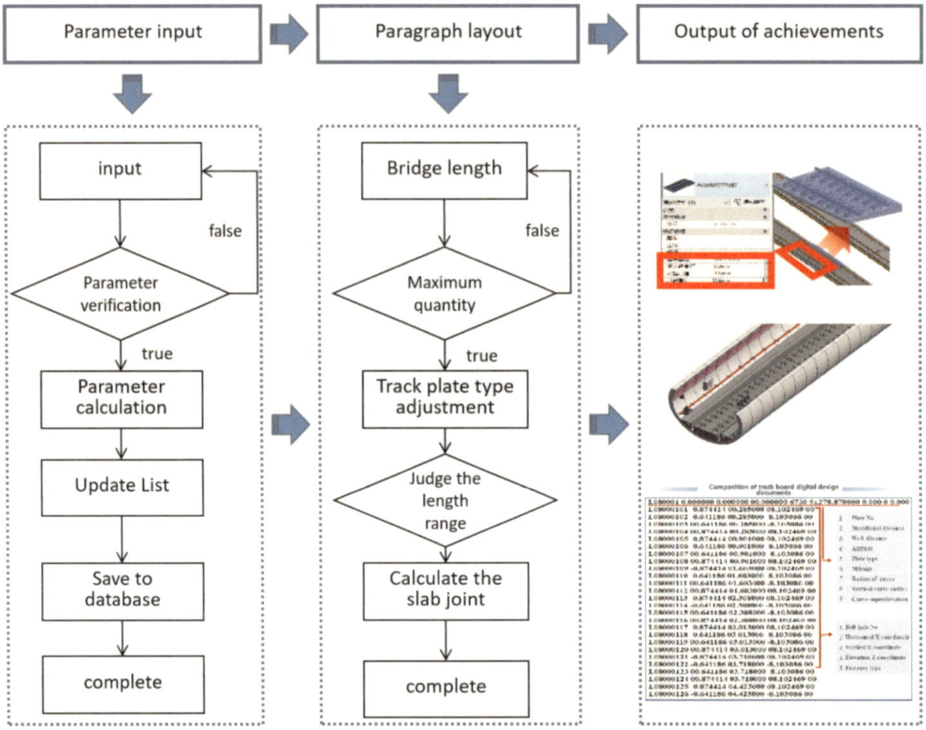

Figure 4.
Schematic diagram of track layout algorithm.

route. As shown in the figure, take the calculation of any point coordinates. The mileage of the starting and ending points (HZ/ZH) of the curve and the coordinates of the intersection point (JD) are known. The azimuth of the line conductor is expressed in A, the radius of the circular curve is expressed in R, and the length of the transition curve is expressed in l_0. The coordinates of the straight-line segment, the transition curve segment (front/back), and the circular curve segment are calculated section by section to form the track design data file.

The horizontal and vertical relative coordinates of each calculation point within the straight line are fixed, and the longitudinal coordinates are the cumulative value of the fastener spacings. The coordinates and azimuth of the main point and control point of the plane curve are calculated through the plane design data of the line, including the coordinates of the intersection point, the curve radius and the length of the transition curve. The elevation of the main point and control point of the vertical curve is calculated through the design data of the line profile, including the elevation of gradient change point, slope, grade length and radius of the vertical curve. At the same time, the coordinates of track slab rough laying position point and base corner point are calculated.

Each point of the transition curve is a gradually changing three-dimensional coordinate, the plane takes the starting point of the plate as the coordinate origin, and the vertical takes the midpoint of the rail top line of the starting point of the plate as the coordinate origin.

The horizontal coordinate of each point within the circle curve has an offset value, the vertical coordinate is fixed, and the longitudinal coordinate is the cumulative value of fastener spacings.

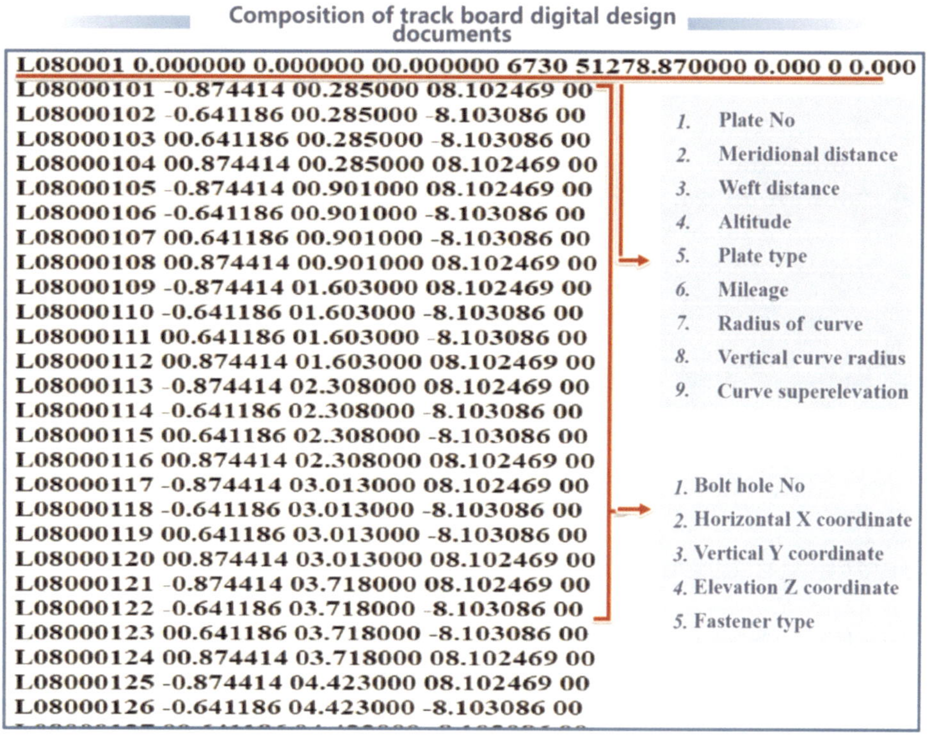

Figure 5.
Generated digital results.

After calculation, the system will automatically generate the digital results guiding the construction survey, as shown in the **Figure 5**.

Description of relevant figures of track slab data file

Line 1: Slab number; Coordinates of the starting point of the slab, X coordinate perpendicular to the line direction; Y coordinate along the line direction; Elevation Z coordinate; Type of track slab; Mileage; Radius of plane circular curve; Vertical curve radius; Superelevation;

Line 2: Slab number and control point; X coordinate perpendicular to the line direction; Y coordinate along the line direction; Elevation Z coordinate; Type of fastener;

...

The last line: Slab number; Coordinates of the ending point of the slab, X coordinate perpendicular to the line direction; Y coordinate along the line direction; Elevation Z coordinate; Type of track slab; Mileage; Radius of plane circular curve; Vertical curve radius; Superelevation.

2.3 Three dimensional model building and updating technology of ballastless track

Based on the results of (2.1) and (2.2) in this section, the 3D visualization building technology of ballastless track is developed, and the real-time dynamic updating of 3D model during construction is realized.

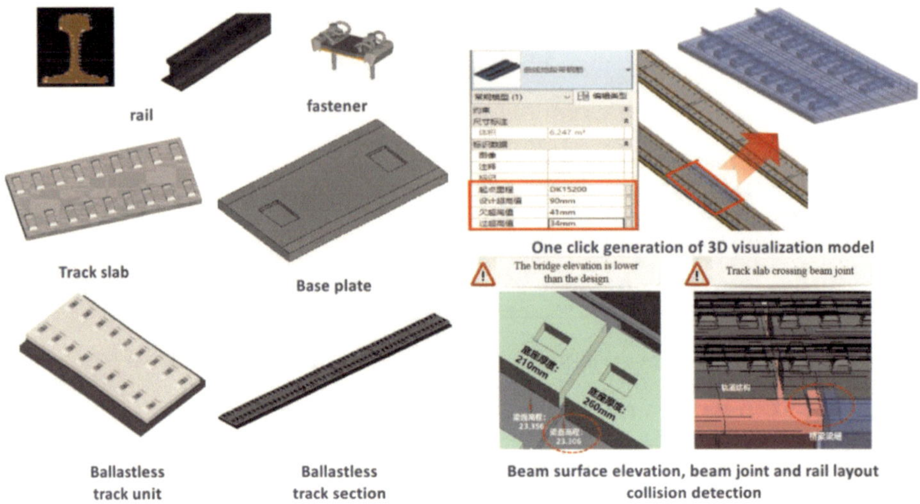

Figure 6.
3D model building and updating technology of ballastless track.

The track structure is composed of steel rails, fasteners, track slab, etc. There are many types of track components. To avoid repeated creation of models, a component library was developed. Unified standards and formats were used for creating component models, such as component classification, naming, material standards, modeling accuracy, etc.; Developing parametric modules to build models of the same type but with different sizes, such as track slabs and bases; Realizing the unified management of component models to facilitate the reuse of components at all stages. Based on the unified coding system, the detailed design model of track unit was constructed. Create component models with fixed dimensions such as rails, fasteners and sleepers, by inputting design parameters, parameterized design of different types of track slabs and different plane section bases can be achieved. For example, ballastless track structures (track slabs, etc.) can be rapidly designed according to the information of non-standard slabs and curve superelevation value generated by layout design. The creation of composite model is realized by inputting elevation, layout design, etc. Combined with the component model and the track engineering design data, the track model can be quickly created and updated according to the change of relevant professional information. By means of human-computer interaction, track components creation, data model generation, 3D digital model layout, etc. are all completed through the system interface (**Figure 6**).

3. Data driven construction measurement and control technology for ballastless track of high-speed railway

Using new generation information technologies such as big data, mobile Internet, edge computing, etc., based on the digital design results of track engineering, and data driven mode, the intelligent construction measurement and control mode of "data perception, real-time transmission, dynamic analysis" was established by using the data automatic acquisition and measurement of total station, track inspection trolley, measurement and control terminal, etc.

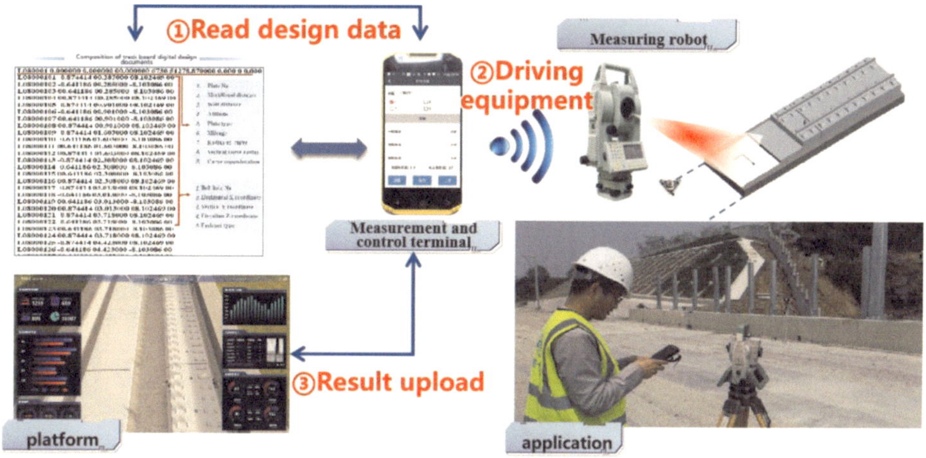

Figure 7.
Intelligent measurement and control scheme based on data drive.

The portable intelligent measurement and control terminal, fine tuning robot and rail geometry and position information acquisition device were developed. The data interface of the whole process of design and construction has been opened up, and the automatic measurement and control of key processes of ballastless track, such as base laying and positioning, track slab fine adjustment, track panel fine adjustment and rail fine adjustment, has been realized, which improves efficiency and reduces manual inputs. It realizes the interaction and flow of "design, measurement and control, feedback" of construction information flow. The GeoCom communication technology is used to open the control interface with the measuring robot, realize the wireless communication and automatic control with the measuring robot, realize the automatic calculation of fine adjustment deviation, and complete the wireless upload of fine adjustment results to the management and control platform (**Figure** 7).

3.1 Base laying and positioning based on data drive

Through the CP III control network and digital design results, the station setting of the measuring robot is completed. The terminal is used to control the measuring robot wirelessly to realize automatic measurement of base plate lofting. At the same time, the elevation measurement data of the subgrade surface, beam surface and other offline foundations are obtained, and the elevation deviation is obtained through real-time comparison with the digital design results. This determines the influence of beam surface elevation before the construction of the base, the base thickness adjustment scheme, automatically generates the formwork height of the base, and realizes the accurate control of the plane position, elevation and thickness of the base (**Figure** 8).

3.2 Fine adjustment of track slab based on data drive

The fine adjustment of track slab is basically a precise measurement in the installation stage, that is, the key point measurement or setting out of the track slab under to acheive the design theoretical position. During the preparation for fine adjustment, the process uses curve elements, slope information, long and short chains, other basic information and dynamic calculation modules to calculate the three-dimensional

Figure 8.
Schematic diagram of base laying and positioning process based on data drive.

coordinates of the rail support platform based on the positional mileage of the track slab. A measuring frame accurately matched with a bolt hole is placed on the rail support platform. A prism for the identification by the measuring robot is installed in the center of the frame. The intelligent measuring robot is used to collect the three-dimensional coordinates of the prism at the corresponding point and send them to the mobile terminal. The measurement and adjustment software automatically analyzes, compares and calculates between the measured coordinates and the design coordinates in real time to form the geometric position deviation under the three-dimensional space of the track slab at the rail support platform. The fine adjustment robot is controlled according to the deviation to complete automatic adjustment (**Figure 9**).

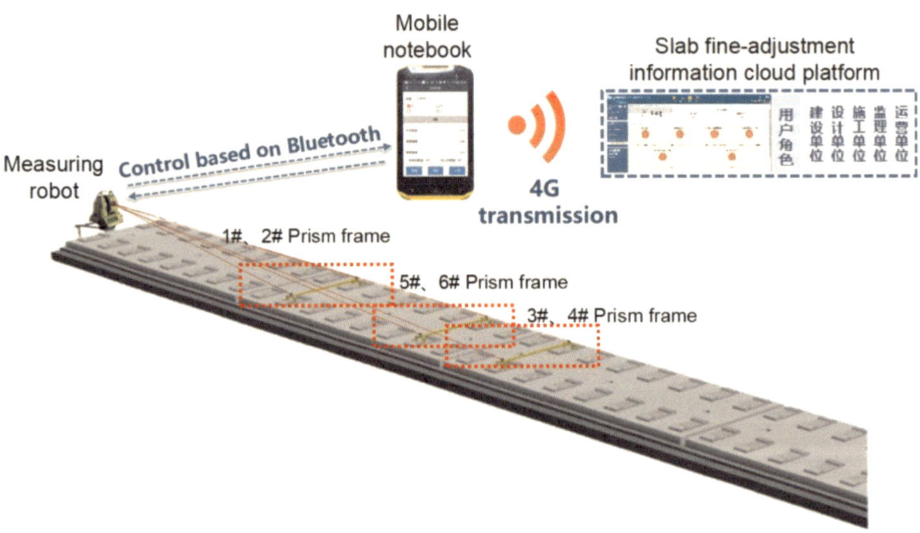

Figure 9.
Schematic diagram of track slab fine adjustment scheme based on data drive.

The measurement and control is mainly realized by a GeoCom interface, and a communication unit is composed of the request of the mobile terminal and the response of the measuring robot. The GeoCom interface is a dynamic link library that encapsulates multiple secondary development functions. When developing the software, the above encapsulation functions can be used to realize the calibration, rotation control, automatic target recognition, accurate ranging, angle measurement, etc. of the measuring robot. It is developed based on an Android platform and uses ASCII protocol to realize control communication with the measuring robot. The Bluetooth communication module of the mobile terminal is used to establish the information channel with the measuring robot and send ASCII commands through GeoCom interface technology. The measuring robot receives ASCII character commands, and returns a response string after protocol analysis. It can then realize the basic measurement and control operations such as equipment initialization, instrument connection, station setting, parameter configuration, verification, measurement, upload, port release, etc.

With the SQLite database as the data organization and storage carrier, the wireless control of the measuring robot is realized through ASCII command mode, and the automatic measurement of track slab fine adjustment is realized. The main functions of the software include the following functions: engineering information configuration of track slab fine adjustment operation, parameter setting of track slab fine adjustment operation system, layout data interface configuration, calibration of fine adjustment standard frame, track slab overlapping orientation, track slab fine adjustment measurement, fine adjustment result data upload, etc. (**Figure 10**).

The measured deviation is fed back to the developed automatic fine track slab adjustment robot in real time, and the electric control system is combined with the servo motor to automatically execute the fine adjustment command without manual adjustment. The fine adjustment deviation and adjustment time are reduced, and the precision of fine adjustment is guaranteed. Compared with manual fine adjustment, the comprehensive efficiency is increased by more than 3 times (**Figure 11**).

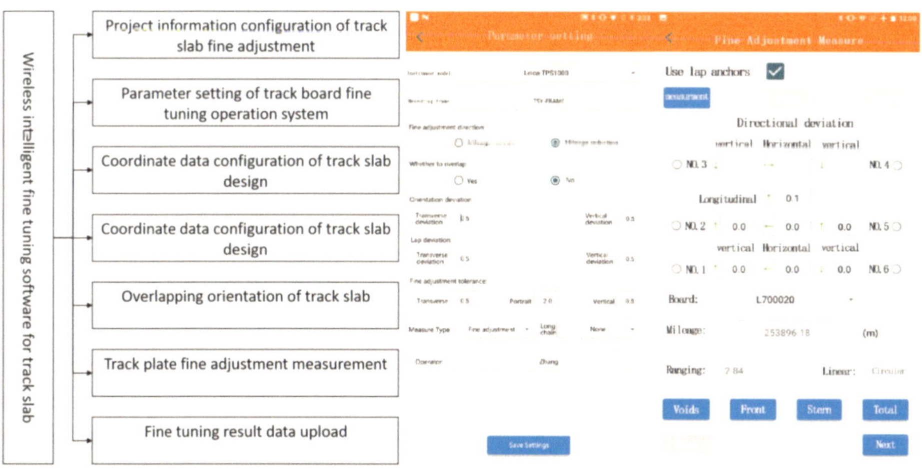

Figure 10.
Functional structure of wireless intelligent fine adjustment software for track slab.

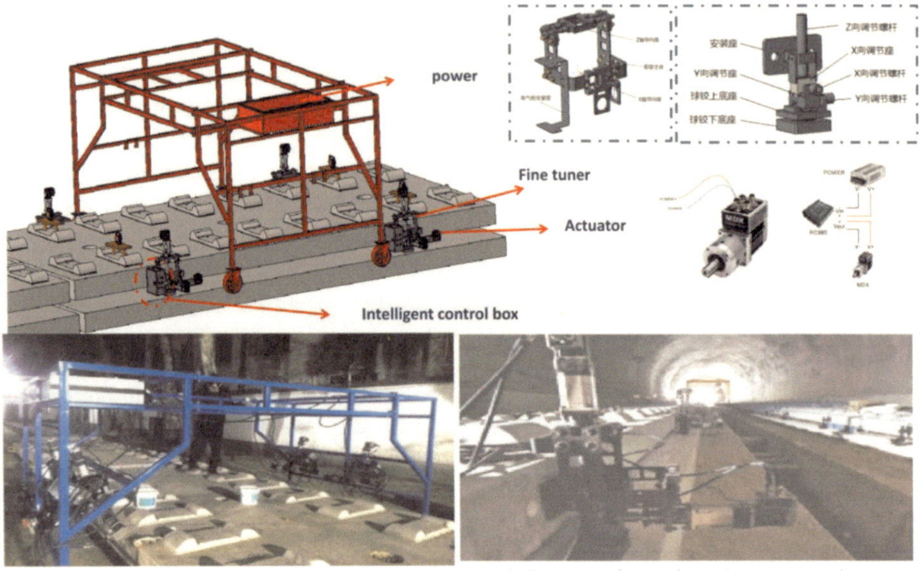

Figure 11.
Schematic diagram of track slab fine adjustment robot scheme.

3.3 Track panel and rail fine adjustment based on data drive

Based on the geometric dimension data of the track in the design results, the automatic measurement of the track centerline, the three-dimensional coordinates of the left and right rails, and the calculation of the lateral and vertical deviations of the track is determined. Based on the developed track inspection trolley information equipment, the automatic collection of track geometry and position data is determined. The thickness distribution of track bed slab is analyzed in real time by measuring the rail top elevation information and combining the elevation data of base slab/bearing layer; determine the year-on-year analysis of the fine adjustment data of the rail panel, the retest data after pouring, and the fine adjustment data of the rail, so as to provide support for obtaining the variation law of the construction deviation of the rail panel, improving the construction control process, and also provide a pre-foundation for the subsequent fine adjustment of the rail (**Figure 12**).

4. High speed railway ballastless track construction intelligent management and control technology

The intelligent construction control technology of ballastless track is composed of four parts: automatic data collection, intelligent transmission, big data storage, and the whole process supervision of the control platform. The information equipment interconnected with the management and control platform, such as the intelligent measurement and control terminal for base positioning and track slab fine adjustment, and the collection device for track geometry and position, are used to realize the automatic upload of relevant information such as construction process measurement and laying. Using big data storage technology, a distributed data warehouse is built on the server to realize the orderly, regular and structured storage of massive amounts of

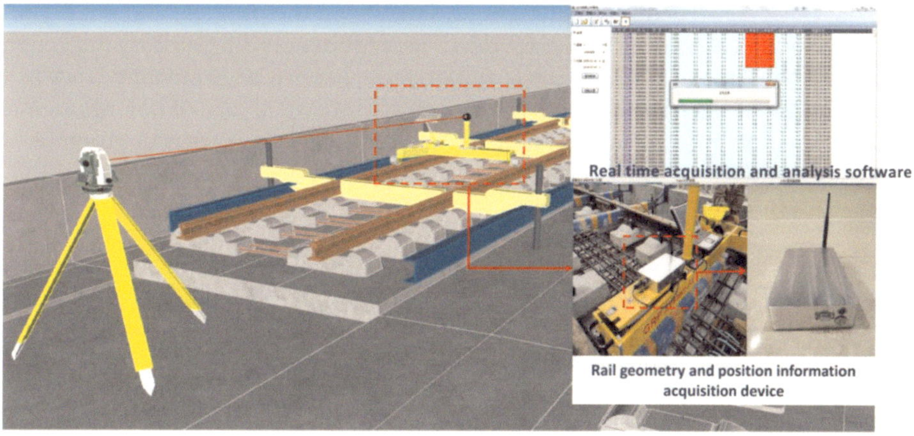

Figure 12.
Data driven rail panel and rail fine adjustment.

data. Through the construction intelligent management and control platform, the massive amount of data in the data warehouse are screened, extracted, analyzed and processed. In combination with the relevant requirements of the slab track construction acceptance standards, the data is judged to exceed the limit. Through the early warning cloud system, various ways of early warning and prediction are realized, so as to realize the interconnection between the construction site and the remote platform, the interconnection between the construction workers and the management personnel, and the control of the process flow, as well as realize the whole process management of quality (**Figure 13**).

The intelligent management and control technology takes the intelligent management and control platform as the core, and takes "design data achievements" and "dynamic data files of the whole construction process" as the data basis. The core of management and control is the progress and accuracy of the whole process of

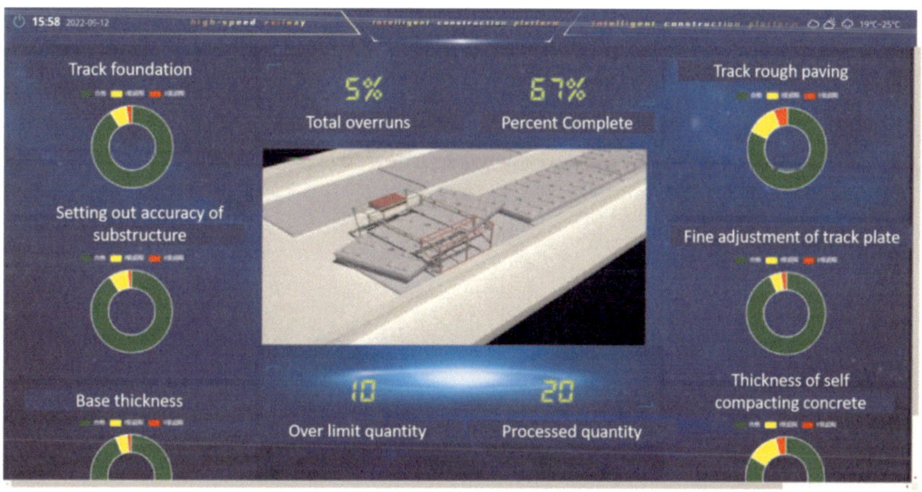

Figure 13.
Control platform for ballastless track construction.

ballastless track construction. Analyze, judge and count the progress and quality of track construction according to the set tolerance requirements by quickly analyzing the on-site construction survey data. It will provide detailed and convenient statistical information for project construction managers, achieve the dynamic management of track construction process on site, improve management efficiency and promote construction quality.

The main functions are as follows:

1. Construction precision control: elevation deviation of the offline foundation surface, plane linetype of the base slab/bearing layer, elevation and thickness control, concrete thickness and elevation control of the track bed slab, precision control of the rail panel and long steel rail laying, etc.

2. Track progress management: Combined with the track 3D visualization design model, the visualization progress management is achieved.

3. Report statistics management: Automatically generate a progress report and precision quality report.

4. Early warning and forecasting: Limit management, early warning value setting, automatic early warning of construction deviation, etc.

5. Online approval and management of inspection batch for concrete, reinforcement and testing.

6. Digital asset management such as full-line 3D model and construction real-time information model.

7. Organization and structure management: Assign authority functions to the management personnel of construction units, supervision units and construction units.

4.1 Schedule management

Prior to the construction of ballastless track, engineering plans are made for the upcoming process, such as track slab rough laying plan, track slab fine adjustment plan, track slab retest plan, etc. Generating a progress Report to display detailed progress information of each process (**Figure 14**).

4.2 Construction quality control

According to the fine adjustment result data uploaded at the construction site, the control platform automatically calculates the qualified rate and the percentage by which the limit of each indicator parameter is exceeded according to the allowable limit set by the user for the fine adjustment of the track board. For example, the platform will display the statistical results in the form of charts and provide the function of querying detailed data of fine adjustment. The fine adjustment data which exceeds the limit will be marked in red in the table for processing. Statistical indicators and parameters include:

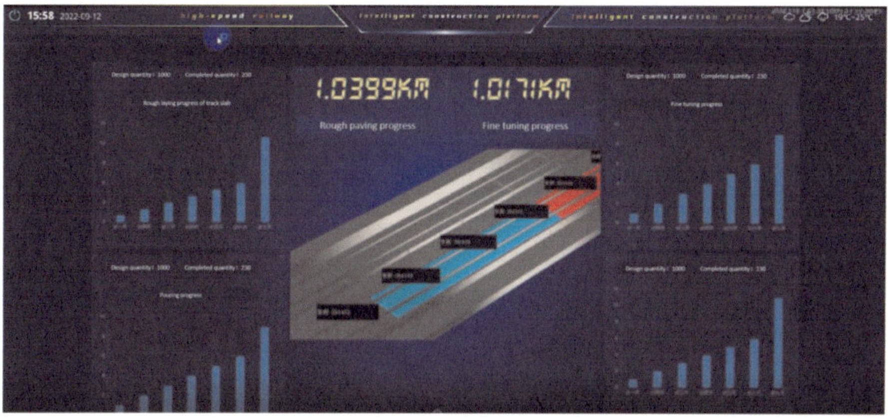

Figure 14.
Project schedule management.

1. The deviation between the actual elevation of the rail seat and the designed elevation after the fine adjustment of the track slab;

2. The deviation between the center line position and the design center line position of the rail seat after the fine adjustment of the track slab;

3. Elevation deviation of adjacent track slab rail seat after fine adjustment of track slab;

4. Center line deviation of adjacent track slab rail seat after fine adjustment of track slab.

After the track plate is perfused with self-compacting concrete, the rail seat is measured again. According to the acceptance standard, the deviation of the height and center line of the rail seat is controlled, and the qualified quantity is counted automatically in the platform and fed back to the management personnel along with with the early warning mechanism.

Track slab retest progress and accuracy pass rate statistics, including elevation of rail seat, center line of rail seat, elevation of adjacent slab rail seat, center line deviation of adjacent center line of rail seat deviation are produced (**Figure 15**).

4.3 Early warning and forecast function

The platform automatically calculates the percentage of qualified rate and limit exceeded for each indicator parameter. Data exceeding the limit will be warned in real time. The administrator of each unit conducts on-site treatment in time according to the over-limit situations and eliminates the alarm information. This can achieve real-time control of construction quality (**Figure 16**).

4.4 Progress and accuracy report

According to the data upload status of each Construction section of the whole line, the platform will conduct statistical work according to the set fixed time and a final

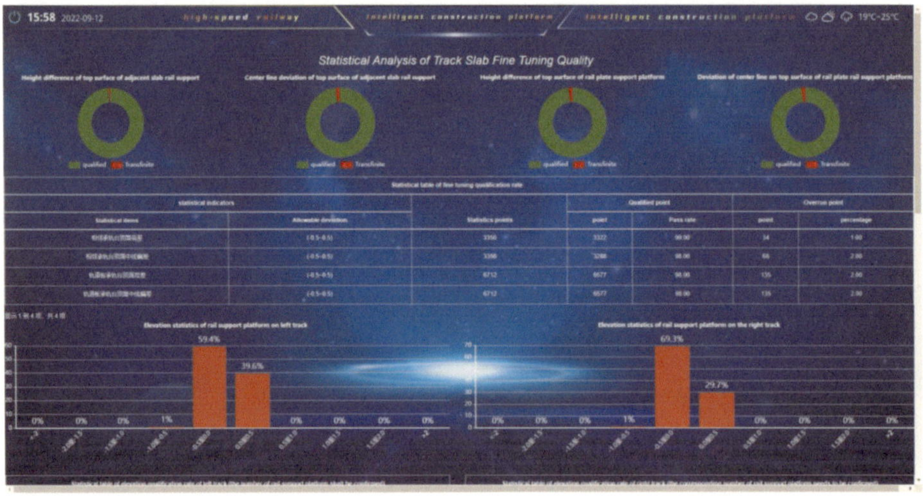

Figure 15.
Construction quality control.

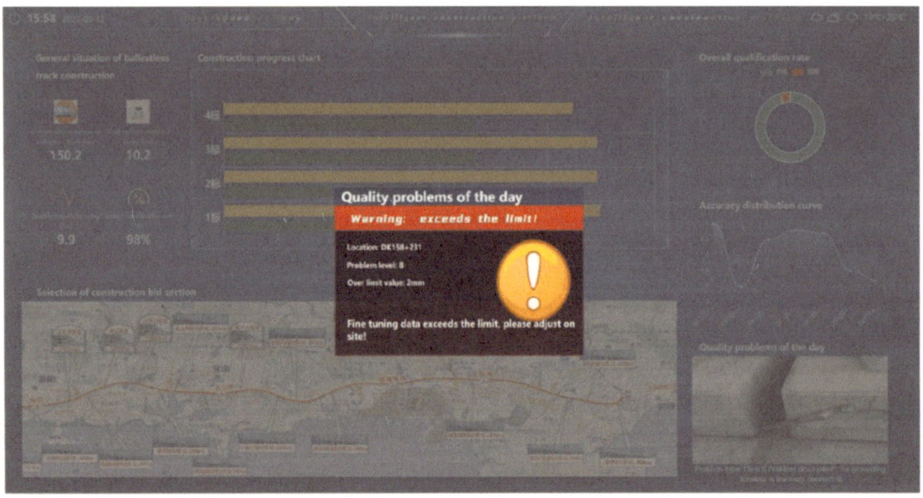

Figure 16.
Early warning and forecasting.

statistical statement is presented. Reports can be divided into daily reports, weekly reports and monthly reports according to different time segments. They can also be divided into base, track plate and other different procedures according to the statistical content of the report. Reports can also be produced for the precision statistics of all processes (**Figure 17**).

5. Conclusion

This chapter mainly describes the intelligent construction technology for ballastless track of high-speed railways from three aspects: construction dynamic adjustment technology, construction measurement, and control technology based on

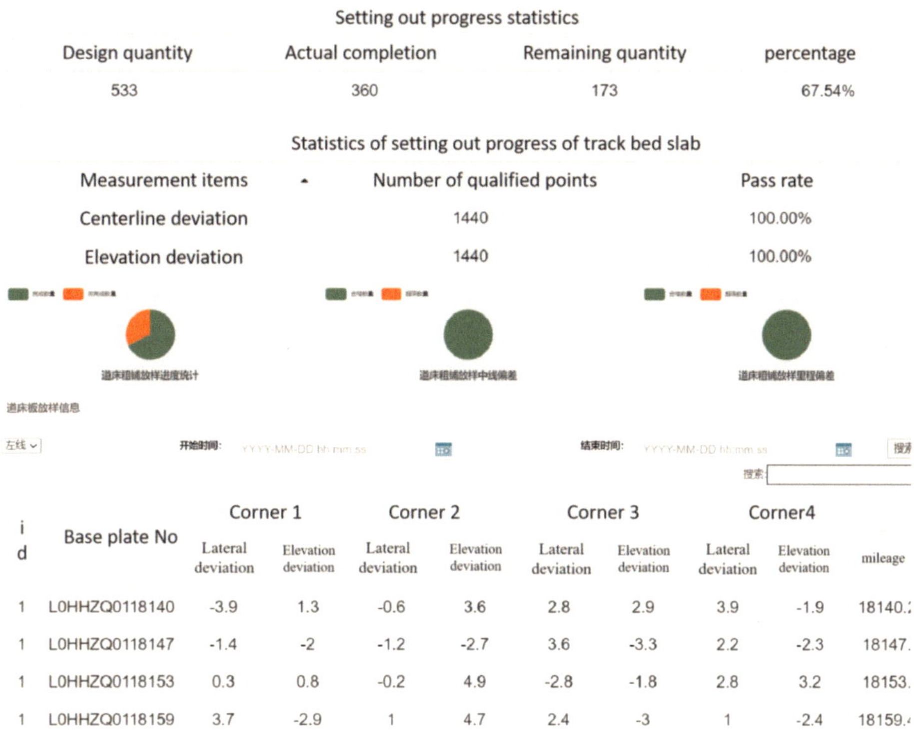

Figure 17.
Progress and accuracy report.

data driven and intelligent construction control technology. Through the application of the above technical measures, the traditional manual calculation, measurement, adjustment and control is replaced by a new intelligent mode. At present, after more than 10 years of research, this technology has advanced to form a complete technical system, which has been applied in more than 11 high-speed railway projects such as Zheng-Xu, Chang-Gan, Shang-He-Hang and Fu-Xia. The submillimeter precision control of ballastless track is realized, the efficiency and intelligence level of construction are improved, and significant social and economic benefits are achieved.

Author details

Qiuyi Li and Luyao Li*
Hubei Key Laboratory of Railway Track Safety Service, China Railway Siyuan Survey and Design Group Co. Ltd., Wuhan, China

*Address all correspondence to: 006149@crfsdi.com

References

[1] TB10082-2017.Code for Design of Railway Track

[2] Meng W, Weibin L, Yong Z. Technology of CRTS III ballastless track system. China Railway. 2017;**8**:11-15

[3] Hedao W. Integrated information management system for CRTS III slab ballastless track Constructionof high speed railway. Railway Engineering. 2019;**59**:107-110

[4] Wan M. Research on the railway track 3 D digital design system on BIM. Journal of Railway Engineering Society. 2021;**38**:90-96

[5] Shuangan X. Discussion on fine adjustment method of CRTS III ballastless track slab. High Speed Railway Technology. 2017;**8**:15-20

[6] Gui L, Yuelei H. Calculation model of track geometry based on three-dimensional coordinate measurement. Urban Mass Transit. 2017;**20**:15-18

[7] Shunhua C, Huijie W, Lihui R. ISO2631 comfort measuring instrument for rail vehicle based on android system. Railway Computer Application. 2020;**29**: 65-70

[8] Weicai W, Chenhui Y, Chengshan P, el. Control measurement and coordinate transfer in automatic monitoring of total station. Geospatial Information. 202;**19**: 66-70

[9] Bo W. Development of CRTS III slab ballastless track layout design system for double track railway. Geomatics & Spatial Information Technology. 2021; **44**:198-201

[10] Jinbo H. Research on measurement and control method for automatic data acquisition of measurement robot under android platform. Journal of Water Resources and Architectural Engineering. 2019;**17**:246-250

Chapter 3

Tunnels

Ebu Bekir Aygar

Abstract

Tunnelling is an indispensable part of infrastructure services. Some of the factors affecting the tunnel design can be listed as the characteristics of the soil or rock mass, tunnel excavation method, tunnel dimensions and the structures located on the tunnel route. Field studies (geological-geotechnical evaluations) during the design phase are one of the most important factors in tunnel design. All studies to be carried out after this stage are based on the determined geological model. The determination of the tunnel support system is made with detailed analytical and numerical solutions. In this study, empirical, analytical and numerical solutions, geotechnical measurements, swelling and squeezing phenomena in tunnels, tunnel support capacity equations, weak rock in tunnelling and tunnel support types are explained briefly.

Keywords: tunnel, support, rock mass classification systems, tunnel excavation, squeezing, swelling, TBM, NATM, geotechnical measurements

1. Introduction

Today, the number of tunnel projects has increased significantly due to the developing infrastructure projects. Tunnels are used in most large highways, railways, water, metro and mining projects. Depending on the increasing number of infrastructure projects, the total number being constructed is gradually increasing. As a consequence of this increase, tunnel design and construction works are also gradually increasing. Project design studies carried out after the determination of project routes are of great importance. It is extremely important to determine the rock and ground support classes, excavation methodology and support elements within the project design studies. Although there are different tunnel project approaches today, all designs are based on the principles of rock and soil mechanics; for this reason, the correct determination of soil and rock parameters is the main factor in determining tunnel design. In addition, the correct realisation of the geological model of the route through which the tunnel will pass is another factor that will directly affect the tunnel design. In addition to all of these, surface structures also affect the design. While deformation is not allowed in tunnels to be excavated under towns and cities, a certain amount of deformation is allowed in mountain tunnels. The design of the support system is affected by natural structures such as streams on the surface. This design effort can be divided into empirical, analytical and numerical methods for the project work to be done. Although these three methods are used separately most of the time, they should be evaluated together for a healthy tunnel project. During the tunnel excavations, tunnel projects should be compared with predicted and encountered geological conditions. If necessary, tunnel projects should be revised according to the encountered geological conditions. It should be well known that the predicted geological conditions and the actual geological conditions encountered will never be the same. For this reason, geotechnical measurements such as geological face maps, convergence measurements and deformation measurements should be made during the tunnel excavation in order to make the necessary revisions and updates. If necessary, the support system and rock mass parameters should be revised with a back-analysis. In addition, laboratory tests made from the boreholes taken during the excavation and rock mass parameters should be re-evaluated.

2. Rock mass classification system

The most important factor when designing a tunnel is to define the geological conditions that the tunnel will pass through. Based on this definition, rock mass classification systems are made by determining the strength parameters of the rock mass and the geological conditions. With these methods, the units that the tunnel will pass through are classified providing detailed information before the tunnel project commences. Rock mass classification systems have been developed in detail since the 1940s, and this process has been refined by different researchers until the present day. A main one of these methods, developed by Terzaghi [1] in 1946 made a classification for steel supports depending on the rock load. In later processes, Lauffer [2] developed a classification based on unsupported standup time, Rabcewicz [3–5], Rabcewicz and Golser [6] introduced the principles of the New Austrian Tunnelling Method, Deere et al. [7] in 1967, the direction of Rock Quality Designation (RQD), and Wickham et al. [8] developed the Rock Structure Rating system. The Rock Mass Rating (RMR) system, which is one of the most well-known systems today, was developed by Bieniawski [9–11] and was last modified in 1989. Barton et al. [12–14] improved the rock classification system known as the Q system. In 1995, Palmstrom [15] developed the Rock Mass Index (RMI) system. All of these methods emerged as empirical studies in the field.

While making rock mass classifications, uniaxial compressive strength of intact rock, condition of discontinuities (spacing, orientation, roughness, etc.), in situ stresses, groundwater condition, tunnel size and rock quality designation are all evaluated. The biggest problem when working with rock mass classification systems is that they are made according to interpretations that vary from person to person, since they are empirical. While one researcher defines discontinuities in a particular way, another researcher may make different interpretations under the same conditions. In addition, incomplete data collection in studies is an important factor. Since the boreholes, from which the main input parameters are obtained, are often not done properly, problems are created in terms of the accuracy of the data. When drilling, the length of the run, sampling and the application of the correct drilling technique will directly affect the results. For this reason, although all rock mass classification systems contain very important and valuable information when describing the rock mass, the results should always be viewed with suspicion. No rock mass classification system specifies a definite support system for the design, it just gives a range. While designing, these support systems should be considered as an initial step in analytical and numerical solutions, and never be considered absolute truth.

2.1 Rock load theory

Rock Load Theory, developed by Terzaghi [1] for sizing steel rib systems, started to be applied in railway tunnels opened in the United States in 1946. Terzaghi considered the pressure exerted by the loose soil (Hp) on the steel rib on the tunnel (**Figure 1**). For this purpose, Terzaghi [1] divided the rock mass into nine main categories. These categories range from solid-intact rock to swelling ground. He calculated the pressure coming to the steel supports depending on Hp Eq. (1). In **Table 1**, rock mass classes and rock load factors are given.

$$p = Hp * \gamma * H \tag{1}$$

2.2 Stand-up time

Lauffer [2] defined the relationship of stand-up time to rock mass quality depending on the tunnel diameter in unsupported condition. The rock mass is classified from very good rock (A) to very weak rock (G). Here, the unsupported period is 100 years in the A rock class, while it ranges from 1 minute to 10 minutes in the G rock grade (**Figure 2**). This classification is valid for a span of 5 m. Stand-up time indicates the time during which the tunnel remains stable in unsupported conditions. Lauffer applied the stand-up time approach to the Bieniawski RMR rock classification system.

2.3 New Austrian tunnelling method (NATM)

The principles of the New Austrian Tunnelling Method (NATM) were introduced by Rabcewicz [3–5] in the 1960s. The NATM method is based on the principle of increasing the carrying capacity of a mountain with a flexible outer belt by allowing deformations around the tunnel. It is divided into rock classes, from very solid rock to swelling soils. In subsequent years very serious criticisms were made of NATM and this process continues today.

According to NATM principles, the rock is divided into three main groups stable (A), friable (B) and stress failure, squeezing (C). These rock classes are again subdivided into subclasses (**Table 2**).

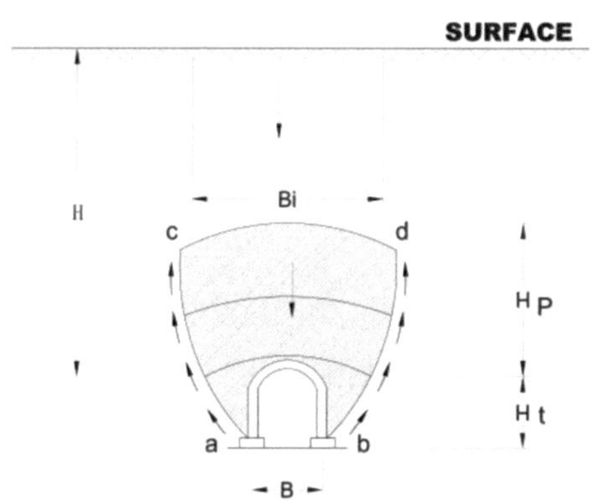

Figure 1.
Rock load concept [1].

Rock class	Rock condition	Rock load factor (Hp)	Remarks
I	Hard and intact	Zero	Light lining is required only if spalling or popping occurs
II	Hard stratified or schistosite	0–0.5B	Light support is mainly for protection against spalling. The load may change erratically from point to point
III	Massive moderately jointed	0–0.25B	No side pressure
IV	Moderately blocky and seamy	0.25B–0.35(B + Ht)	No side pressure
V	Very blocky and seamy	(0.35–1.10)(B + Ht)	Little or no side pressure
VI	Completely crushed but chemically intact	1.10(B + Ht)	Considerable side pressure. Softening effects of seepage toward the bottom of the tunnel require either continuous support for the lower ends of ribs or circular ribs.
VII	Squeezing rock moderate depth	(1.10–2.10)(B + Ht)	Heavy side pressure invert struts are required. Circular ribs are recommended.
VIII	Squeezing rock moderate depth	(2.10–4.50)(B + Ht)	As above
IX	Swelling rocks	Up to 250 ft. (80 m) irrespective of the value of (B + Ht)	Circular ribs are required. In extreme cases, the use of yielding support is recommended.

Table 1.
Rock load in tunnels within various rock classes [1].

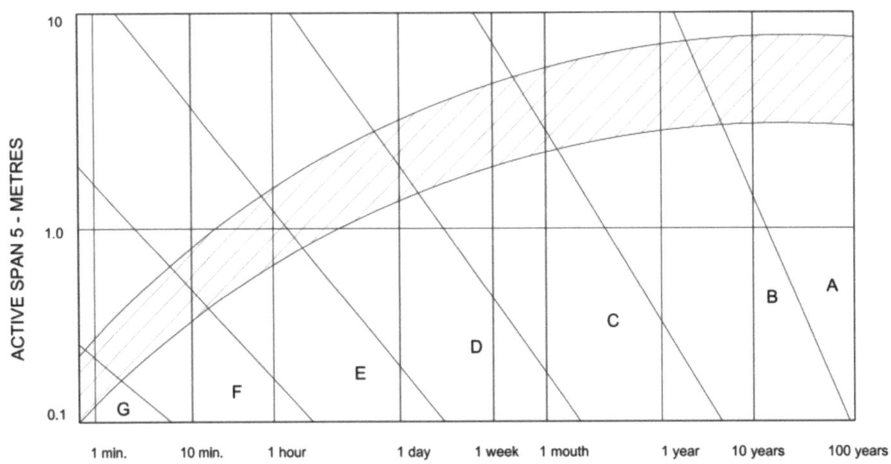

Figure 2.
Relationship between active span and stand-up time and rock mass classes [2].

2.4 Rock quality designation

Deere et al. [7] made a classification according to the core samples obtained from the borehole. Here rock quality is defined by dividing the core samples larger than 10 cm by

Main rock class	Rock classes				
A (stable to overbreak)	A1 (stable)	A2 (friable)			
B (friable)	B1 (friable)	B2 (very friable)	B (ravelling)		
C	C1(rock bursting)	C2 (squeezing)	C3 (heavily squeezing)	C4 (flowing)	C5 (swelling)

Table 2.
NATM rock classes.

RQD	Rock mass quality
<25	Very poor
25–50	Poor
50–75	Fair
75–90	Good
99–100	Excellent

Table 3.
Rock mass quality classification classes [7].

the run size. The RQD value remained as a preliminary approach but has continued to be used as one of the most important data inputs to the other developed rock classification systems. The RQD value is shown in Eq. (2). In the classification made according to the RQD value, the rock mass is divided into five main categories (**Table 3**).

$$RQD = (Total\ length\ core\ pieces > 10\ cm/total\ length\ of\ the\ core) * 100\% \quad (2)$$

2.5 Rock structure rating (RSR)

Wickham et al. [8] developed the RSR system based on their work in small-diameter tunnels. They established a relationship between the parameters determined according to the geological conditions and the construction parameters. RSR value Eq. (3) is also given:

$$\mathrm{RSR} = \mathrm{A} + \mathrm{B} + \mathrm{C} \quad (3)$$

Parameter (A) depends on the rock hardness, geological structure and rock type origin.

Parameter (B) depends on the discontinuity pattern which derives from joint spacing, joint orientation and direction of the tunnel drive.

Parameter (C) depends on the joint condition and water inflow.

2.6 Rock mass rating system (RMR)

Rock Mass Rating System is a method developed by Bieniawski [9–11] between 1973 and 1989 on the basis of mine galleries and road tunnels. Bieniawski classified the rock mass according to six parameters.

a. Uniaxial compressive strength (UCS)

b. Rock quality designation (RQD)

c. Spacing of discontinuities

d. Condition of discontinuities

e. Groundwater conditions

f. Orientation of discontinuities

The RMR system was developed in line with the data obtained from horseshoe tunnels with a diameter of 5–12 m. In this method, there is a rating value for each parameter. The support system is determined according to the RMR score calculated according to these determined values (**Table 4**).

2.7 Tunnelling quality index (Q)

Barton et al. [12–14] proposed a Tunnelling Quality Index (Q) system for underground structures. The Q method was developed in Scandinavia in 212 tunnels

Rock mass class	Excavation	Rock bolt (20 mm diameter, fully grouted)	Shotcrete	Steel sets
I—Very good rock RMR: 81–100	Full face, 3 m advance	Generally, no support is required except spot bolting		
II—Good rock RMR: 61–80	Full face, 1–1.5 m advance. Complete support 20 m from face	Locally, bolts in crown 3 m long, spaced 2.5 m with occasional wire mesh	50 mm in crown where required	None
III—Fair rock RMR: 41–60	Top heading and bench. 1.5–3 m advance in top heading. Commence support after each blast. Complete support 10 m from face	Systematic bolts 4 m long, spaced 1.5–2 m in crown and walls with wire mesh in crown.	50–100 mm in crown and 30 mm in sides.	None
IV—Poor rock RMR: 21–40	Top heading and bench 1–1.5 m advance in top heading. Instal support concurrently with excavation, 10 m from face	Systematic bolts 4–5 m long, spaced 1–1.5 m in crown and walls with wire mesh	100–150 mm in crown and 100 mm in sides	Light to medium ribs spaced 1.5 m where required
V—Very poor rock RMR: <20	Multiple drifts 0.5–1.5 m advance in the top heading. Instal support concurrently with excavation. Shotcrete as soon as possible after blasting	Systematic bolts 5–6 m long, spaced 1–1.5 m in the crown and the walls with wire mesh. Bolt invert	150–200 mm in the crown, 150 mm in sides and 50 mm on the face	Medium to heavy ribs spaced 0.75 m with steel lagging and forepoling if required. Close the invert

Table 4.
Guidelines for excavation and support of 10 m span rock tunnels in accordance with the RMR system [11].

excavated in solid rock. The Q value is logarithmic, ranging from 0.001 to 100. According to the Q value, the rock mass is divided into nine main categories between exceptionally good and exceptionally poor. The Q value depends on a total of six parameters. These are;

RQD: Rock quality designation.
Jn: Joint set number.
Jr: Joint roughness number.
Ja: Joint alteration number.
Jw: Joint water reduction factor.
SRF: Stress reduction factor.

According to these parameters, the Q value is calculated with Eq. (4).

$$Q = (RQD/Jn) * (Jr/Ja)/(Jw/SRF) \tag{4}$$

The (RQD)/Jn) value used in the equation corresponds to the rock mass geometry and the block/wedge size. The Jr/Ja value refers to the inter-block shear strength condition and the Jw/SRF value to the active stress incorporation water pressures and flows condition.

In addition to the Q value, the equivalent dimension (De) value is needed for the tunnel support details. This value is determined by dividing the tunnel diameter by the excavation support ratio. The details of the support system are determined by crossing these two values in the chart in **Figure 3** and **Table 5**.

2.8 Section conclusion

The main purpose of all rock mass classification systems developed empirically is to describe the rock mass. There are many parameters as inputs in all studies. While making the evaluations, the comments that each engineer will make in the field may be different.

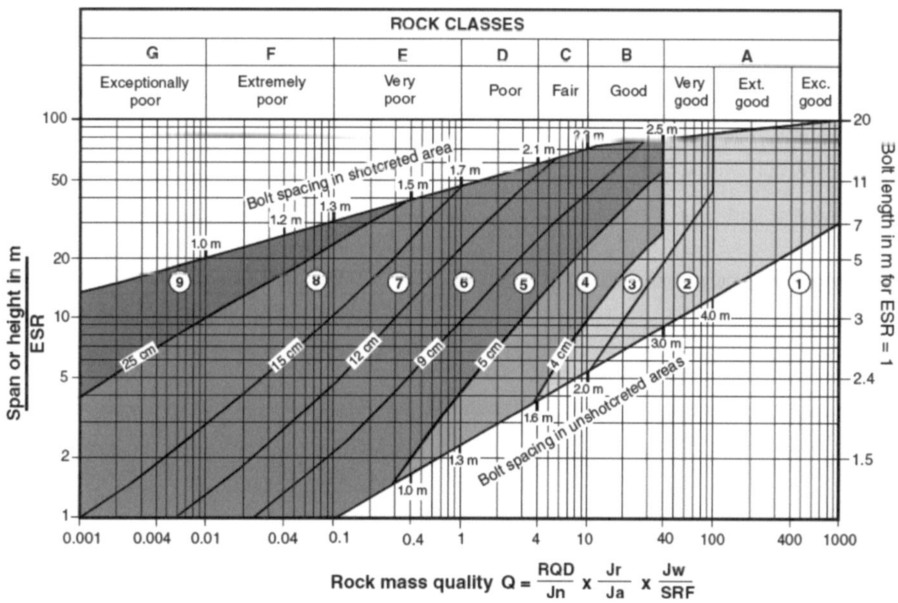

Figure 3.
Q excavation and support system chart [14].

1. Unsupported	6) Fibre-reinforced shotcrete and bolting, 9–12 cm
2. Spot bolting	7. Fibre-reinforced shotcrete and bolting, 12–15 cm
3. Systematic bolting	8. Fibre-reinforced shotcrete, >15 cm Reinforced ribs of shotcrete and bolting
4. Systematic bolting (and unreinforced shotcrete, 4–10 cm.	9. Cast concrete lining
5. Fibre-reinforced shotcrete and bolting, 5–9 cm	

Table 5.
Q support system details [14].

Since these evaluations are empirical, they directly affect the determination of the rock class. Another important factor in rock classification is the evaluation of boreholes, which is one of the basic input parameters and perhaps the most important one. Consideration should be given to the sampling and run lengths required while drilling. Errors made during these processes will directly affect the design. Since the wrong calculation of the RQD value will be a direct input parameter in the rock mass parameters, borehole quality is extremely important. The number and location of drillings is another important factor. The insufficient number of boreholes will significantly increase the uncertainty in determining the tunnel rock classes. The uniaxial compression test or point load test, also appears as direct input parameters. During laboratory experiments, sampling, transportation or errors seriously affect the rock mass classification systems.

It is unwise to accept the results of the rock classification system as invariably correct, even when there are great uncertainties in determining the rock parameters.

Each of the researchers, on the other hand, aimed to make rock mass classifications according to the support systems that were successful in the field. Each method has its own limitations. Terzaghi [1] only offered suggestions for steel supports for tunnels excavated in solid rock. Bieniawski determined the intervals as 20 units in the determination of the RMR system. This situation often brings with it uncertainties. When the RMR value is calculated as 41, the rock class is medium rock, and when it is 39, it is weak rock. According to this situation, one should be very careful in recommending support. As stated, in cases where the rock mass parameters are purely empirical and relative, specifying and applying the support elements directly according to the determined RMR values may lead to wrong results.

Since similar limitations are valid for all rock mass classification systems, taking the rock mass classification systems as invariable correct will give extremely inaccurate results. Rock mass classification systems are, however, extremely important for the preliminary evaluation and preliminary design of the unit that the tunnel will pass through. In light of these obtained data, it is necessary to dimension the support systems with analytical and numerical methods.

Because the support systems determined according to all rock mass classification systems are given within a range a clear and precise sizing cannot be given. For this reason, these approaches are only a guide in terms of tunnel design.

3. Closed-form solutions-rock support interaction

It is often difficult to determine how the rock mass is deformed during tunnel excavation to determine the interaction of rock support systems because there are many uncertainties in defining the rock. Most of the time, such approaches give results on a homogeneous medium, but in a complex situation, they should be evaluated with numerical analysis.

In addition to empirical methods, closed-form solutions are used in the determination of tunnel support systems. Closed-form solutions calculate the plastic behaviour of the rock mass and determine the deformations that occur according to the support pressures. The interaction of the ground and support systems is revealed by drawing the ground interaction curve and support interaction curves in the tunnel.

The assumptions made in the analysis are;

- The tunnel is circular
- In situ stresses are considered hydrostatic
- The rock mass is homogeneous and isotropic.

4. Analysis of tunnel behaviour during construction

Figure 4 shows the deformation vectors and the plastic zone in the tunnel propagation direction in weak rocks. **Figure 5** gives a summary of the deformations that occur in the tunnel. The elastic deformations that occur in the tunnel start at a distance of two tunnel diameters in front of the tunnel and reach the maximum level after two tunnel diameters behind the tunnel face [16]. The maximum displacement occurring in the tunnel face is one-third of the total displacements (**Figure 5**).

The unsupported deformation situation for the rock-soil interaction is given in **Figure 6**. If the uniaxial compressive strength of the rock mass is $\sigma_{cm} > 2p_o$ ($p_i = 0$), the displacements are elastic and continue linearly. If a failure occurs, the displacements are plastic and are curved in **Figure 6**.

Ground reaction curve or characteristic line depends on the convergence occurring in the tunnel with the internal support pressure.

Here, the tunnel radius is taken as r_0, the hydrostatic pressure p_0 and the support pressure as p_i (**Figure 7**).

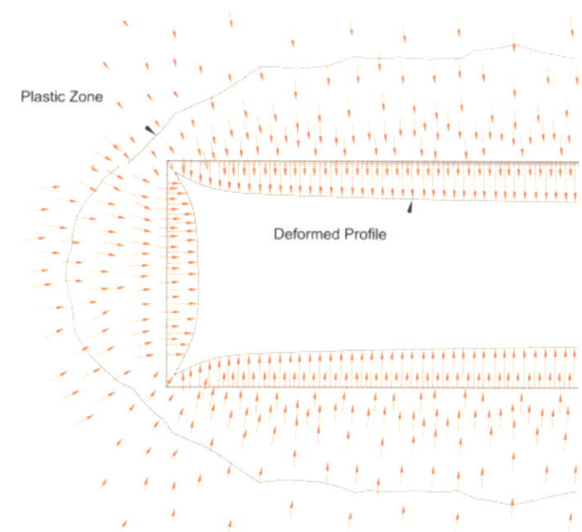

Figure 4.
Vertical section through a three-dimensional finite element model of the failure and deformation of the rock mass surrounding the face of an advancing circular tunnel [16].

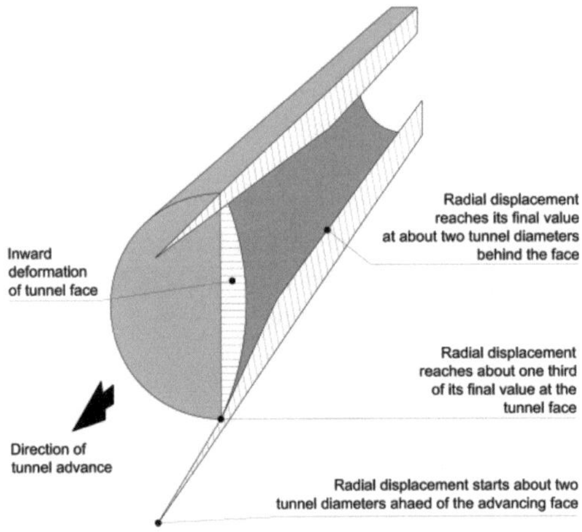

Figure 5.
Radial displacements around the tunnel [17].

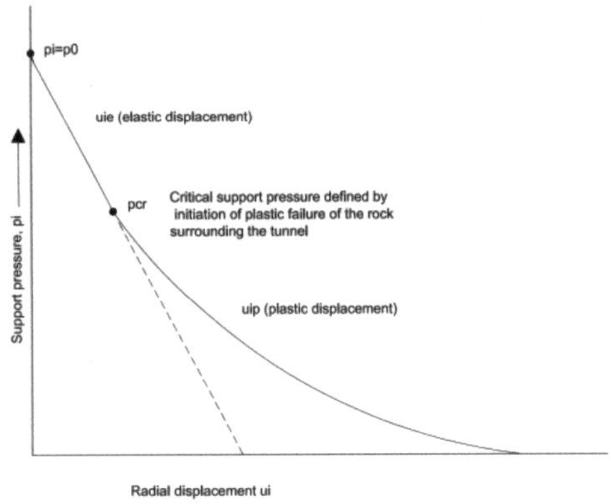

Figure 6.
Graphical representation of relationships between support pressure and radial displacement of tunnel walls [16].

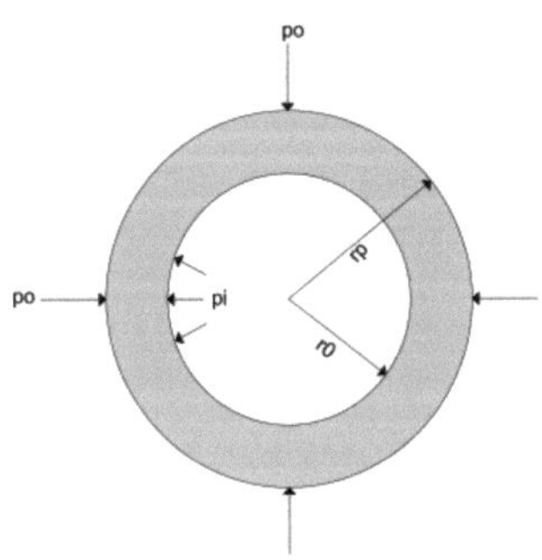

Figure 7.
Plastic zone surrounding a circular tunnel.

If the support pressure p_i is less than the critical pressure p_{cr}, the rock mass will fail. If the pi pressure is greater than p_{cr}, no failure occurs around the tunnel and the rock mass behaves elastically. The critical support pressure is given in Eq. (5):

$$Pcr = \frac{2p0 - \sigma cm}{1 + k} \tag{5}$$

Elastic displacement is given in Eq. (6):

$$uie = \frac{r0(1 + \vartheta)(p0 - pi)}{Em} \tag{6}$$

Here E_m is the rock mass deformation modulus and ν is Poisson's ratio.

If the support pressure (p_i) is less than the critical pressure p_{cr}, failure occurs around the tunnel and a plastic zone is formed. In this case, plastic deformation and plastic zone radius are defined by Eq. (7). The resulting plastic deformation is given in Eq. (8):

$$rp = ro\left[\left(\frac{2(p0(k-1) + \sigma cm)}{(1+k)((k-1)pi + \sigma cm)}\right]^{\frac{1}{k-1}} \tag{7}$$

$$uip = \left(\frac{ro(1+\vartheta)}{Em}\right)\left[2(1-\vartheta)(p0 - pcr)\left(\frac{rp}{r0}\right)^2 - (1-2\vartheta)(p0 - pi)\right] \tag{8}$$

The graph of radial displacement with Pi to the support pressure drawn with the help of the given equations is given in **Figure 8**. Here,

if $p_i = p_o$, no deformation occurs,

$p_i > p_c$ elastic deformation occurs,

$p_i < p_{cr}$ise plastic deformation occurs.

After the support installation, the deformations continue elastically. Maximum elastic deformation is defined as u_{sm} and maximum support pressure is defined as p_{sm}.

Support interaction analysis depends on three main parameters. These are

- Deformations that occur before support are made
- Support stiffness
- Support capacity

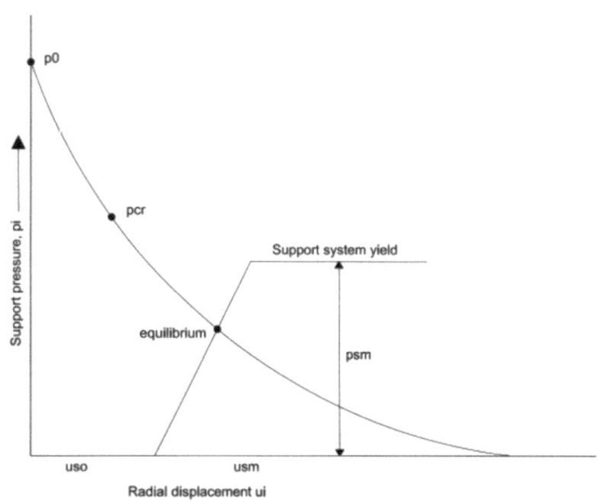

Figure 8.
Response of support system to tunnel wall displacement resulting in the establishment of equilibrium [16].

Steel set	
σys is the yield strength of the steel (MPa) Es is the Young's modulus of the steel (MPa) As is the cross-sectional area of the section (m^2) sl is the set spacing along the tunnel axis (m) ro is the radius of the tunnel (m) $Pssmax$ is the maximum support pressure Kss is the stiffness	$Pssmax = \frac{As * \sigma ys}{sl * lro}$ (9) $Kssmax = \frac{Es * As}{sl * lro^2}$ (10)
Rock bolts	
d_b is the rockbolt or cable diameter (m) l is the free length of the bolt or cable (m) E_s is the Young's modulus of the bolt or cable (MPa) sc is the circumferential bolt spacing (m) sl is the longitudinal bolt spacing (m) Tbf is the ultimate bolt or cable load $Psbmax$ is the maximum support pressure Kss is the stiffness	$Psbmax = \frac{Tbf}{sl * sc}$ (11) $Ksb = Es * \pi * \frac{db^2}{4lslsc}$ (12)
Concrete or shotcrete	
σ_{cc} is the uniaxial compressive of the concrete or shotcrete (MPa) E_c is the Young's modulus of the concrete or shotcrete (MPa) υ is the Poisson ratio of the concrete or shotcrete tc is the thickness of the lining (m) ro is the radius of the tunnel (m) $Pscmax$ is the maximum support pressure Kss is the stiffness	$Pscmax = \frac{\sigma cc}{s} * \left[1 - \frac{(ro-tc)^2}{ro^2}\right]$ (13) $Ksc = \left(Ec * \frac{ro^2-(r0-tc)^2}{2*\left(1-\vartheta^2\right)*(r0-tc)*ro^2}\right)$ (14)

Table 6.
Support capacity equations [18, 19].

In the support reaction curve, elastic deformations occur in the tunnel after the tunnel excavation. The support reaction curve has reached equilibrium if the ground reaction curvature crosses the curve before deformations in the rock mass increase substantially. However, if the deformations that occur develop very quickly and the reinforcements are insufficient, failure occurs and balance cannot be achieved.

Equations for the stiffness and capacity of supports have been published by Hoek and Brown [18] and Brady and Brown [19]. The equations are given in **Table 6**.

5. Numerical methods

In tunnel design, empirical solutions and analytical solutions do not always give exact results. Numerical analysis and solutions are required for complex geological structures and complex underground structures. The Finite Element Method, Finite Difference Method and Discrete Element Methods are used as common numerical analysis methods. Numerical analysis methods are used both in 2D and 3D. Numerical programs have been increasingly used in tunnel design in recent years. In numerical programs, in-situ conditions, geological units and boundary conditions can be reflected exactly. In addition, while creating the models, the tunnel geometry, excavation stages and support details can be entered into the model exactly.

In numerical methods, the stresses and strains occurring in the ground can be calculated and the cross-sectional effects on the support systems can also be determined. Thus, much more accurate results can be obtained compared to empirical and analytical solutions for design.

6. Weak rock tunnelling-soil tunnelling

In the evaluations of rock mass classification systems, support details for weak rocks are usually detailed as cast lining or ring closure of the inverted section with rigid lining. However, in tunnels defined as weak rock or soil, dimensioning support details is often not possible according to rock mass classification systems. In these sections, the dimensioning of support systems can be detailed as a result of analytical solutions and numerical solutions. According to ISRM classification systems, sections with a value of less than 1 MPa are considered soil. In addition, special support system solutions emerge in weak rocks, especially in units with weak strength such as clay, claystone and schist. The squeezing mechanism that develops under high overburden affects the long-term performance of the support systems. For this reason, both squeezing and swelling potential are the most important factors in tunnels excavated in weak rocks and soils. Two different tunnel support system approaches are used for tunnels excavated in weak rocks and soils [19–22]. These are called the passive method and the active method. While the passive method is based on the principle of increasing the support pressure by allowing deformations in the tunnel, in the active method, the support systems are dimensioned without allowing such deformations. However, for the passive approach in squeezing ground, the long-term deformations that may occur in real projects were often unsuccessful and an active approach was adopted in the revisions made. For this reason, it is very important to determine and evaluate the squeezing mechanism, which is one of the most important factors in weak ground.

6.1 Squeezing in tunnels

In the evaluations for the squeezing mechanism, Jethwa et al. [23], Sakurai [24], Singh et al. [25], Goel et al. [26], Hoek and Marinos [27], Aydan et al. [28] approaches are quite common. In these approaches, the uniaxial compressive strength of the rock mass, unit weight and overburden height appear as the main factors.

Singh et al. [25] defined according to the Q value. If the H value determined according to Eq. (15) is greater than $350 * Q^{1/3}$, squeezing is expected, while if the H value is less than $350 * Q^{1/3}$, squeezing is not expected (**Figure 9**):

$$H = 350 * Q^{1/3} \tag{15}$$

Goel et al. [26], on the other hand, defined the squeezing state similarly according to the Q value. He stated the calculated value of the Q value according to the stress-free conditions as N (rock mass number). If the H value (Eq. (16)) calculated according to this value is greater than the thickness of the overburden height, squeezing will occur, and if it is small, there will be no squeezing. In **Figure 10**, Goel et al. [26] the squeezing condition is shown according to the N value:

$$H = \left(275N^{0.33}\right)B^{-1} \text{ (m)} \tag{16}$$

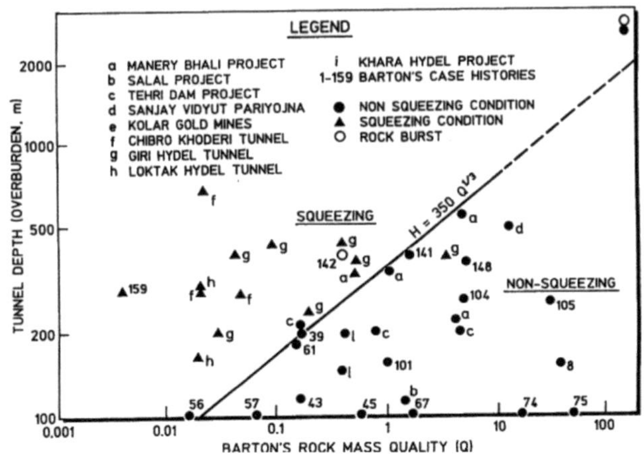

Figure 9.
Squeezing conditions regarding Q and H values [25].

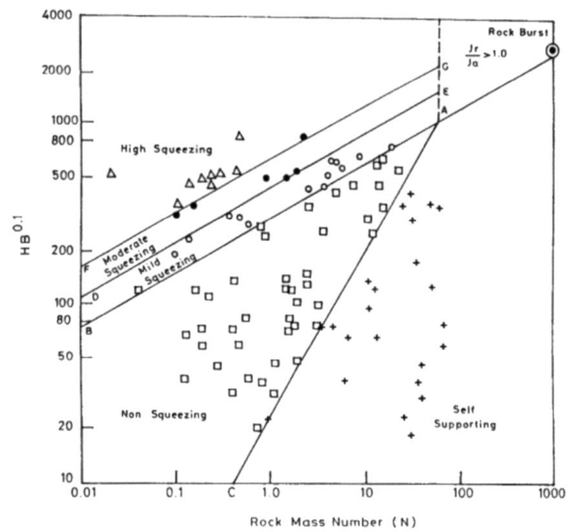

Figure 10.
Squeezing conditions according to Goel et al. [26].

Sakurai [24] classified the compression mechanism according to the strain value calculated based on the uniaxial compressive strength of the rock mass. He proposed Eq. (17) to determine the strain value:

$$\varepsilon pc = 1.073 \sigma cm^{-0.318} \tag{17}$$

The relationship between the uniaxial compressive strength of the rock mass depending on the strain value is given in **Figure 11**.

Jethwa et al. [23] classified the compression according to the N coefficient. Nc coefficient (competency factor) is given in Eq. (18):

$$Nc = \sigma cm / Po \tag{18}$$

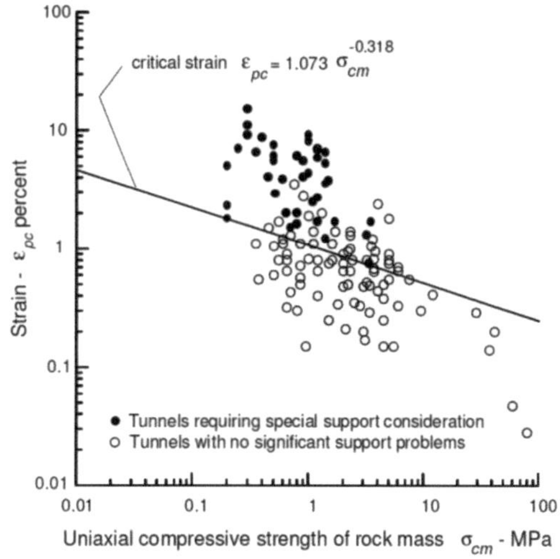

Figure 11.
Relationship between uniaxial compressive strength of strain and rock mass [24].

P_o is the in situ stress and σ_{cm} is the uniaxial compressive strength of the rock mass. The compression status according to the calculated Nc value is given in **Table 7**.

Hoek and Marinos [27] defined the squeezing according to the strain value depending on the relationship between the uniaxial compressive strength of the rock and the in-situ stress. The strain value is given in Eq. (19):

$$\varepsilon = 0.2 * (\sigma cm/Po)^{-2} \tag{19}$$

The squeezing mechanism is given in **Figure 12** depending on the calculated strain value and the ratio of the compressive strength of the rock mass to the in situ stress.

While evaluating the squeezing mechanism, the critical factors are the height of the overburden, the unit weight and the uniaxial compressive strength of the rock mass.

6.2 Swelling in tunnels

Swelling soils can cause failures in the support systems due to unexpected loads. Most of the time, it can cause failures and swell even in the sections where inner lining and invert concrete are completed in tunnels. For this reason, the swelling potential of the ground should be evaluated during tunnel design. Support systems should be designed to carry these new loads that may occur. Einstein and Bischef [29] suggested a design procedure for swelling soils. First of all, the process can be listed as

Degree of squeezing	Ranges (N)
High	<0.4
Moderate	0.4–0.8
Slightly	0.8–2
Non-squeezing	>2

Table 7.
Squeezing degree according to Jethwa et al. [23].

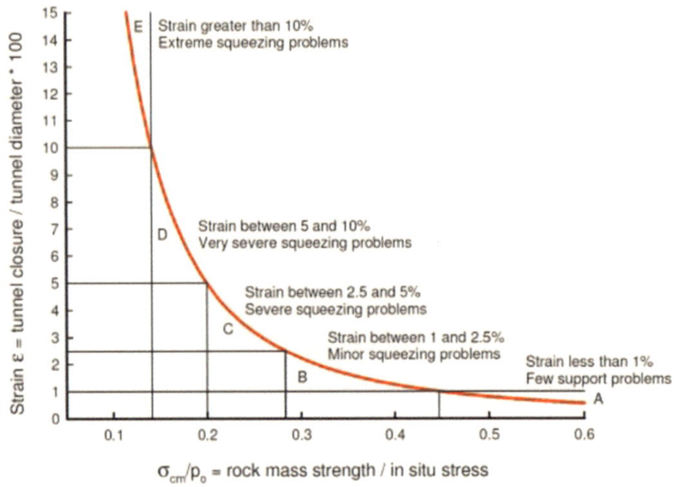

	Strain ε %	Geotechnical issues	Support types
A	Less than 1	Few stability problems and very simple tunnel support design methods can be used. Tunnel support recommendations based upon rock mass classifications provide an adequate basis for design.	Very simple tunnelling conditions, with rockbolts and shotcrete typically used for support.
B	1 to 2.5	Convergence confinement methods are used to predict the formation of a 'plastic' zone in the rock mass surrounding a tunnel and of the interaction between the progressive development of this zone and different types of support.	Minor squeezing problems which are generally dealt with by rockbolts and shotcrete; sometimes with light steel sets or lattice girders are added for additional security.
C	2.5 to 5	Two-dimensional finite element analysis, incorporating support elements and excavation sequence, are normally used for this type of problem. Face stability is generally not a major problem.	Severe squeezing problems requiring rapid installation of support and careful control of construction quality. Heavy steel sets embedded in shotcrete are generally required.
D	5 to 10	The design of the tunnel is dominated by face stability issues and, while two-dimensional finite analyses are generally carried out, some estimates of the effects of forepoling and face reinforcement are required.	Very severe squeezing and face stability problems. Forepoling and face reinforcement with steel sets embedded in shotcrete are usually necessary.
E	More than 10	Severe face instability as well as squeezing of the tunnel make this an extremely difficult three-dimensional problem for which no effective design methods are currently available. Most solutions are based on experience.	Extreme squeezing problems. Forepoling and face reinforcement are usually applied and yielding support may be required in extreme cases.

Figure 12.
Approximate relationship between strain and the degree of difficulty associated with tunnelling through rock [27].

determining the primitive stress state of the current situation, determining the swelling soils around the tunnel and performing the swelling tests (odometer). In addition, they suggested closing the ring with invert, draining the water in the tunnel and closing the surface with steel-wire shotcrete (SFRS) after excavation.

Komornik and David [30, 31] proposed Eq. (20) for the analytical determination of the pressure that will occur due to the swelling:

$$\log ps = 2.132 + 0.0208\omega L + 0.000665\gamma d - 0.0269\omega n \tag{20}$$

where.

ps = selling pressure (kg/cm^2) at zero swelling strain.

ωL=liquid limit (%).

γd = natural dry density (kg/m^3) and.
ωn=natural moisture content (%).

7. Tunnel support details for classical tunnelling

Support systems in tunnels are generally divided into two main parts.

a. Outer lining

- Shotcrete
- Steel rib
- Bolt
- Forepoling/Umbrella

b. Inner lining

The outer lining used in the tunnel is made to ensure the stability of the tunnel after the tunnel excavation.

Shotcrete: Shotcrete is used in tunnels as a basic carrier element. Shotcrete applied immediately after the excavation not only prevents the ground from loosening by preventing atmospheric effects but also has the feature of carrying the loads coming from the ground. Shotcrete application is divided into dry and wet in tunnels.

Steel rib: With the increase in the thickness of shotcrete in tunnels, it acts as reinforcement in the concrete and can resist the first deformations that will occur after excavation in the tunnel. Steel rib can be divided into I type, H type, lattice girder and TH type (sliding rib).

Rock bolt: The rock bolt is an integral part of the external support as a support system. It is a support system that is necessary both to prevent block slipping due to discontinuities in rocks and to connect the plastic zone to the elastic zone in weak rocks. Rock bolts are used passively in tunnels. It is divided into SN, PG, swellex and

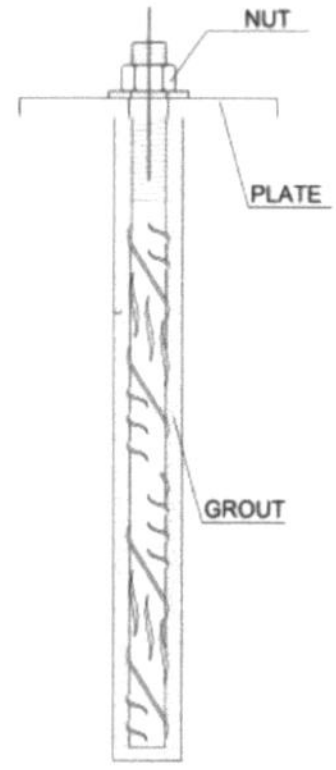

Figure 13.
SN-type rock bolt.

self-drilling bolt (IBO). Rock bolt diameters are generally used between 28 mm and 51 mm depending on the soil type.

In SN-type bolt applications, firstly the hole is drilled, then the borehole is filled with injection and finally, the bolt is placed into the hole (**Figure 13**).

In PG-type bolts, after the hole is drilled, the PG bolt is driven into the hole. In the last stage, the grout is pumped into the hole using the grout pipe (**Figure 14**).

Swellex-type bolts are used in tunnels opened in rocks with water ingress. After drilling the hole, the swellex bolt is placed. Afterwards, high pressure is applied from the bolt mouth, allowing the bolt to swell and hold onto the rock (**Figure 15**).

IBO (injection boring outside) is known as self-drilling bolts. In the application, the hole is drilled with the help of the bit attached to the bolt and the bolt remains in the borehole. In long bolts, bolts connected with coupling are added every 3 or 4 m and placed in the hole together with the drill (**Figure 16**). The injection process is completed by filling the entire drilling through the IBO bolts.

Forepoling/umbrella: Forepolings are not basically considered as a carrier element. The main purpose of the forepoling is to provide the stability of the ceiling in order to prevent slips that may occur during the excavation (**Figure 17**). Forepoling diameters and types vary depending on the ground.

Inner Lining: The inner lining concrete is made after the completion of the outer supports of the tunnel in order to give the tunnel its final form and to provide an

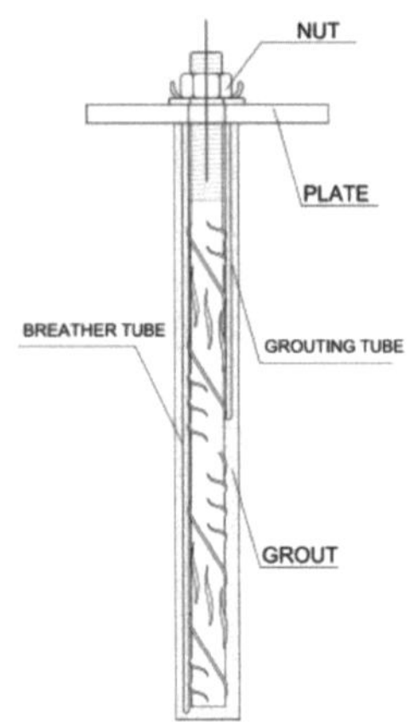

Figure 14.
PG type bolts.

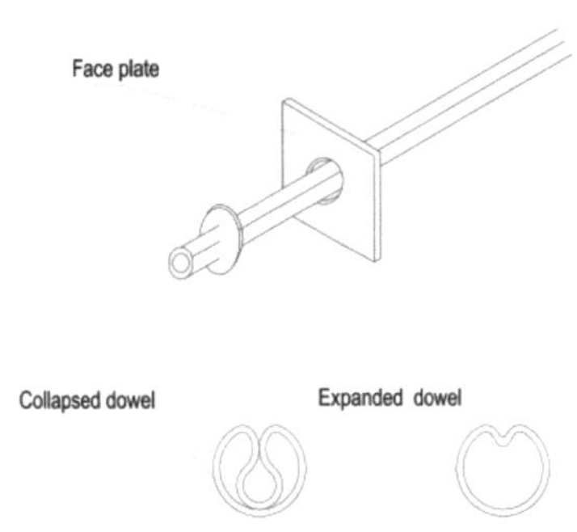

Figure 15.
Swellexs type rock bolt.

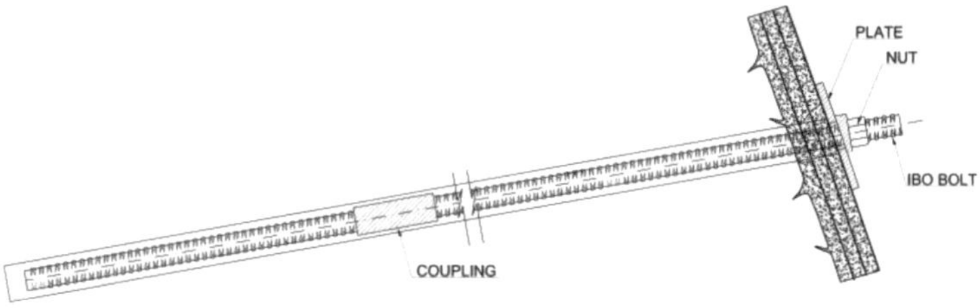

Figure 16.
IBO type bolts.

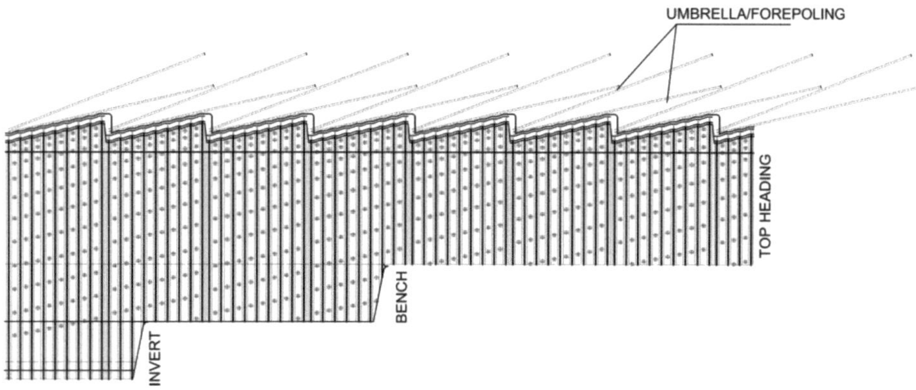

Figure 17.
Umbrella type forepoles.

architectural view. In general, inner-lining concrete is not considered as a load-bearing element. However, in weak or very weak rock conditions, it has recently been designed as a carrier element. The inner lining concrete is a necessary structure for the placement of drainage pipes required for tunnel drainage, as well as for the placement of ventilation fans and electro-mechanical devices to be used in the tunnel (**Figure 18**).

8. Tunnel excavation systems

Circular tunnel cross-section is always the most accurate method for tunnel stability and stress distribution (**Figure 19**). However, in most cases, this is not possible. Horseshoe section is often applied in mine galleries and water tunnels (**Figure 20**).

As a tunnel excavation method, it is preferred that it is completed as a full section, and the stress distributions are created once, in terms of tunnel stability. However, in large-diameter tunnels, the excavation process is carried out gradually. While tunnel excavation is divided into top heading and bench in solid rock, invert excavation is also carried out in weak rocks (**Figure 21**).

In solid rocks, the distance between the top heading and the bench can be longer due to the low deformations. This distance can be up to 70–100 m in solid rocks. On weak grounds, the top heading, bench and invert distance should be minimum to complete the ring. The distance between the top heading and the bench is between 15 and 25 m in weak ground conditions. Temporary invert is performed to prevent

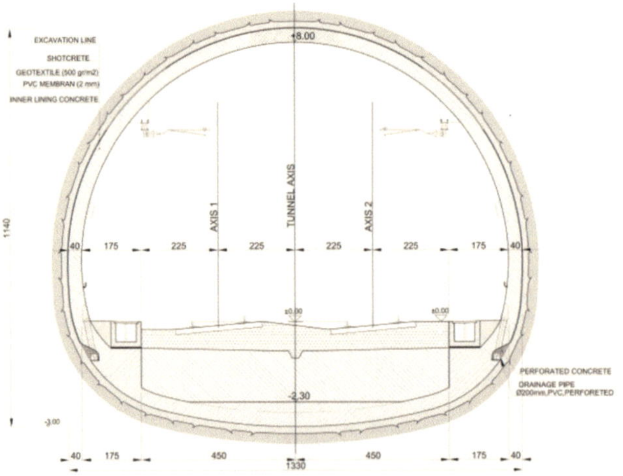

(a)

(b)

Figure 18.
Tunnel inner lining section (a) tunnel section (b) application in site.

deformations that may occur in the top heading in very weak soils. Thus, deformations are limited by closing the ring in the top heading.

In addition, the ADCEO-RS (The Analysis of Controlled Deformation in Rocks and Soils) method also suggests a near-full cross-section excavation in both rocks and weak soils [32, 33]. In this method, the tunnel face and ceiling section are reinforced with fibre bolts, and the tunnel excavation can be done as a full section (**Figure 22**). Thus, the tunnel is excavated close to the circular, and the ring is closed immediately.

Tunnel excavation in long tunnels (L > 5 km) can be conducted using a Tunnel Boring Machine (TBM). It is generally preferred in metro tunnels and water tunnels (**Figure 23**). The TBM machine carries out the excavation with mechanical excavation and segments or supports are placed right behind it.

The type and selection of TBM completely depend on the type of soil or rock. While open TBM is preferred for solid rocks, earth pressure balanced (EPB) type TBMs are preferred for very weak rocks. TBM types are summarised in **Table 8** [34].

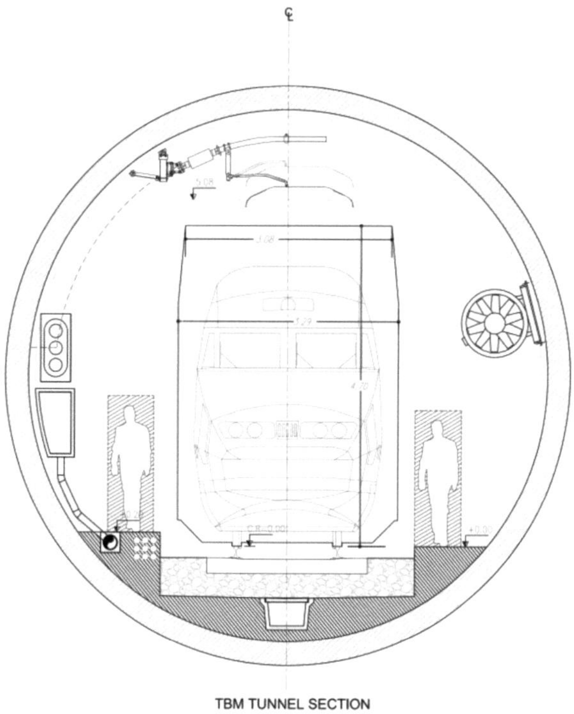

Figure 19.
Circular tunnel section.

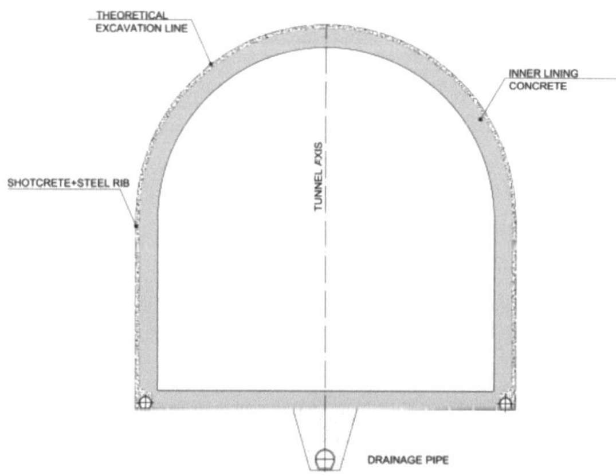

Figure 20.
Horseshoe tunnel section.

9. Geotechnical measurements

Studies carried out during the design phase can never give a definitive result. Because the research studies for tunnel design are limited and the acceptances to be made require that tunnel projects always be controlled with the measurements made during the tunnel excavation. This is because not all the works carried out during the tunnel design phase

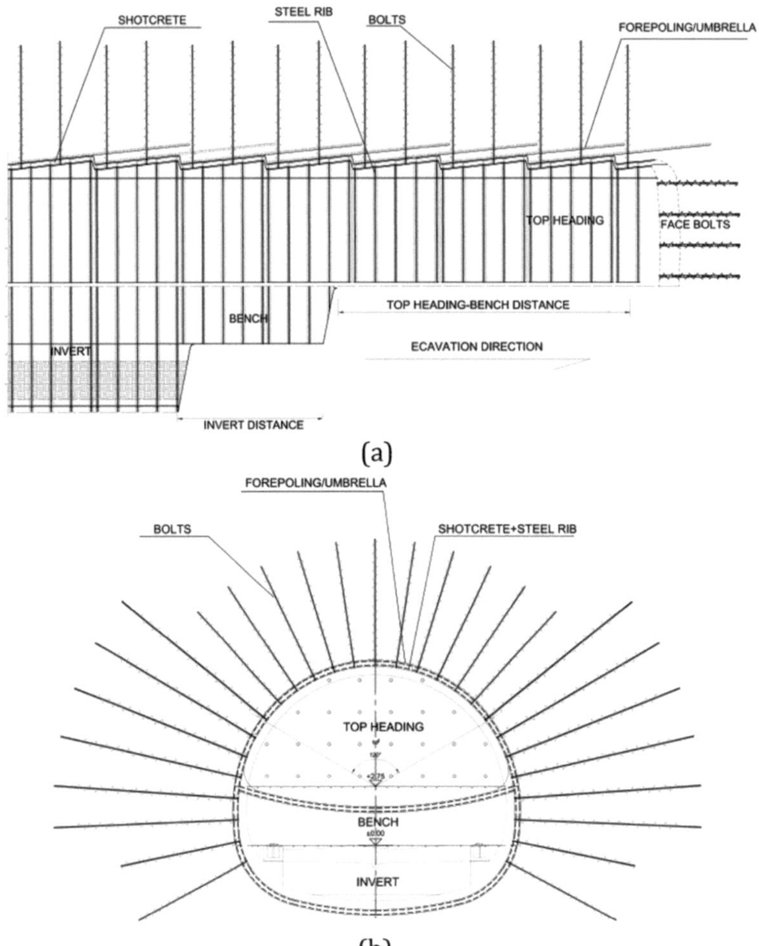

Figure 21.
Tunnel excavation section (a) longitudinal section (b) front view.

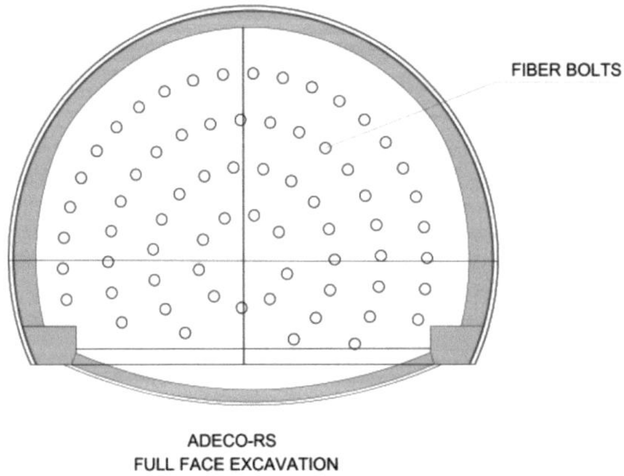

Figure 22.
ADECO-RS full-face excavation method.

Figure 23.
Hard rock TBM section for water tunnels.

Face support	Full face excavation	Partial face excavation
None or passive	Open hard rock TBM or Main Beam TBM	Digger shield
	Single shield hard rock TBM	Auger or road header
	Double shield Hard Rock TBM	
Active	Earth Pressure Balanced TBM (EPB-TBM)	Digger shield with compressed air
	Slurry TBM	Auger or road header with compressed air
	Compressed Air Shields	
Combination	Dual or Multimode-TBMs, combining for example hard rock excavation with EPB-and/or Slurry mode	

Table 8.
Most common TBM with or without face support [34].

reflect the tunnel behaviour exactly. Every tunnel project requires revision during tunnel excavation. For this purpose, geotechnical measurements made during the tunnel excavation are of great importance and it is an indispensable condition of tunnelling.

A correct programming of measurements during the tunnel excavation will both prevent uncertainties in the tunnel project and contribute to the formation of a safer and more economical outcome. The geotechnical measurements to be made in the tunnel are:

a. Deformation measurements.

b. Convergence measurements.

c. Extensometer measurements.

d. Inclinometer measurements.

e. Stress cell-pressure measurements.

f. Strain gauge measurements.

g. Piezometric measurements.

h. Geotechnical face maps.

9.1 Deformation measurements

During the tunnel excavation, deformation measurements should be carried out every 10 or 20 m for solid rocks and every 2 m for weak rocks. Deformation measurements are made by electronically reading targets placed along the tunnel section with a theodolite. These measurements provide extremely important information regarding tunnel behaviour. According to the results of the measurements, the deformations in the tunnel can be determined and the support systems and the excavation sequence can be revised. The measurement frequency can be adjusted on-site according to the deformation rate occurring in the tunnel. If the measurements made during the first week after the excavation remain constant, they are continued weekly, while in cases where the deformation increases daily, two readings can be made daily (**Figure 24**).

9.2 Convergence measurements

During tunnel excavation, convergence measurements should be made with tape extensometers (**Figure 25**) to see the closures in the tunnel. In addition to deformation measurements, convergence measurements are also made to determine on which side and how much the closure is.

9.3 Extensometer measurements

Extensometer measurements are made in the tunnel in order to see the plastic zone thickness around the tunnel (**Figure 26**). They play an important role in determining the length of the bolts by detecting the loosened zone thickness around the tunnel. In

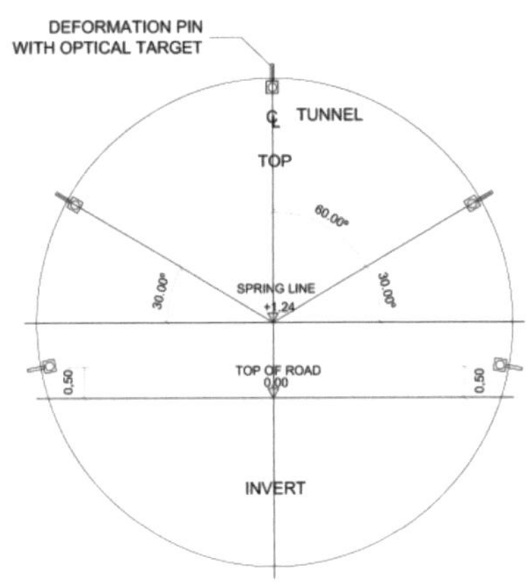

Figure 24.
Deformation target points in circular tunnel.

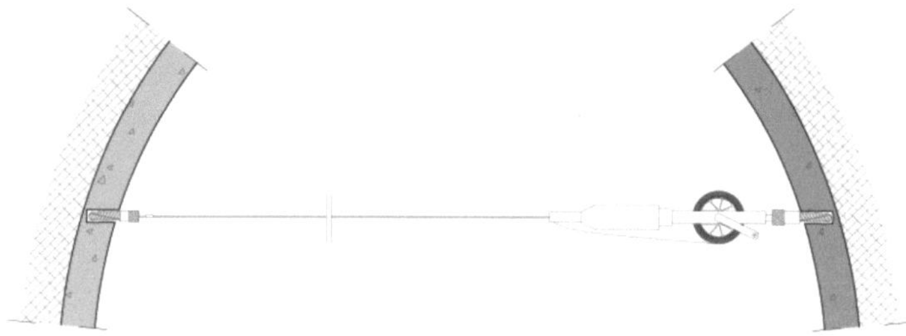

Figure 25.
Tape the extensometer section in the tunnel.

addition, they give information about the magnitude of the loads that will come to the outer lining depending on the loosened zone thickness. Extensometer lengths are determined according to the tunnel diameter. Extensometers can be applied individually as well as in single, double or triple types. Measurements are started immediately after the placement of extensometers and are made daily or weekly depending on the frequency of movement.

9.4 Inclinometer measurements

Inclinometer measurements are made from the surface to see possible movements in the portal sections or on the tunnel route (**Figure 27**). The inclinometer is selected according to the thickness of the land movement or possible loose zone in the drilled boreholes and generally goes down to the level of the tunnel. By means of inclinometers, possible movements in the portal section are determined and the necessary information is provided for the portal design, and it can be determined whether possible movements on the tunnel route affect the tunnel structure.

9.5 Stress cell

Stress cells are placed on the ground or concrete after excavation to see the stresses acting on the supports. In addition, stress cells are placed in the inner lining of concrete and the stresses that occur in the lining are determined. With the determination of the incoming loads, necessary revisions can be made in the shotcrete or inner lining concrete design/implementation. Because, when the assumptions made in tunnel lining calculations often differ with the site conditions, it is extremely important to determine the incoming loads on site. The lining thicknesses can be determined by making back analyses by means of the loads determined in the field. Stresses can be measured by placing the stress cell in both the radial and tangential directions (**Figure 28**).

9.6 Strain gauge measurements

Strain gauges are placed in shotcrete or interior lining concrete to see the strains in the lining and to detect the incoming stresses (**Figure 28**).

9.7 Piezometric measurements

Piezometric measurements can be carried out to measure the water pressure or the water level around the tunnel. These measurements are of great importance for inner lining design, especially in the long term.

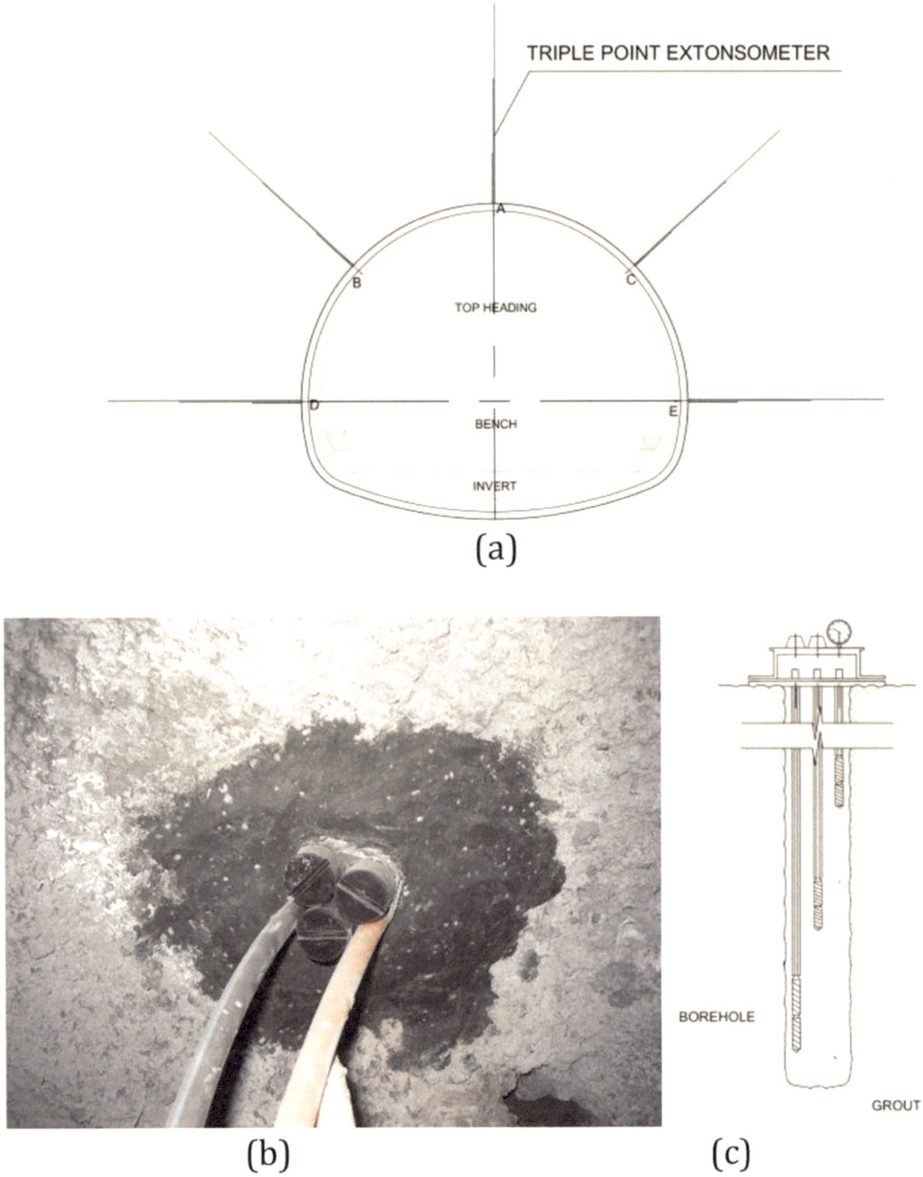

Figure 26.
Extensometer measurements in tunnel (a) installation section in tunnel, (b) in tunnel application and (c) triple type extensometer.

9.8 Geotechnical face maps

After the tunnel excavation, the geological units should be defined and perimetric maps should be made. The changes in geological conditions with these maps are extremely important in terms of evaluating the changes in geological conditions between those envisaged in the project and the geological conditions actually encountered. If there is a difference between the encountered geological conditions and the predicted geological conditions, the tunnel design may need to be revised according to the new conditions.

(a) (b)

Figure 27.
Inclinometer typical section (a), measurements on site (b).

Figure 28.
Stress cell and strain gauge application on site.

9.9 Conclusion

Acceptances and geotechnical investigations carried out during the tunnel design phase are limited. All tunnel projects are made with these deficiencies and in a sense, they are in the form of a preliminary project. It should never be possible to go with uncertainties and to implement the same support systems made in the design phase. Geotechnical measurements and project data should be continuously evaluated and revised according to the encountered geological conditions.

10. Conclusion

Empirical, analytical and numerical approaches should be used together in tunnel design. Empirical methods provide preliminary information for the rock mass and support design during the design phase. The details of support systems should be complemented by numerical analysis.

Each tunnel reaches its final design status during the process of tunnel excavation. All the acceptances made during the design phase should be considered as preliminary assumptions. Based on the geological-geotechnical research carried out during the design phase, the tunnel behaviour cannot be fully revealed. For this reason, measurements such as geological and geotechnical evaluations, deformation measurements, convergence measurements and pressure cells should be done to revise the tunnel design according to the encountered geological conditions.

During the tunnel excavation, geological-geotechnical measurements, face maps and horizontal boreholes, if necessary, should be carried out continuously.

TBM selection and design are completely dependent on geological conditions. The wrong TBM selection can be made because of the geological profile that is not determined correctly. This situation causes very serious project problems in the tunnel.

While determining the tunnel excavation method, tunnel opening and ground conditions are some of the most important factors. In large-diameter tunnels, the tunnel is usually divided into top heading, bench and invert if necessary. It is obligatory to close the ring (top heading, bench and invert) as soon as possible when excavating in weak rock conditions and soils. Otherwise, serious deformations can be observed in the tunnel. The most correct approach for tunnel section is circular or near-circular type section.

Acknowledgements

The author would like to thank Fugro Sial Geosciences Engineering Ltd.

Conflict of interest

The author declares no conflict of interest.

DOI: http://dx.doi.org/10.5772/intechopen.109608

Author details

Ebu Bekir Aygar
Fugro Sial Geosciences Consulting and Engineering Ltd., Ankara, Türkiye

*Address all correspondence to: eaygar@gmail.com

References

[1] Terzaghi KV. Rock defects and loads on tunnel supports. In: Proctor RV, White TL, editors. Rock Tunneling with Steel Supports. Youngstown: Commercial Shearing and Stamping Company; 1946

[2] Lauffer H. Gebirgsklassifizierung für den Stollenbau. Geol. Bauwesen. 1958; **24**(1):46-51

[3] Rabcewicz L. The New Austrian tunnelling method, Part one. Mint: Water Power. 1964;**1964**:453-457

[4] Rabcewicz L. The New Austrian tunnelling method, Part two. Mint: Water Power. 1964;**1964**:511-515

[5] Rabcewicz L. The New Austrian tunnelling method, Part three. Mint: Water Power. 1965;**1965**:19-24

[6] Rabcewicz L, Golser J. Principles of dimensioning the supporting system for the "New Austrian Tunnelling Method". Mint: Water Power. 1973;**1973**:88-93

[7] Deere DU, Hendron AJ, Patton FD, Cording EJ. Design of surface and near surface construction in rock. In: 8th U.S. Symposium on Rock Mechanics: Failure and Breakage of Rock. New York: Society of Mining Engineers, American Institute of Mining, Metallurgical, and Petroleum Engineers; 1967

[8] Wickham GE, Tiedemann HR, Skinner EH. Support determination based on geologic predictions. In: Lane KS, editor. North American Rapid Excavation and Tunneling Conference. Chicago, New York: Society of Mining Engineers of the American Institute of Mining, Metallurgical and Petroleum Engineers; 1972. pp. 43-64

[9] Bieniawski ZT. Engineering classification of jointed rock masses. Transaction of the South African Institution of Civil Engineers. 1973;**15**: 335-344

[10] Bieniawski ZT. Rock mass classification in rock engineering. In Exploration for rock engineering. In: Bieniawski ZT, editor. Proceedings of the Symposium. Cape Town: Balkema; 1976. pp. 97-106

[11] Bieniawski ZT. Engineering Rock Mass Classifications: A Complete Manual for Engineers and Geologists in Mining, Civil, and Petroleum Engineering. New York: Wiley; 1989. p. 251

[12] Barton NR, Lien R, Lunde J. Engineering classification of rock masses for the design of tunnel support. Mint: Rock Mechanics. 1974;**6**:189-239

[13] Barton N, Løset F, Lien R, Lunde J. Application of the Q-system in design decisions. In: Bergman M, editor. Subsurface S. New York: Pergamon; 1980. pp. 553-561

[14] Barton N. Some new Q-value correlations to assist in site characterization and tunnel design. Mint. International Journal of Rock Mechanics and Mining Sciences. 2002;**39**:185-216

[15] Palmstrom A. RMi—A Rock Mass Characterization System for Rock Engineering Purposes. Norway: University of Oslo; 1995

[16] Hoek E. Support for very weak rock associated with faults and shear zones. In: Proceedings of International Symposium on Rock Support and Reinforcement Practice in Mining. Kalgoorlie, Australia. 1999. pp. 14-19

[17] Hoek E. Tunnel support in weak rock. In: Symposium of Sedimentary Rock Engineering. Taipei, Taiwan. 1998. pp. 20-22

[18] Hoek E, Brown ET. Underground Excavations in Rock. London: Instn Min. Metall; 1980

[19] Brady BHG, Brown ET. Rock Mechanics for Underground Mining. London: Allen and Unwin; 1985

[20] Aygar EB. Evaluation of new Austrian tunnelling method applied to Bolu tunnel's weak rocks. Mint. Journal of Rock Mechanics and Geotechnical Engineering. 2020;**12**:541-556

[21] Aygar EB, Gokceoglu C. Problems encountered during a railway tunnel excavation in squeezing and swelling materials and possible engineering measures: A case study from Turkey. Mint: Sustainability. 2020;**12**:1166. DOI: 10.3390/su12031166

[22] Aygar EB, Gokceoglu C. Analytical solutions and 3D numerical analyses of a shallow tunnel excavated in weak ground: A case from Turkey. Mint: International Journal of Geo-Engineering. 2021;**2**(1):9. DOI: 10.1186/s40703-021-00142-7

[23] Jethwa JL, Singh B, Singh B. Estimation of ultimate rock pressure for tunnel linings under squeezing rock conditions—A new approach. In: Hudson JA, editor. Design and Performance of Underground Excavations, ISRM Symposium. Cambridge: E.T. Brown and; 1984. pp. 231-238

[24] Sakurai S. Displacement measurements associated with the design of underground openings. In: Proc. Int. Symp. Field Measurements in Geomechanics. Vol. 2. Zurich; 1983. pp. 1163-1178

[25] Singh B, Jethwa JL, Dube AK, Singh B. Correlation between observed support pressure and rock mass quality. Mint: Tunn Undergr Sp Technol Inc Trenchless. 1992;**1992**:59-74. DOI: 10.1016/0886-7798(92)90114-W

[26] Goel RK, Jethwa JL, Paithankar AG. Indian experiences with Q and RMR systems. Mint: Tunn Undergr Sp Technol Inc Trenchless. 1995;**10**:97-109. DOI: 10.1016/0886-7798(94)00069-W

[27] Hoek E, Marinos P. Predicting tunnel squeezing problems in weak heterogeneous rock masses. Tunnels Tunnelng International. 2000;**Part 1–2**: 1-20

[28] Aydan O, Akagi T, Kawamoto T. The squeezing potential of rocks around tunnels; theory and prediction. Mint: Rock Mechanics and Rock Engineering. 1993;**26**(2):137-163

[29] Einstein HH, Bischef N. Design of tunnels in swelling rock, design methods in rock mechanics. In: Proc. 16th Symp. On Rock Mechanics. U.S.A: Univ. of Minnesota; 1995. pp. 185-195

[30] Komornik A, David D. Prediction of swelling pressures of clays. In: Proc. ASCE, SMFE. United States; 1969. pp. 209-255

[31] Luque A, Hegedus S. Handbook of Photovoltaic Science and Engineering. 2nd ed. Chichester: Wiley; 2011. p. 1132. DOI: 10.1002/9780470974704

[32] Lunardi P. Design and constructing tunnels—ADECO-RS approach. Tunnels and Tunnelling International. Special supplement. 2000. pp.3-30

[33] Lunardi P, Bindi R. The evolution of reinforcement of the advance core using fibre glass elements for short and long term stability of tunnels under difficult stress-strain conditions: Design, technologies and operating methods. In:

Progress in Tunnelling after 2000, AITES-ITA. Vol. 2. Milano, Italy: World Tunnel Congress; 2001. pp.309-322

[34] ITA, International tunnelling and underground space association, Recommendations on the development process for mined tunnels, Working Group 14 Mechanized Tunnelling, Working Group 19 Conventional Tunnelling. France. 2016. Available from: https://about.ita-aites.org/publications/wg-publications/1468/recommendations-on-the-development-process-for-mined-tunnels

Section 2

Rolling Stock

Chapter 4

Experimental Research of New Design Solutions for Fencing Refrigerated Wagon Bodies and Containers

Rustam V. Rahimov and Bakhrom A. Abdullaev

Abstract

In this chapter, in order to select the directions of development and develop scientifically based technical solutions to improve the thermal parameters of the bodies of refrigerated wagons and containers for the transportation of perishable goods in the conditions of the Republic of Uzbekistan, experimental studies of new design solutions for fencing refrigerator bodies and containers have been carried out. For conducting experimental research a method of experimental determination of the body heat transfer coefficient using a closed thermally insulated chamber in the form of a parallelepiped with a replaceable upper face (cover) has been developed. As a result of comprehensive research using analytical calculations and field experiments, it was found that a promising option for thermal body fencing is a technical solution where polyurethane foam and "Corundum" are used as thermal insulation. The use of these technical solutions in the fences of refrigerated wagons bodies and containers will reduce the heat transfer coefficient by up to 20% and, accordingly, reduce the thickness of the fence by 20–30%, which will lead to an increase in the internal useful volume of the body, a decrease in its mass and consumption of materials used, and an improvement in the thermal state of the body.

Keywords: refrigerated wagon, refrigerated container, body fencing, heat transfer coefficient, thermal insulation, thermal insulation material, corundum material, test chamber, climatic chamber

1. Introduction

Effective functioning of railway transport in the Republic of Uzbekistan plays a significant role in creating conditions for modernization, transition to an innovative path of development and sustainable growth of the national economy of the country [1, 2]. By their geographical location, the railways of Uzbekistan are an integral part of the Eurasian Railway Network and are also the largest transport and logistics facility of national importance [3]. Throughout the existence and functioning of railway transport works in close relationship with other sectors of the national economy of the

Republic of Uzbekistan. One of its main tasks is to meet the needs of the economy by providing timely cargo and passenger transportation. The work in this direction is carried out in various aspects and is aimed at ensuring the safe maximum possible throughput and carrying capacity of the railway network of the Republic of Uzbekistan [4, 5].

The location and peculiarity of the climate of the Republic of Uzbekistan – aridity, a large amount of solar heat, continentality contribute to the mass cultivation of fruit and vegetable products and allow sufficiently high yields of fruits, berries and vegetable products, to provide not only domestic needs but also to supply products to the markets of the near and far abroad [6–12].

For the export of fruits and vegetables, a sufficiently large number of vehicles equipped with thermal insulation and refrigeration and heating installations are needed. However, at present, the fleet of refrigerated rolling stock, as well as the entire fleet of freight wagons of the railways of Uzbekistan, is experiencing a shortage of serviceable vehicles [4, 13–21]. The current state of the fleet of refrigerated rolling stock does not allow to supply of a large number of perishable goods to other countries, which negatively affects the development of perishable traffic in the Republic of Uzbekistan.

The thermal engineering properties of the thermal insulation material play an essential role in the design of the thermal fencing of the bodies of refrigerated vehicles [21, 22]. Various materials are used for the thermal insulation of wagons and containers: mineral wool, foams and polyurethanes, characterized by low values of thermal conductivity coefficients [14, 18, 20, 23, 24]. However, due to the presence of thermal bridges, moisture and deterioration of thermal insulation properties during operation in refrigerated wagons and containers, the thickness of body fences (walls, roof and floor) usually exceeds 160–200 mm and reaches 250 mm, which leads to a decrease in the internal useful volume of the body and an increase in the weight of the container. The reason for this is the imperfection of technical solutions for the construction of body fencing structures and thermal insulation materials. During operation, the insulation of the body fence ages and wears out due to the effects of vibration, temperature and humidity changes. Deterioration of the insulation quality of fences leads to fuel overspending and an increase in financial costs, and the volume of work increases with planned types of repairs.

At the same time, new materials are being used in related branches of technology [22, 25, 26], having the lowest coefficient of thermal conductivity and greater strength, which makes it possible to create solid bearing structures of smaller size and weight. Therefore, the development of new technical solutions and the search for promising thermal insulation materials for fencing bodies of refrigerated wagons and containers are relevant. In this regard, the main purpose of this research is to select areas of development and develop scientifically sound technical solutions to improve the thermal parameters of the bodies of refrigerated wagons and containers for the transportation of perishable goods in the conditions of the Republic of Uzbekistan.

2. Analysis of thermal insulation materials for fencing bodies of refrigerated wagons and containers

In recent years, the most commonly used insulation is made of polyurethane foam, which is applied to finished surfaces by spraying [21, 22]. **Table 1** shows some characteristics of the materials used.

Name of the material	Coefficient of thermal conductivity, λ, W/m·°C	Specific weight, ρ, kg/m³
Foam PCV-1	0.035	70–100
Foam PCV-2	0.047	100–130
Foam resin phenolformal FRP-1	0.058	70
Foam resin phenolformal FRP-2	0.058	100
Mineral plate	0.075–0.08	300–350
Mipor	0.035–0.045	12–15
Expanded polystyrene PSBS	0.038–0.041	25–40
Polyurethane foam PPU-36	0.04–0.046	25–40
Fiber glass	0.058	170
Wood	0.14–0.23	500–600
Dry air	0.023	—

Table 1.
Characteristics of insulation materials.

To ensure a low value of the heat transfer coefficient, the dimensions of the thickness of the body fence with these materials are in the range of 160–240 mm. The fences of the vehicle body with the specified materials do not protect against the formation of fogging of the body fence elements and the formation of moisture inside the body, which is harmful to the transported goods.

When repairing refrigerated wagons and containers, it is necessary to replace almost all the elements of the body fence due to moisture damage [13, 21].

The consumption of materials for body fencing is quite high, which undoubtedly affects the cost of vehicles. The search for materials for thermal insulation of the body of vehicles that could become ideal continues [27–29].

To protect building structures from moisture and heat preservation indoors at low temperatures, the thermal insulation material "Corundum" [30], created in 2008, and

Indicators	Traditional thermal insulation		Corundum	Manufactured by LLC "Regent Baltika" and LLC "BALTMASH"		
	Mipora TU 6-05-1112-92	Foamed polystyrene		Foamed polystyrene reinforced in volume with fiberglass (RZhD – 1)	Honeycomb panel and foamed polystyrene reinforced in volume	Honeycomb panel
Thermal conductivity, W/°C	0.034-0.052	0.025–0.03	0.001	0.026	0.007	0.032
Labor costs for installation, person·h/m2	10	1–2	1–2		0.5–1	
Service life, years	Up to 5	From 10	From 10		30–50	

Table 2.
Comparison of technical characteristics of traditional and new thermal insulation.

produced by Russian manufacturers, is used. This material is a one-component polymer suspension, to which the smallest ceramic heat-sealed granules are added, which are characterized by the physical principle of reflection and heat transfer. The content of such granules (microspheres) in the suspension reaches 70%. Depending on the intended purpose, pigmenting fire-resistant, anticorrosive and inhibitory components are added to the composition. All this makes it possible to obtain thermal insulation with high adhesion and elasticity, which does not change under the influence of external factors: humidity, ultraviolet rays, temperature changes, mold and fungi. The basis of corundum is an Aqueous-F acrylic solution, to which the above components are added [31]. In addition, a number of new materials have appeared. Basic information about these materials is given in **Table 2**. Therefore, the task arises to assess the possibility of using these materials in the construction of fencing of wagons and containers [32].

3. Methods of conducting experimental research

The methods of measuring the coefficients of thermal conductivity and heat transfer of material are based on the equation of Fourier's law [33].

$$\frac{Q}{F} = \frac{\lambda}{\delta}(T_1 - T_2), \tag{1}$$

The determination of the heat transfer coefficients was carried out in works [27, 28, 34–41]. The main difficulty is that part of the heat flux created by heating is dissipated into the environment, and measuring the heat flux through the structure under study presents a known difficulty. Therefore, it was proposed to use a closed chamber in the form of a parallelepiped with a replaceable top face (cover). It is known that the average heat transfer coefficient of a closed chamber can be calculated by the expression:

$$K_{av} = \frac{4K_{sw} \cdot F_{sw} + K_{fl} \cdot F_{fl} + K_{\text{cov}} \cdot F_{\text{cov}}}{4F_{sw} + F_{fl} + F_{\text{cov}}}, \tag{2}$$

To determine the heat transfer coefficient of the tested structures, it is necessary to know the heat transfer coefficient of the auxiliary elements of the chamber.

Since the bottom and side wall of the test chamber are made of one of the materials, their heat transfer coefficients are equal to $K_{fl} = K_{sw} = K_0$,

$$K_{av} = K_0 \frac{F_0}{F_0 + F_{\text{cov}}} + K_{\text{cov}} \frac{F_{\text{cov}}}{F_0 + F_{\text{cov}}}, \tag{3}$$

With a known thermal conductivity coefficient of the material λ, the dimensions of the chamber, from formula (3), according to the experimentally determined average heat transfer coefficient of the chamber, it is possible to find the heat transfer coefficient of the investigated technical solution:

$$K_{\text{cov}} = \frac{K_{av} \cdot (F_0 + F_{\text{cov}})}{F_{\text{cov}}} - \frac{K_0 \cdot F_0}{F_{\text{cov}}}. \tag{4}$$

The values of the thermal conductivity coefficient λ of the material may differ depending on its structure. Therefore, to improve accuracy, it was proposed to make two graduated covers from the same material from the same delivery but with different thicknesses: 50 and 100 mm. When the thickness is reduced up to 2 times, the heat transfer coefficient of the cover becomes 1.87 times less.

With the help of the camera calibration data using covers of different thicknesses, we obtain a system of two equations for a more accurate determination of the heat transfer coefficient of the camera:

$$\begin{cases} K_{av}^{\delta=100} = \dfrac{K_0F_0 + K_{cov100}F_{cov}}{F_0 + F_{\kappa p}} \\ K_{av}^{\delta=50} = \dfrac{K_0F_0 + 1.87K_{cov100}F_{cov}}{F_0 + F_{cov}} \end{cases}. \tag{5}$$

By solving the system of Eqs. (7) concerning two unknowns, we obtain the refined values of the heat transfer coefficient K_0 of the walls and floor of the chamber:

$$K_0 = \frac{(F_0 + F_{cov}) \cdot \left(1.87F_{cov} \cdot K_{cov}^{\delta=100} - K_{av}^{\delta=50} \cdot F_{cov}\right)}{F_0 - F_0 \cdot F_{cov}}. \tag{6}$$

Thus, the method of experimental determination of the heat transfer coefficient of the tested technical solution, made in the form of a cover to a heat-insulated chamber (HIC), is as follows [32]:

- the area of the enclosing surfaces of the chamber and the cover is calculated;
- the chamber is installed in a heat-insulated room or a refrigerating chamber;
- the temperature difference is measured outside and inside the chamber;
- the air inside the chamber is heated by an electrical device connected to an electricity meter;
- upon reaching a conditionally stationary mode with a constant temperature difference outside and inside the chamber, the average heat transfer coefficient is calculated;
- using the obtained average heat transfer coefficient, the heat transfer coefficient of the lower enclosures of chamber K_0 is calculated according to the formula (5) (performed only when the chamber is calibrated);
- with the help of the value of the coefficient K_{av}, the heat transfer coefficient of the investigated technical solution of thermal insulation is found according to (4).

To study the effect of the external temperature, the required temperature of the refrigerating chamber is set, and the experiments are repeated in the same order.

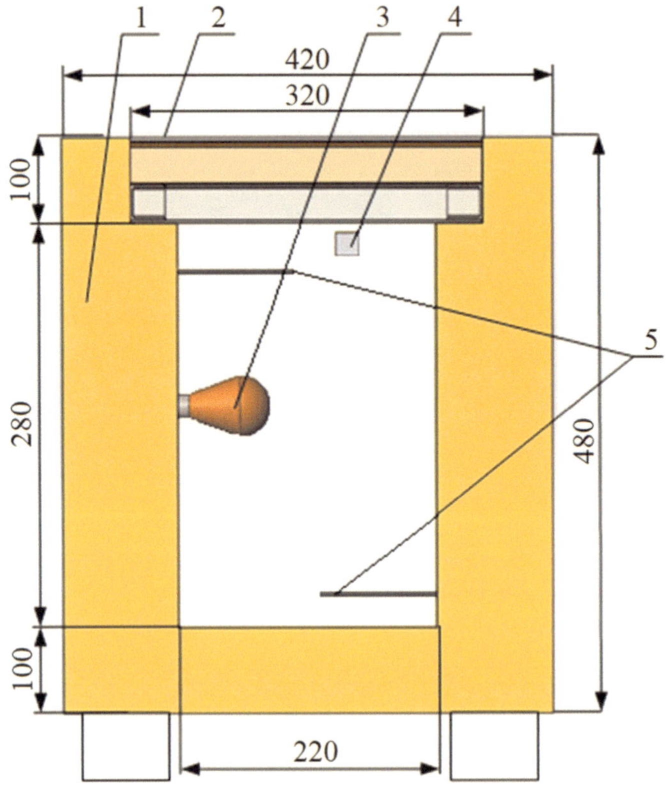

Figure 1.
Test chamber TIC-1 for conducting experiments to determine the heat-shielding properties of thermal insulation: 1 – Expanded polystyrene; 2 – Test sample (cover) of the chamber; 3 – Heater; 4 – Test chamber thermostat; 5 – Thermocouples.

4. Experimental setup and calibration

A general view of the thermally insulated chamber (TIC-1) in **Figure 1** and the geometric mean areas of the chamber elements are given in **Table 3**.

The test chamber is made of 100 mm thick expanded polystyrene, the inner dimension of which is 220 × 220 × 280 mm. The top of the test chamber (cover) is a test piece of thermal insulation (fence). Samples for the calibration of the test chamber were designed in the form of a cover measuring 320 × 320 mm and a thickness of 50 and 100 mm.

Chamber elements names	The area of the elements outside, m^2	The area of the elements inside, m^2	The average area of elements, m^2
Side wall, F_{sw}	0.8064	0.2464	0.4457
Floor, F_{fl}	0.1764	0.0484	0.092
Cover, F_{cov}	0.1024	0.1024	0.1024
	Total average chamber area, $\sum F$		0.6401

Table 3.
Geometric average areas of the chamber elements.

The geometric mean areas of the fence elements of the test chamber (**Table 3**) were determined by the formula:

$$F_{av} = \sqrt{F_{out} \cdot F_{in}}, \tag{7}$$

The TIC-1 chamber is equipped with a heating element (electric lamp 25 W), connected through a meter to an electric source, and MS-227R4 thermocouples, as well as a W1209 thermostat for temperature control. The heating element is installed so that it evenly heats the inside of the test chamber.

The test chamber was placed in a special climatic chamber of the (CHALLENGE CH600C) series equipped with the necessary measurement and monitoring means, where the temperature can be maintained in the range from −75 to +180°C. When calibrating, the test chamber was placed inside the climatic chamber, and the temperature was maintained at 40°C inside and −20°C outside. The correlation between

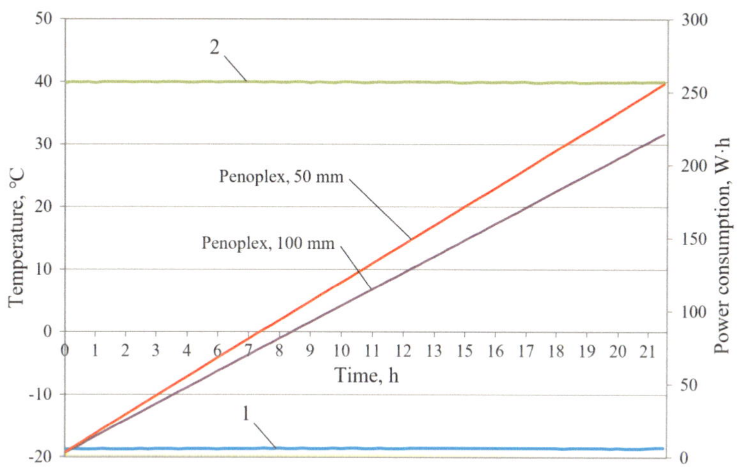

Figure 2.
Dependence of power consumption on the time of the experiment at a temperature outside −20°C (2) and inside the test chamber +40°C (3).

Test parameters	δ = 100 mm	δ = 50 mm
Holding time of samples, h	22	
Electricity consumption, W·h	220	255
Internal temperature of the test chamber, °C	40.1	40.486
Internal temperature of the climatic chamber, °C	−19.9	−19.52
Temperature difference, °C	60	59.71
Average heat transfer coefficient of the chamber, W/m^2·K	0.260	0.304
Average heat transfer coefficient of the graduation cover K_{cov}, W/m^2·K	0.263	0.537
Average heat transfer coefficient of side walls and floor K_{sw} and K_{fl}, W/m^2·K	0.259	

Table 4.
Experimental data and results of determining the heat transfer coefficient of the TIC-1 test chamber.

electricity consumption and temperature inside and outside is shown in **Figure 2**. It can be seen from it that the temperature remained practically constant, and the power consumption linearly depended on the time of the experiment. The results of processing the experimental data are shown in **Table 4**.

5. Heat-protective fence design

To assess the effectiveness of various technical solutions for the fence, the following samples were made (**Table 5**).

- Sample 1 is a typical fragment of the thermal insulation of refrigerated wagons and containers. It consists of a steel sheet with a thickness of 2 mm, a U-shaped frame with dimensions of 30 × 30 × 2 mm, expanded polystyrene with a thickness of 60 mm, an inner lining of plywood with a thickness of 4 mm, an aluminum sheet with a thickness of 2 mm;

Sample No.	Schemes of sample covers for testing	General view of the sample
1	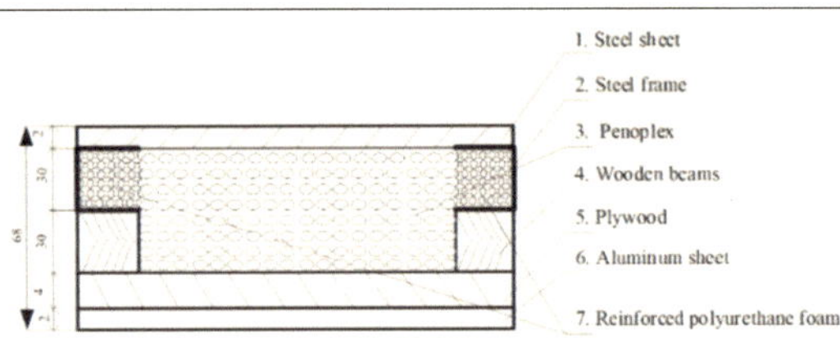	
2	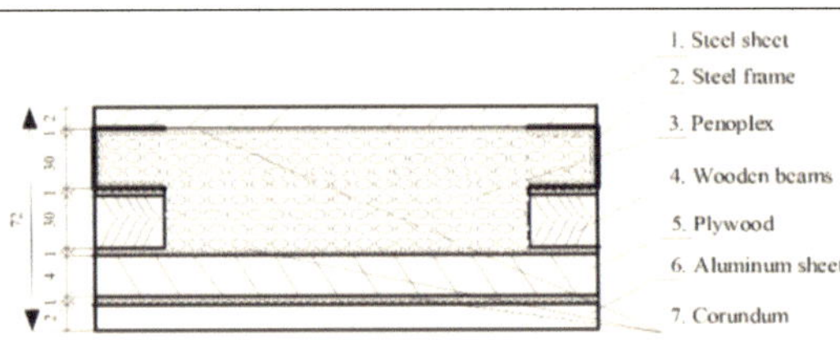	
3	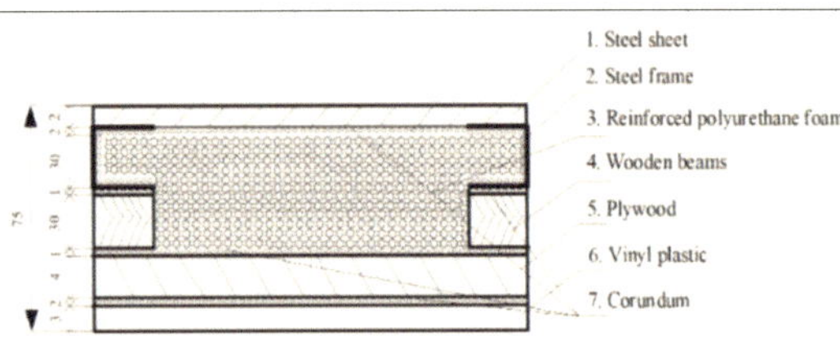	
4	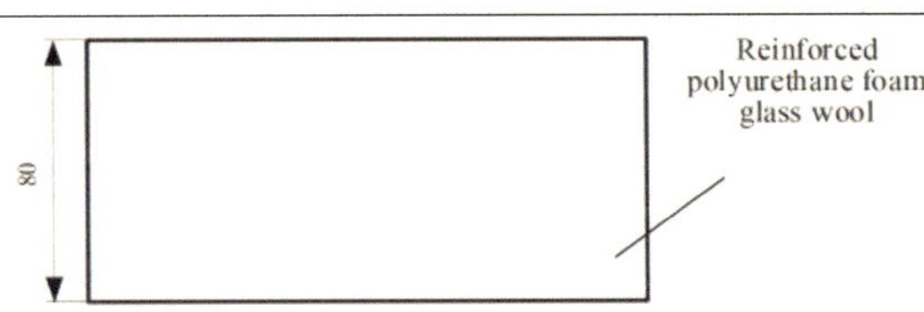	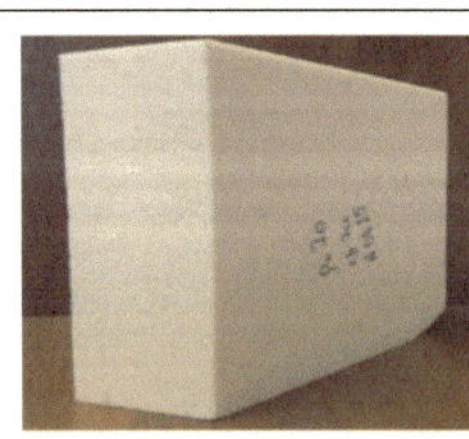

Sample No.	Schemes of sample covers for testing	General view of the sample
5		
6	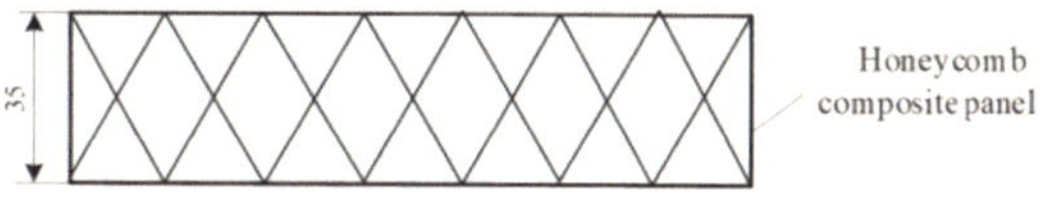	

Table 5.
Technical solutions for the experimental determination of the heat transfer coefficient.

- Sample 2 is similar to Sample 1, but layers of "Corundum" 1 mm thick are applied between the steel sheet and expanded polystyrene, expanded polystyrene and plywood, plywood and aluminum sheet, as well as along the perimeter of the U-shaped frame;

- Sample 3 is similar to Sample 1, but instead of an aluminum sheet with a thickness of 2 mm, a sheet of vinyl plastic with a thickness of 3 mm is installed; a layer of "Corundum" with a thickness of 2 mm is applied between the plywood and vinyl plastic, and layers of "Corundum" with a thickness of 1 mm are applied between expanded polystyrene and plywood, as well as along the perimeter of the U-shaped frame;

- Sample 4 – in this sample, reinforced polyurethane foam with fiberglass, manufactured by "Regent Baltika", with a thickness of 80 mm, was used as a fragment of thermal insulation;

- Sample 5 – a combination of reinforced polyurethane with a thickness of 100 mm and honeycomb insulation with a thickness of 35 mm was used as a promising fragment of thermal insulation;

- Sample 6 – a honeycomb panel, 35 mm thick, was used as a fragment of thermal insulation.

6. Results of experimental studies of thermophysical properties of fences

In the first stage, Samples 1–3 were tested in the CHALLENGE CH600C (**Figure 3**) climatic chamber. The holding time of the samples was 22 hours; inside the TIC-1

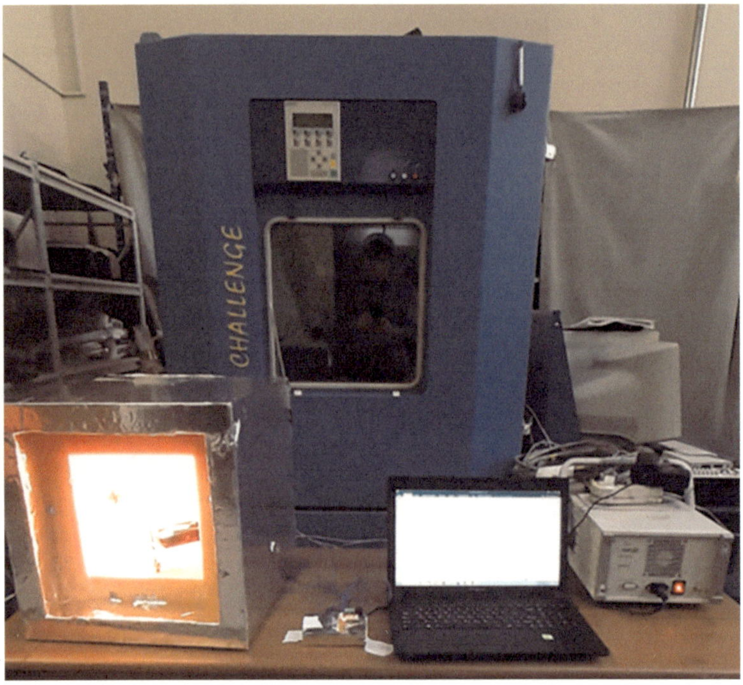

Figure 3.
General view of the CHALLENGE CH600C climatic chamber and the TIC-1 test chamber.

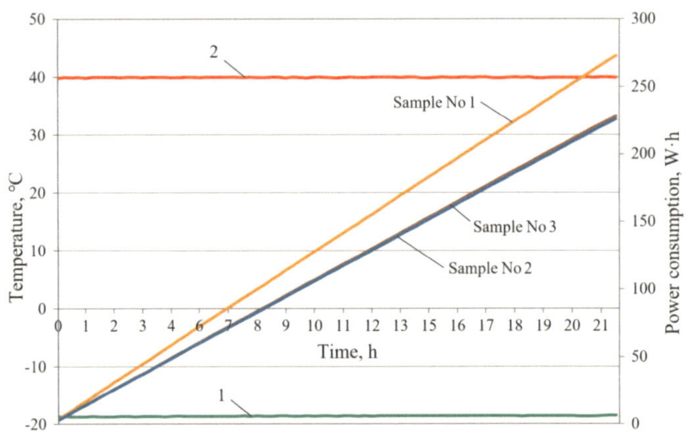

Figure 4.
Dependence of power consumption at the time of the experiment at a temperature outside −20°C (2) and inside the test chamber +40°C (3).

chamber, the temperature was maintained at 40°C, outside −20°C (**Figure 4**). The average heat transfer coefficients of the cover and chamber are given in **Table 6**. The thermal insulation properties of Samples 2 and 3, containing thermal insulation "Corundum", were almost 2 times better. In Sample 3, the thickness of the layers of "Corundum" was 1.5 times thicker, but the heat transfer coefficient of the sample

Test parameters	Sample no.		
	1	2	3
Holding time of samples, h		22	
Power consumption, W h	273	228	226
Internal temperature of the test chamber, °C	40.009	40.066	40.1
Internal temperature of the climatic chamber, °C	−19.73	−19.49	-19.56
Temperature difference, °C	59.73	59.56	59.66
Average heat transfer coefficient of the chamber, W/m^2 K	0.325	0.272	0.270
Average heat transfer coefficient of the cover, W/m^2 K	0.673	0.341	0.332

Table 6.
Data obtained from experimental studies using a climatic chamber.

decreased by less than 3% compared to Sample 2. This requires further research to select the optimal thickness of insulation of the "Corundum" type.

In the second stage, the determination of the heat transfer coefficients was carried out at the testing ground of the Scientific-Implementation Center "Wagons" in a room at an outside air temperature of 18–20°C without using a climatic chamber. Inside the test chamber, heating was carried out to 60–70°C. In this case, the power of the heating element – an electric lamp (25 W) – was reduced by means of a power regulator to 8.8 W.

Six prototype chamber covers were tested for 18 hours each. All parameters – power consumption, air temperature outside and inside the chamber and test duration were recorded. As a result of the experiment, the dependences of the external and internal temperatures and the average heat transfer coefficient on the experiment's time were determined, as shown in **Figure 4**. The results of the statistical analysis of the data obtained in experimental studies are shown in **Table 7**.

In **Figure 5a**, it can be seen that the air temperature in the test chamber increases during 5–6 hours, and then it stabilizes at the level of 58–75°C due to the transfer of heat from the test chamber to the outside. From **Figure 5b**, it follows: at the beginning of the experiment, during 5 h, the heat transfer coefficient of the test chamber drops sharply, which can be explained by an increase in the temperature difference between the outside and inside air $(t_{in} - t_{ex})$. Then the heat transfer coefficient is stabilized and

Test parameters	Sample No.					
	1	2	3	4	5	6
Holding time of samples, h	160	160	160	160	160	160
Power consumption, W h	19.2	19.2	19.3	18.9	17.4	17.4
External temperature, °C	62	70.2	71	70	72.4	58.2
Internal temperature of the test chamber, °C	42.8	51	51.7	51.1	55	40.8
Temperature difference, °C	0.324	0.272	0.269	0.271	0.252	0.340
Average heat transfer coefficient of the chamber, W/m^2 K	0.664	0.341	0.322	0.332	0.220	0.761

Table 7.
Data obtained in experimental studies without a climatic chamber.

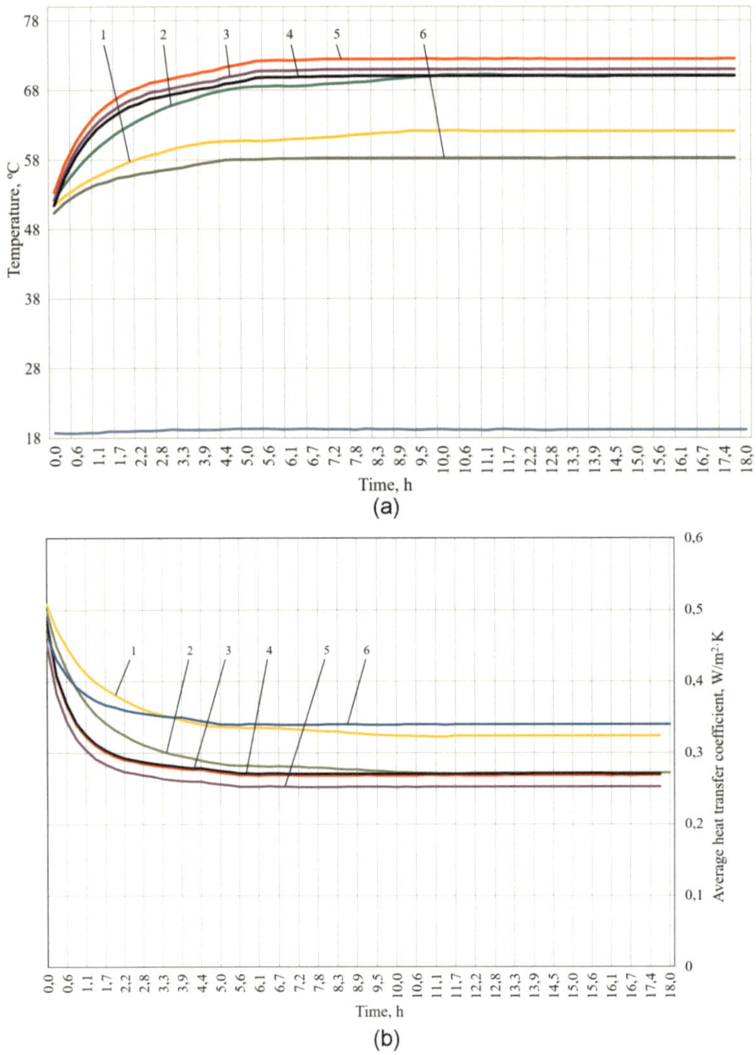

Figure 5.
Dependences of the internal temperature (a) and the heat transfer coefficient of the test chamber (b) on the time of the experiment: 1 – Sample No. 1, 2 – Sample No. 2, 3 – Sample No. 3, 4 – Sample No. 4, 5 – Sample No. 5, 6 – Sample No. 6, 7 – Outside the test chamber.

takes on a constant value. The heat transfer coefficients of the chamber covers are shown in **Figure 6**.

The heat transfer coefficients with and without a climatic chamber practically did not change. Their difference does not exceed 3%, which indicates a small dependence of the thermal conductivity coefficients of Samples 1–3 on temperature.

Repeated tests of Samples 1–3 confirmed (**Figure 6**) that the application of thermal insulation "Corundum" reduced the heat transfer coefficient almost twice. Thus, using "Corundum" as a layer of thermal insulation, it is possible to reduce the side wall thickness of refrigerated wagons and containers. For real wagons and containers, it is possible to reduce the insulation thickness by 20–30%. Insulation Samples 5 manufactured by "Regent Baltika" have the lowest heat transfer coefficients, but their thickness was significantly greater (135 mm) than Samples 1–3. Their

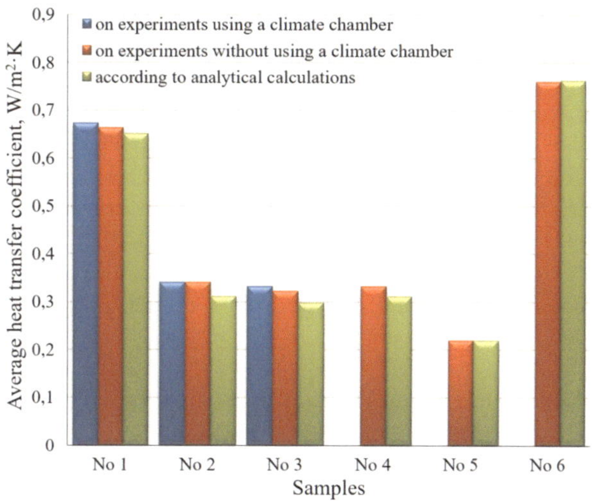

Figure 6.
Average heat transfer coefficients of cover samples No. 1–6.

application is promising, but it is necessary to develop technical solutions for their implementation.

7. Conclusions and recommendations

This chapter presents a set of theoretical and experimental research on the choice of development directions and the development of scientifically sound technical solutions to improve the thermal parameters of the bodies of refrigerated wagons and containers for the transportation of perishable traffic in the conditions of the Republic of Uzbekistan.

1. Based on the analysis of the designs of vehicles for the transportation of perishable goods and their thermal insulation materials, the main problems and prospects for the development of transportation of perishable goods in the conditions of the Republic of Uzbekistan are determined.

2. A method of experimental determination of the body heat transfer coefficient using a closed thermally insulated chamber in the form of a parallelepiped with a replaceable upper face (cover) has been developed.

3. A heat-insulated chamber has been developed that allows experimentally determining the coefficients of heat transfer and thermal conductivity of model thermal insulation samples with sufficient accuracy (error no more than 3%).

4. As a result of comprehensive research using analytical calculations and field experiments, it was found that a promising option for thermal body fencing is a technical solution where polyurethane foam and “Corundum” are used as thermal insulation.

5. The minimum value of the heat transfer coefficient from the materials studied was provided by the option of RZhD-1 thermal insulation (manufactured by "Regent Baltika"), which combines reinforced polyurethane and a honeycomb structure on a paper base.

6. The use of these technical solutions in the fences of refrigerated wagons bodies and containers will reduce the heat transfer coefficient by up to 20% and, accordingly, reduce the thickness of the fence by 20–30%, which will lead to an increase in the internal useful volume of the body, a decrease in its mass and consumption of materials used, and an improvement in the thermal state of the body.

7. The comparison of the calculation data using analytical formulas with the experimental results showed satisfactory accuracy in determining the average value of the heat transfer coefficient (the error is not more than 9%).

Thus, as a result of comprehensive studies using analytical calculations and field experiments, it was found that a promising option for thermal body fencing is a technical solution where polyurethane foam and "Corundum" are used as thermal insulation. The minimum value of the heat transfer coefficient from the studied materials was provided by the option of thermal insulation of RZhD-1 (produced by "Regent Baltika"), combining reinforced polyurethane and a honeycomb structure on a paper base. The use of these technical solutions in the fences of the bodies of refrigerated wagons and containers will reduce the heat transfer coefficient to 20% and, accordingly, reduce the thickness of the fence by 20–30%, which will lead to an increase in the internal useful volume of the body, reduce its weight and consumption of materials used, improve the thermal condition of the body. Theoretically and experimentally, the possibility of piecewise continuous recording of vertical and lateral forces from the wheel/rail interaction was achieved by measuring the stresses in two rail sections on a significant part of the sleeper gap.

Author details

Rustam V. Rahimov* and Bakhrom A. Abdullaev
Tashkent State Transport University, Tashkent, Uzbekistan

*Address all correspondence to: rakhimovrv@yandex.ru

References

[1] Resolution of the President of the Republic of Uzbekistan No. PP-1623 dated 04.10.2011 "About the Program of Priority Measures to Expand Production Volumes and Develop the Production of New Types of Competitive Products" [Electronic source]. Available from: https://lex.uz/acts/2427043 [Accessed: December 12, 2022]

[2] Resolution of the President of the Republic of Uzbekistan dated 06.03.2015. No. PP-2313 "About the Program of Development and Modernization of Engineering and Communication and Road Transport Infrastructure for 2015-2019" [Electronic source]. Available from: http://lex.uz/docs/3134726 [Accessed: December 12, 2022]

[3] Rasulov MK. Problems of increasing the competitiveness of national railway corridors. In: Rasulov MK, Ibragimov UN, Rahimov RV, editors. Scientific Works of the Republican Scientific and Technical Conference with the Participation of Foreign Scientists "Resource-Saving Technologies in Railway Transport". Tashkent: TashIIT; 2013. pp. 14-17

[4] Rahimov RV. Analysis of the state and prospects of the development of the freight wagon fleet of the Republic of Uzbekistan. Non-Ferrous Metals. 2018; **44**(1):7-11

[5] Rahimov RV. Uzbekistan on the path of revival of the Great Silk Road. Wagons and Wagon Economy. 2022; **3**(71):25-27

[6] Bakiyeva IA. Production and export of fruit and vegetable products in the Republic of Uzbekistan. Young Scientist. 2015;**21**:348-350

[7] Ibragimov NN. Development of a container design for the transportation of fruit and vegetable products. Young Scientist. 2015;**21**:168-173

[8] Muminov A. Physical geography of Uzbekistan. In: Rakhimbekov. Tashkent: Mekhnat; 2001. p. 224

[9] Ibragimov NN. Research of design and operating conditions of containers for fruit and vegetables. Proceedings VIII International Scientific Conference "Transport Problems 2016". Katowice: Silesian University of Technology Faculty of Transport. 2016:174-181

[10] Rahimov RV. Development of improved technical means for transportation fruits and vegetables. European Science Review. 2016;**1-2**: 175-177. DOI: 10.29013/ESR-16-1.2-175-177

[11] Rahimov RV. Development of the design and operating conditions of containers for the transportation of fruit and vegetable products. Bulletin of Transport of the Volga Region. 2016; **2**(56):75-81

[12] Abdullayev BA. The effect of solar radiation on the heat flows of refrigerated wagons and containers. Journal of the St. Petersburg University of Railway Transport. 2019;**16**(1):18-24

[13] Abdullayev BA. Analysis and prospects for the development of transportation of perishable goods in the Republic of Uzbekistan. Bulletin of Transport of the Volga Region. 2019; **3**(75):63-69

[14] Abdullayev BA. Analytical review of vehicle designs for the transportation of perishable goods. In: Materials of the XIII International Scientific and Technical

Conference “Rolling Stock of the XXI Century: Ideas, Requirements, Projects”. St. Petersburg: FSBEI HE SPSTU; 2018. pp. 109-111

[15] Rahimov RV. Wagon Park of JSC “Railways of Uzbekistan”: Current state and prospects for development. Bulletin of Tash STU. 2018;**2**(103): 136-142

[16] Rahimov RV. Analysis of the state and prospects of development of the freight wagon park of JSC “railways of Uzbekistan”. In: Scientific Works of the Republican Scientific and Technical Conference with the Participation of Foreign Scientists “Resource-Saving Technologies in Railway Transport”. Tashkent: TashIIT; 2017. pp. 91-94

[17] Rahimov RV. The choice of directions for the development of the railway wagon park of Uzbekistan. Transport of the Russian Federation. 2018;**1**(74):71-74

[18] Rahimov RV. The state and prospects of development of the railway wagon park of Uzbekistan. In: Materials of the XIII International Scientific and Technical Conference “Rolling stock of the XXI century: Ideas, requirements, projects”. Saint-Petersburg: St. Petersburg: FSBEI HE SPSTU; 2018. pp. 124-128

[19] Rahimov R.V. Ways to Increase the Transit Potential of the Republic of Uzbekistan. Bulletion of TashIIT. 2017. No 1. pp. 120-127

[20] Ruzmetov Ya O. Prospects for the development of wagon building in the Republic of Uzbekistan. In: Collection of Scientific Papers of the VIII All-Russian Scientific and Practical Conference “Problems and Prospects of Wagon Building Development”. Bryansk: Bryansk: BSTU; 2019. pp. 147-150

[21] Rahimov RV. Analysis and prospects for the development of performance cargo transportation in the Republic of Uzbekistan. In: Proceedings of the 1st International Conference on Problems and Perspectives of Modern Science. Vol. 2432.030066. AIP Conference Proceedings; 2022. pp. 1-6. DOI: 10.1063/5.0089484

[22] Abdullayev BA. Selection of thermal insulation materials for refrigerated wagons and containers. In: Materials of the XIV International Scientific and Technical Conference “Rolling Stock of the XXI Century: Ideas, Requirements, Projects”. St. Petersburg: FSBEI HE SPSTU; 2019. pp. 50-53

[23] Abdullayev BA. Analysis and Prospects for the Development of Transportation of Perishable Goods in the Republic of Uzbekistan. Collection of Abstracts of the National Scientific and Technical Conference “Future Prospects in the Educational Process”. St. Petersburg: FSBEI HE SPSTU; 2020. pp. 10-13

[24] Abdullayev BA. Analysis of vehicle designs for the transportation of perishable goods. In: Collection of Abstracts of the National Scientific and Technical Conference “Future Prospects in the Educational Process”. St. Petersburg: FSBEI HE SPSTU; 2018. pp. 15-17

[25] Boronenko YP. Strategic objectives of wagon builders in the development of heavy traffic. Transport of the Russian Federation. 2013;**5**(48):68-73

[26] Kiselev IG. Thermal Engineering on Railway Rolling Stock: Educational Aid/I. G. Kiselev. Moscow: Transport; 2008. p. 277

[27] Abdullayev BA. Testing of new thermal insulation materials «regent

Baltika». In: Materials of the XV International Scientific and Technical Conference "Rolling Stock of the XXI Century: Ideas, Requirements, Projects". St. Petersburg: FSBEI HE SPSTU; 2021. pp. 89-93

[28] Abdullayev BA. Influence of Solar Radiation on Heat Flows of Refrigerated Wagons and Containers. Toshkent: International Conference on Problems and Perspectives of Modern Science. ICPPMS; 2021

[29] Zaynitdinov OI. Development of new polymer composite materials for the flooring of rail carriage. International Journal of Engineering and Technology. 2020;**9**(2):378-381. DOI: 10.14419/ijet.v9i2.30519

[30] VGASU Technical Report. 2011 "Investigation of thermal characteristics of ultrathin thermal insulation "Corundum"". Available from: https://stroi34.ru/wp-content/uploads/2017/12/Ispytanie-Korund-VolgGASU-teplofizika [Accessed: December 12, 2022]

[31] Technical Conditions TU – 5760-001-83663241–2008. Liquid Ceramic Thermal Insulation Coatings of the Series "Corundum". Comp: A.S. Platov. Volgograd; 2008. p. 11

[32] Abdullayev BA. Experimental researches of new design solutions for fencing the bodies of refrigerated wagons and containers. Bulletin of the St. Petersburg University of Railway Transport. 2020;**17**(4): 498-513

[33] Kitayev BN. Thermal Effect of Solar Radiation on Wagons. Moscow: Transzheldorizdat; 1962. p. 32

[34] Anisimov PS. Testing of Wagons: Monograph/P.S. Anisimov. Moscow: Marshrut; 2004. p. 197

[35] Golubin AA. Analysis of Methods of Experimental Determination of the Heat Transfer Coefficient. Issues of Railway Transport Development in a Market Economy: A Collection of Articles by Young Scientists and Graduate Students. Moscow: Intext; 2007. pp. 108-113

[36] Golubin AA. Influence of measurement errors in determining the heat transfer coefficient of the enclosing structures of an insulated wagon. Bulletin of VNIIZHT. 2019;**78**(2): 100-104

[37] Meister AO. Determination of the average heat transfer coefficient of passenger wagon bodies. VI scientific and practical conferences. "Problems and Prospects of Wagon Building Development" Bryansk: BSTU. 2014: 93-95

[38] Naumenko SN. Accuracy of determining the heat transfer coefficient. In: Semechkin AE, editor. Railway Transport at the Present Stage: Tasks and Ways to Solve Them: Coll. Moscow: Intext; 2008. pp. 76-78

[39] Naumenko SN. Evaluation of the accuracy of determining the heat transfer coefficient of the isothermal wagon body under depot conditions. In: Coll. of Reports of Participants of Merged Science RAS Session. Moscow: MIIT; 2008. pp. 189-192

[40] RD 24.050.15-89. Methodology for Determining the Average Heat Transfer Coefficient of Enclosing Structures of Passenger Wagon Bodies. Moscow: Appr. by the Ministry of Heavy, Energy and Transport Engineering; 1989. p. 20

[41] Sokolov MM. Measurements and Control during the Repair and Operation of Wagons. Moscow: Transport; 1991. p. 158

Chapter 5

The Influence of Inclined Barriers on Airflow Over a High Speed Train under Crosswind Condition

Masoud Mohebbi, Yuan Ma and Rasul Mohebbi

Abstract

During the last decade the problem of crosswind has developed into an important subject amongst the topics in railway engineering. When high speed trains are exposed to extreme weather conditions such as intense lateral winds, storms and tornadoes, lateral loads acting on the train can cause overturning of the train. This present work analyzed the aerodynamic mechanism of a high-speed train with and without two inclined barriers. A three-dimensional numerical model of a train-barrier-crosswind system is adopted to investigate the effects of inclined angles of barriers on the flow patterns and the aerodynamic coefficients. This perusal surveys the design criteria indispensable for barriers that are installed alongside the tracks to protect the passing trains under strong side winds. By using a numerical code based on Lattice Boltzmann Method (LBM) it is attempted to initially investigate the behavior of airflow behind the barriers. Finally, it is found that the presence of the barriers has a great impact on decreasing the intensity of the air flow above the train. This study's findings could be utilized as a reference for practical usage of barriers in railway transportation.

Keywords: train aerodynamic, crosswind, high speed train, barrier, LBM

1. Introduction

The aerodynamic behaviour caused by the effect of crosswinds is one of the most serious challenges concerning the safety of high-speed trains [1–3]. The existence of crosswind would lead to the high-speed train being accompanied by a rather complicated flow field, that fluctuates both temporally and spatially. As a consequence, the train's aerodynamic properties are affected, and the running safety would be imperilled, especially when the train is through a bridge, the crosswinds become more complicated due to the bridge structure [4, 5]. Thus, the effects of crosswinds on aerodynamic behaviour are crucial.

With the fast growth of high-speed trains, the effect of crosswinds has become more and more prominent [6, 7]. To reduce the effects of crosswind, the typical windproof is widely used. It includes wind barriers and anti-wind open-cut tunnels. Wind barriers are simple and convenient devices, that are utilized in high-speed tracks at strong crosswinds conditions. In recent years, many researchers have investigated the impact of wind barriers on the aerodynamic characteristics of a high-speed train. Deng et al. [8] numerically studied the windproof performance of wind barriers in the wind-vehicle-bridge system. They found that the wind barrier is exceedingly important and significantly affects the aerodynamic coefficient, flow structure, and traveling safety. Guo et al. [9] assessed the impact of wind barriers on the traveling safety of a high-speed train to crosswinds. They determined that the existence of the wind barriers causes negative effects on the bridge. Gu et al. [10] experimentally and numerically studied the aerodynamic characteristics of a train with various lengths of vertical wind barriers. They found that when the wind barrier length varies, the impact on the train's head is more obvious than on the tail.

Liu et al. [11] numerically studied the aerodynamic behaviour of a high-speed train running through a windbreak region while being buffeted by crosswinds. They found that when the train entered the region under the crosswind, the aerodynamic coefficients change suddenly. Zou et al. [12] performed a numerical simulation to evaluate the effect of wind barriers of a high-speed train on a bridge. In their work, two vertical wind barriers were placed on either sides of the railway. Xiang et al. [13] executed a wind tunnel testing to evaluate the aerodynamic load of a high-speed train. Their findings revealed that a wind barrier of a specific height increases lift. However, it is still indistinct how the inclined angle of the barriers affects the flow pattern around the train and barriers, which benefits the development of high-speed trains. Therefore, the present work seeks to explore the mechanisms of the impacts of barriers inclined angles on the high-speed train and explain the relationships between the barriers with different inclined angles and the train.

For investigating the effects of barriers on the train aerodynamic mechanism, there are four approaches, including analytical method, numerical simulation, field measurement, and wind tunnel test. In the work of Yang et al. [14], the Finite Volume Method (FVM) in ANSYS FLUENT software was used to solve the 3D unstable incompressible Navier-Stokes equations. Catanzaro et al. [15] compared the CFD results and wind-tunnel tests of a high-speed train in a crosswind. They found that the results of the stationary model become more different from the moving model and the environment has a major impact on the train's incoming flow. Wang et al. [16] conducted an experiment work to study the influences of crosswinds on the aerodynamic properties of a high-speed train.

As a recognized and powerful numerical method, the Lattice Boltzmann approach has been widely utilised to simulate fluid flow and heat transfer problems [17, 18]. The Lattice Boltzmann equation was created and developed as a computational alternative to the solving the Navier-Stokes equations of continuum fluid physics [19]. Due to the advantages of LBM, such as its ability to dealing with complicated boundaries, parallelize the algorithm, and incorporating microscopic interactions, it has also been used to model the aerodynamic behaviours of the high-speed train. Mohebbi and Rezvani [20] utilized LBM to investigate the consequences of windbreaks geometry on two-dimensional airflow past a high-speed train. They determined that the performance of windbreak is significantly dependent on its height and edge angle. The LBM was also used by Wang et al. [21] to predict the aerodynamic behaviour of a high-speed train. They concluded that LBM has many advantages compared with the traditional CFD method. In the previous work, the authors [22] assessed the impact of porous shelters alongside a high-speed track on the vehicle's aerodynamic behaviour and the modelling was conducted utilizing the lattice Boltzmann method. The authors have proved that the LBM codes with smaller lattices can provide a reasonable accuracy result.

To the best of the authors' knowledge, the effects of two inclined barriers on the aerodynamic mechanism of a train have never been studied. In the present work, the German Intercity Express (ICE3) high-speed train was focused and a three-dimensional numerical model of the train-barrier-crosswind system is adopted to investigate the effect of two inclined barriers on the train's aerodynamic mechanism through the Lattice Boltzmann method. The effect of barrier inclined angle and direction on the velocity, pressure, turbulence intensity, streamlines, and aerodynamic coefficients are investigated.

2. The model description

Figure 1 shows the geometry under consideration in the present work. The height of the ICE3 high-speed train (H_{Train}) is 3.5 m. Two barriers are set on the right and left sides, respectively. The distance between the two barriers ($L_{Barrier\ Distance}$) is 14 m. The spacing between the barrier and the boundary of the calculation domain is $10H_{Train}$, which is the same as the height of the computational domain. The wind blows parallel to ground level from the right to the left, with a uniform velocity of 80 m/s. The height and width of the calculation domain are $10H_{Train}$ and $2H_{Train}$, respectively. Besides, the forces and momentums concerning the coordinate system's directions (x, y, z) are specified by the European Standard EN14067-6 (**Figure 1**).

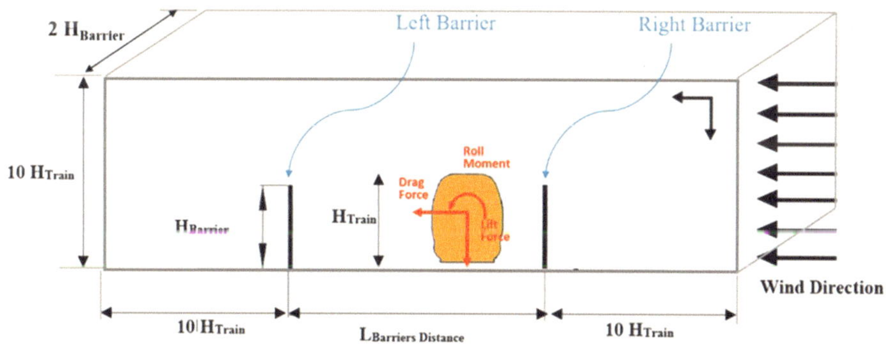

Figure 1.
The computational domain and the aerodynamic coefficients definitions.

Figure 2.
The model of considered geometry.

Type I: θ = 0°	Type II: θ = +2.5°	Type III: θ = +5.0°
Type IV: θ = +7.5°	Type V: θ = +10.0°	Type VI: θ = −2.5°
Type VII: θ = −5.00°	Type VIII: θ = −7.5°	Type IX: θ = −10.0°

Table 1.
The barrier types with different inclined angles.

In the present work, the effects of barriers are focused on. The heights of the two barriers ($H_{Barrier}$) are the same and $H_{Barrier}$ = 3.0 m. As shown in **Figure 2**, the effect of barrier inclined angle (θ) is investigated in the present work. According to the different barrier inclined angles, nine types of cases are simulated (**Table 1**).

3. The numerical method

The Lattice Boltzmann Method is used in this paper to model fluid flow past a high-speed train with two barriers. The velocity of the two-dimensional nine-speed (D2Q9) model in multiple directions can be defined as [23]:

$$e_i = \begin{cases} (0,0) & (i=0) \\ \cos\left(\frac{i\pi}{2}-\frac{\pi}{2}\right), \sin\left(\frac{i\pi}{2}-\frac{\pi}{2}\right).c & (i=1,\ldots,4) \\ \sqrt{2}\left(\cos\left(\frac{i\pi}{2}-\frac{9\pi}{4}\right)\right), \sin\left(\frac{i\pi}{2}-\frac{9\pi}{4}\right).c & (i=5,\ldots,8) \end{cases} \tag{1}$$

where $c = \Delta x/\Delta t$ is the velocity of lattice, Δx is the lattice space, Δt is the time step and i is the different direction.

The governing equation in the lattice Boltzmann method is:

$$f_i(x+e_i\Delta t, t+\Delta t) = f_i(x,t) + \frac{\Delta t}{\tau_v}\left[f_i^{eq}(x,t) - f_i(x,t)\right] \tag{2}$$

where f_i is the distribution function and f_i^{eq} is the equilibrium distribution function, which can be calculated according to:

$$f_i^{eq} = w_i\rho\left[1 + \frac{e_i u}{c_s^2} + \frac{1}{2}\frac{(e_i u)^2}{c_s^4} - \frac{1}{2}\frac{u^2}{c_s^2}\right] \tag{3}$$

where w_i are the weights, $w_0 = 4/9$, $w_{1-4} = 1/9$, $w_{5-8} = 1/36$.

The distribution functions can be used to obtain the macroscopic variables,

$$\rho = \sum_i f_i \tag{4}$$

$$\rho u = \sum_i e_i f_i \tag{5}$$

With the multi-scaling expansion, the mass and the moment equations can be obtained:

$$\frac{\partial \rho}{\partial t} + \nabla \bullet (\rho u) = 0 \tag{6}$$

$$\frac{\partial \rho}{\partial t} + \nabla \bullet (\rho u u) = -\nabla P + \vartheta\left[\nabla^2(\rho u) + \nabla(\nabla \bullet (\rho u))\right] \tag{7}$$

More information on the present numerical method can be found in the authors' previous paper [19].

In this study, Wall Modelled Large Eddy Simulation (WMLES) [24] approach was used to consider the turbulence. LES has been proved as a compliant numerical approach in computing and simulating unsteady turbulent flows. WMLES takes the wall models into account and its primary idea is that the near-wall turbulence length scales grow linearly with the wall distance, leading to the smaller and smaller eddies as the wall is approached.

The aerodynamic coefficients in terms of non-dimensional characteristics are described by EN 14067-1 as follows:

$$\text{Forces} : F = \frac{\rho U_\infty^2}{2} \bullet A.C \tag{8}$$

$$\text{Momentums} : M = \frac{\rho U_\infty^2}{2} \bullet A \bullet l.C \tag{9}$$

$$\text{Pressure} : P - P_\infty = \frac{\rho U_\infty^2}{2} \bullet C_p \tag{10}$$

U_∞ is the free stream velocity, P is the local static pressure, and P_∞ is the free-stream static pressure.

4. The grid independence and validation

For the validation of the current CFD code based on LBM, the aerodynamic force and moment coefficients are calculated to compare with the experimental wind tunnel works from EN 14067-6:2010, which is performed by Schober et al. [25]. **Table 2** shows the comparison of the aerodynamic coefficient by present code and the standard data. The high-speed train model with no barrier is tested and the drag, lift, and rolling moment coefficients have been calculated for comparison. It is clear to see that the disparity is acceptable, which validated the present code and results.

Mode	No-barrier		
	CFD	Standard	Error percentage CFD relative to standard (%)
Drag coefficient	0.33	0.38	−14
Lift coefficient	−4.86	−5.49	−11
Rolling moment coefficient	3.00	3.35	−11

Table 2.
Verification with the standard EN 14067-6: 2010.

5. The results and discussion

In the present section, the effects of barriers with the inclined angle (nine types) on the flow field and aerodynamic forces and moments are discussed in form of velocity contours, velocity vectors, total pressure contours, turbulence intensity contours, and aerodynamic force.

Figure 3 shows the effects of barriers with the inclined angle on the velocity contours around the train. When there is no barrier, the high flow velocity can be found at the upper windward of the train as the flow approaches the train. A large low-velocity zone occurs on the leeward side, which is the left side due to the inflow direction. Meanwhile, the high-velocity gradient is generated on the top of the train. It is obvious that the flow velocity around the train is rather uneven, leading to a considerable negative influence on the stability of the train. As the barriers are introduced, the flow velocity around the train becomes more even, and the flow pattern changes substantially. When $\theta = 0°$, the right barrier takes the place of the train to withstand the crosswind. As a result, the line of a large velocity gradient is formed

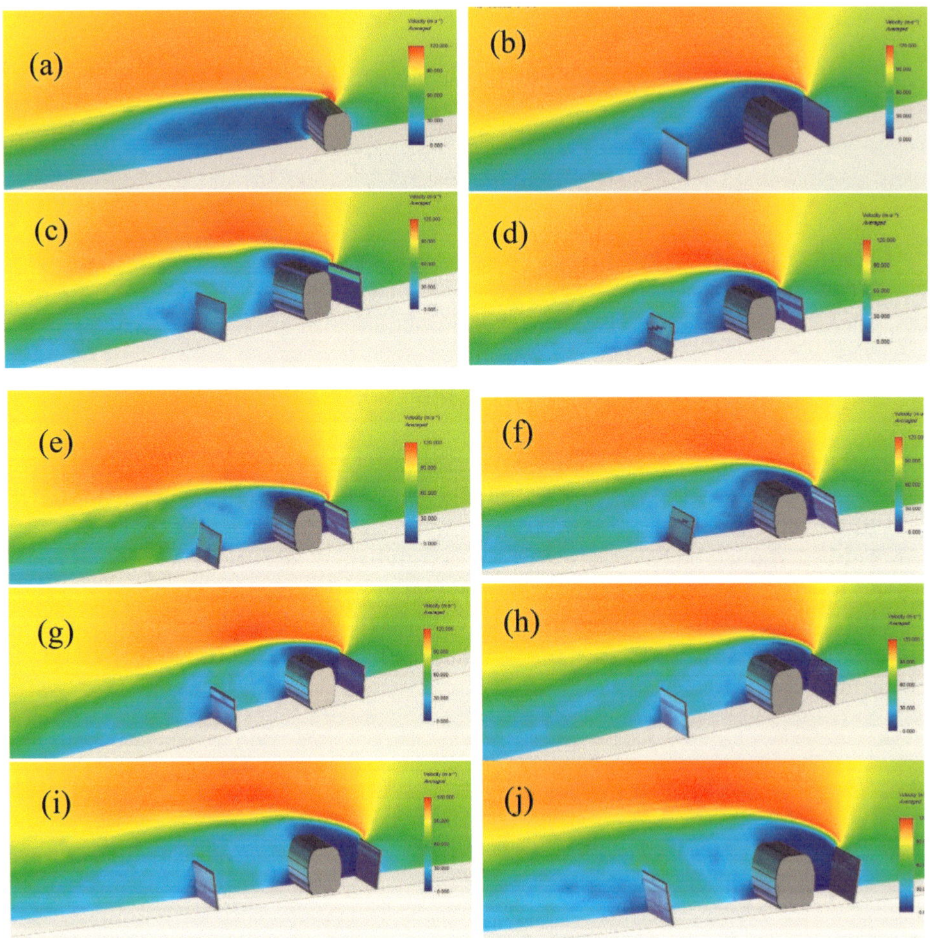

Figure 3.
The velocity contours (a) without barriers, (b) $\theta = 0°$, (c) $\theta = +2.5°$, (d) $\theta = +5.0°$, (e) $\theta = +7.5°$, (f) $\theta = +10.0°$, (g) $\theta = -2.5°$, (h) $\theta = -5.0°$, (i) $\theta = -7.5°$, and (j) $\theta = -10.0°$.

from the top side of the barrier. And the train is surrounded by the low-velocity air caused by the right barrier and the flow velocity around the train is almost even. However, when the angle changes to θ = +2.5°, +5.0°, +7.5°, and +10.0°, the function of the right barrier mentioned above weakens. The area of the low-velocity region caused by the right barrier is reduced and the distribution of flow velocity on the left of the train becomes unequal correspondingly. As shown in **Figure 3g–j**, when the angle becomes a negative value, the flow velocity on the left side of the train increases and becomes larger than that without a barrier. As a result, the velocity unevenness of the flow field around the train is significantly increasing.

Figure 4 shows the velocity vectors around the train and barriers. Without the barriers, the flow separations happen from the train's leeward side and then leave the surface and are involved in the leeward vortex. When the barriers are induced, the flow separations begin from the barrier, and the train is surrounded by vortices. As a result, the difference in velocity between the left and right sides of the train decreases, and the velocity on the windward of the train is lower than that on the leeward. However, when the inclined angle changes from positive to negative, the velocity on the leeward of the train increases significantly.

To investigate the effect of barriers and inclined angles on the pressure of the train, **Figure 5** depicts the total pressure contours. Without the barrier, the pressure on the right side of the train is apparently higher than on the left side because of the

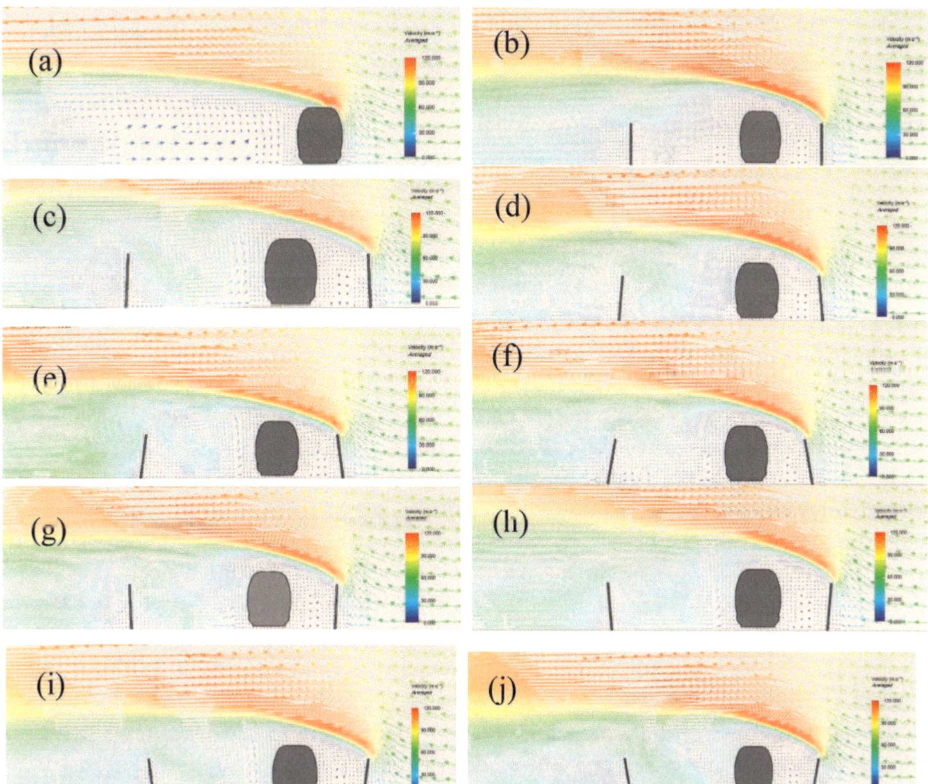

Figure 4.
The velocity vectors (a) without barriers, (b) θ = 0°, (c) θ = +2.5°, (d) θ = +5.0°, (e) θ = +7.5°, (f) θ = +10.0°, (g) θ = −2.5°, (h) θ = −5.0°, (i) θ = −7.5°, and (j) θ = −10.0°.

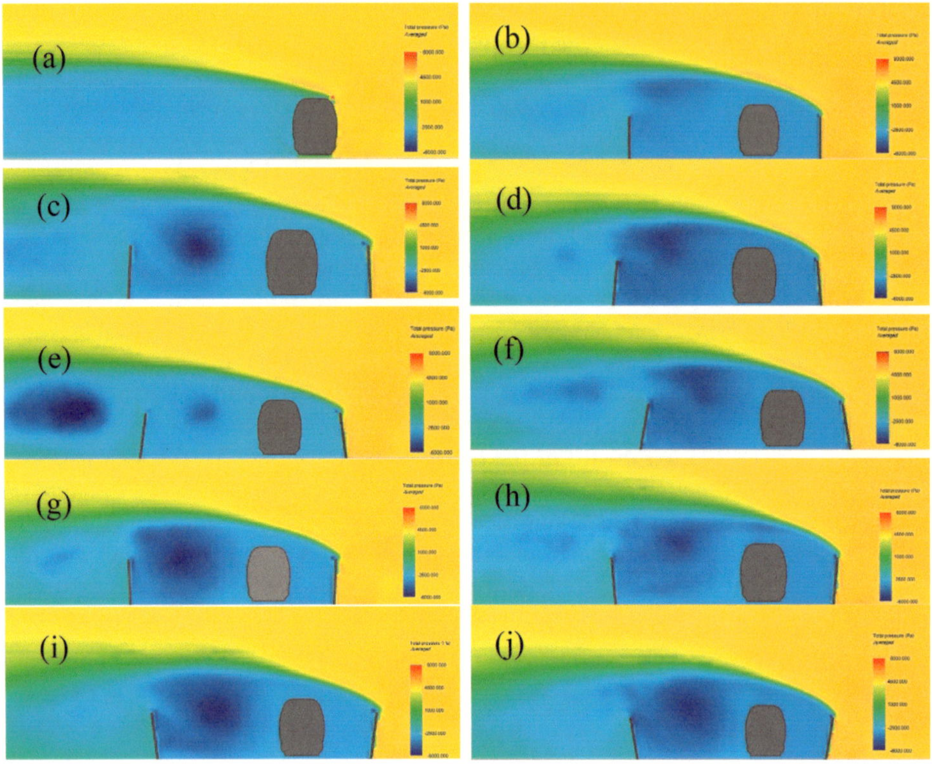

Figure 5.
The total pressure contours (a) without barriers, (b) $\theta = 0°$, (c) $\theta = +2.5°$, (d) $\theta = +5.0°$, (e) $\theta = +7.5°$, (f) $\theta = +10.0°$, (g) $\theta = -2.5°$, (h) $\theta = -5.0°$, (i) $\theta = -7.5°$, and (j) $\theta = -10.0°$.

crosswind. It should be stated that the pressure on the left side of the train is approximately identical. When two barriers with $\theta = 0°$ are placed, a low-pressure zone is formed between two barriers. The pressure gradient in this low-pressure zone is small and the relatively low pressure occurs around the top of the left barrier. When the angle increases to +2.5°, the pressure on the right side of the train increases slightly. However, there is one noticeable lower pressure region between the left barrier and the train, which also leads to the pressure gradient. When the inclined angles are negative value, as seen in **Figure 5g–f**, the area of the lower pressure region increase and occupies almost the whole zone between the left barrier and the train. Moreover, when the inclined angle of barriers varies, the pressure gradient between the train and the right barrier is always small.

To analyse the flow pattern characteristics, the turbulence intensity is calculated to show the turbulence level, which is presented in **Figure 6**. Firstly, the existence of the barriers and the inclined angle affect the flow pattern characteristics. Without the barriers, the high turbulence intensity occurs on the train's leeward side, and the turbulence intensity on the windward side is almost equal to 0. However, when two barriers with $\theta = 0°$ are placed on both sides of the train, the turbulence intensity around the train is affected significantly. Both the windward and leeward sides experience a rise in turbulence intensity. The windward side, which is between the train and the right barrier, has larger turbulence intensity than the leeward side. Besides,

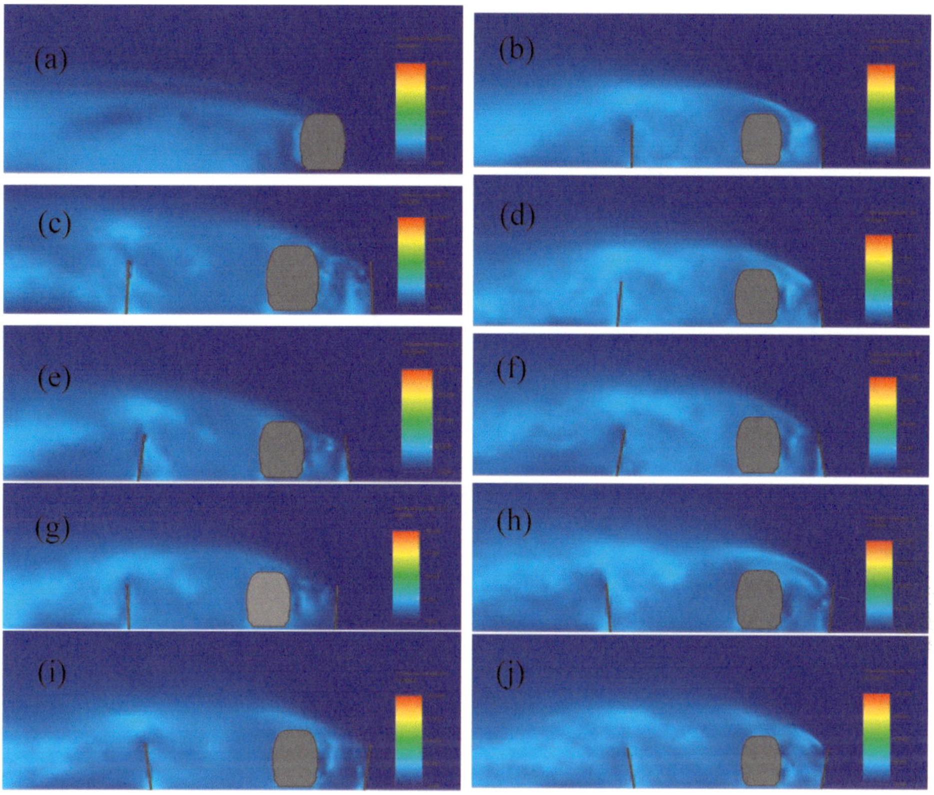

Figure 6.
The turbulence intensity contours (a) without barriers, (b) $\theta = 0°$, (c) $\theta = +2.5°$, (d) $\theta = +5.0°$, (e) $\theta = +7.5°$, (f) $\theta = +10.0°$, (g) $\theta = -2.5°$, (h) $\theta = -5.0°$, (i) $\theta = -7.5°$, and (j) $\theta = -10.0°$.

the turbulence intensity on the right side of the right barrier is adjacent to 0. As the barrier inclined angle increases to +2.5°, the turbulence intensity decreases and the diminishment is more significant between the left barrier and the train, which shows the effect of barriers on the turbulence intensity difference on both sides. However, when the inclined angle increases to +5.0° and +7.5°, the turbulence intensity difference between both sides decreases. That's to say, the difference between the leeward and windward sides of the train becomes small. When the inclined angle increases from +7.5° to +10.0°, the difference increases and the turbulence intensity on the leeward of the train is slightly larger than on the windward. When the inclined angle becomes minus, the turbulence intensity changes slightly.

Figure 7 shows the effect of barrier inclined angle on the drag force coefficient, lift force coefficient, rolling moment coefficient, and lee-rail rolling moment coefficient. Firstly, it can be found that using the barriers with a positive inclined angle cannot decrease the drag coefficient. The barriers with zero or negative angles can decrease the drag coefficient. The barriers with a positive inclined angle cause a positive drag coefficient of the train, and the barriers with a negative angle lead to a negative value. Besides, the absolute value of the drag coefficient of positive angles is larger than those of negative ones. This can be analysed by comparing two types of barriers. The positive angles of barriers lead to the narrower space around the train. It should be

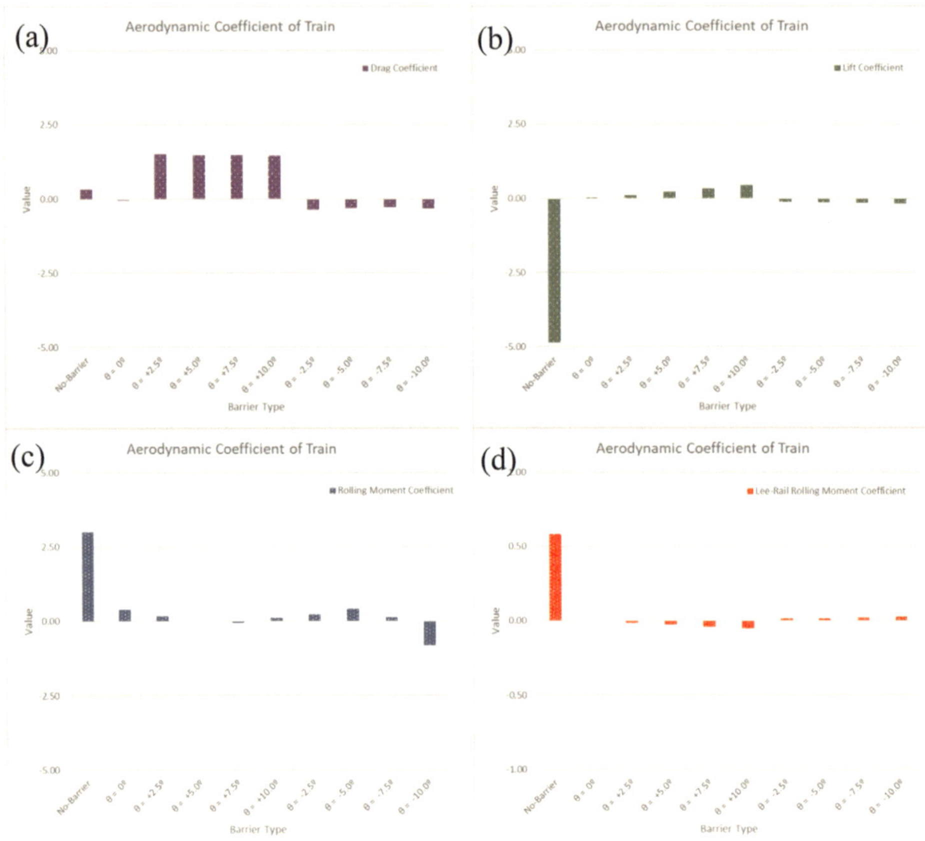

Figure 7.
The aerodynamic coefficients: (a) drag force coefficient, (b) lift force coefficient, (c) rolling moment coefficient, and (d) lee-rail rolling moment coefficient.

stated that the barriers with zero inclined angle have the smallest value of drag coefficient. As for the lift coefficient, the existence of barriers is beneficial to the diminution of the lift coefficient regardless of the inclined angle. The zero inclined angles can cause the smallest lift coefficient. The barriers with positive angles lead to the positive value of the lift coefficient and the negative angles cause the negative value, which is similar to the effect on the drag coefficient. Concerning the rolling moment and the lee-rail rolling moment coefficients, the presence of barriers can decrease them. These parameters are important for the train, which is because they are answerable for the loading and unloading of wheelsets. It is obvious that the existence of barriers with any inclined angles, including zero, can decrease the rolling and lee-rail rolling moment coefficients. As the inclined angle of the barrier become $\theta = +5.0°$, the value of the rolling moment coefficient become the minimum. And then when the angle increases, the value also increases. The effect of inclined angle on the lee-rail rolling moment coefficient is more distinct. The positive inclined angle leads to the negative lee-rail rolling moment coefficient and the negative angle cause the positive coefficient.

To demonstrate the impact of various types of barriers (with barriers or not, barriers inclined angle), the effects of barrier type on aerodynamic coefficients are

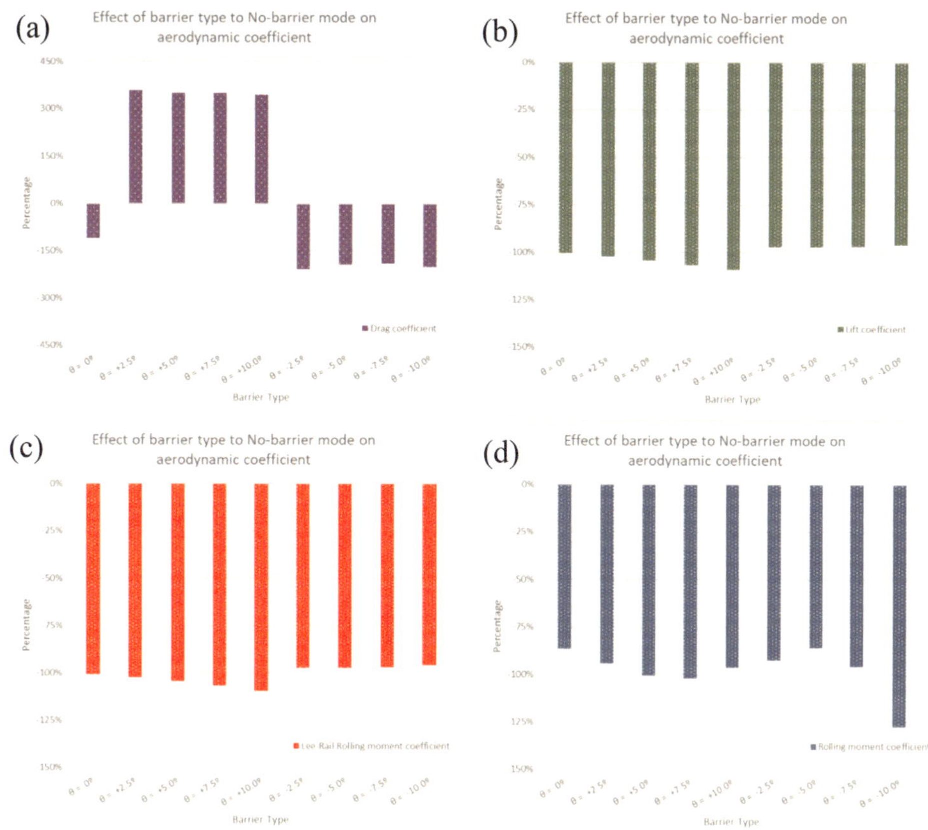

Figure 8.
The effects of barrier type on aerodynamic coefficients, (a) drag force coefficient, (b) lift force coefficient, (c) rolling moment coefficient, and (d) lee-rail rolling moment coefficient.

shown in **Figure 8**. In summary, to reduce the aerodynamic coefficients of the train, the barriers with zero inclined angle are the most optimal choice. It appears that the barriers with a positive inclined angle have an inverse effect on the drag coefficient. But the vertical barrier with zero inclined angles has the same effect as the negative ones. As for the lift coefficient, one can find that the existence of any barriers, including the zero inclined angle barriers, leads to a similar influence on the lift coefficient of the train. The same trend also happens for the lee-rail rolling moment coefficient and rolling moment coefficient.

6. Conclusions

In the present work, a 3D numerical model of a train-barrier-crosswind system is adopted to investigate the influence of two inclined barriers on the aerodynamic mechanism of a high-speed train by utilizing the Lattice Boltzmann Method. The influence of barrier inclination angle and direction on the velocity, pressure, turbulence intensity, streamlines, and aerodynamic coefficients are investigated. It has been discovered that utilizing barriers with a positive inclination angle cannot reduce the drag coefficient. The drag coefficient can be reduced by using barriers with zero or negative angles. When all aerodynamic coefficients are taken into account, barriers with zero or negative inclination angles are the best choice for reduction of its. The value of the rolling moment coefficient reaches a minimum when the inclined angle of the barrier becomes = +5.0°.

Author details

Masoud Mohebbi[1*], Yuan Ma[2] and Rasul Mohebbi[3]

1 Iran University of Science and Technology, Tehran, Iran

2 Hong Kong Polytechnic University, Kowloon, China

3 Damghan University, Damghan, Iran

*Address all correspondence to: mohebbi@rail.iust.ac.ir

References

[1] Schetz JA. Aerodynamics of high-speed trains. Annual Review of Fluid Mechanics. 2001;**33**(1):371-414

[2] He XH, Li H. Review of aerodynamics of high-speed train-bridge system in crosswinds. Journal of Central South University. 2020;**27**(4):1054-1073

[3] Jingping XI, Zhixiang H, Li C. Review of aerodynamic investigations for high speed train. Mechanics in Engineering. 2013;**35**(2):1-2

[4] Barcala MA, Meseguer J. An experimental study of the influence of parapets on the aerodynamic loads under cross wind on a two-dimensional model of a railway vehicle on a bridge. Proceedings of the Institution of Mechanical Engineers, Part F: Journal of Rail and Rapid Transit. 2007;**221**(4): 487-494

[5] Yao Z, Zhang N, Chen X, Zhang C, Xia H, Li X. The effect of moving train on the aerodynamic performances of train-bridge system with a crosswind. Engineering Applications of Computational Fluid Mechanics. 2020; **14**(1):222-235

[6] Zhou L, Liu T, Chen Z, Li W, Guo Z, He X, et al. Comparison study of the effect of bridge-tunnel transition on train aerodynamic performance with or without crosswind. Wind and Structures. 2021;**32**(6):597-612

[7] Guo Z, Liu T, Liu Z, Chen X, Li W. An IDDES study on a train suffering a crosswind with angles of attack on a bridge. Journal of Wind Engineering and Industrial Aerodynamics. 2021;**217**: 104735

[8] Deng E, Yang W, He X, Zhu Z, Wang H, Wang Y, et al. Aerodynamic response of high-speed trains under crosswind in a bridge-tunnel section with or without a wind barrier. Journal of Wind Engineering and Industrial Aerodynamics. 2021;**210**:104502

[9] Guo W, Xia H, Karoumi R, Zhang T, Li X. Aerodynamic effect of wind barriers and running safety of trains on high-speed railway bridges under cross winds. Wind and Structures. 2015;**20**(2): 213-236

[10] Gu H, Liu T, Jiang Z, Guo Z. Experimental and simulation research on the aerodynamic effect on a train with a wind barrier in different lengths. Journal of Wind Engineering and Industrial Aerodynamics. 2021;**214**:104644

[11] Liu T, Chen Z, Zhou X, Zhang J. A CFD analysis of the aerodynamics of a high-speed train passing through a windbreak transition under crosswind. Engineering Applications of Computational Fluid Mechanics. 2018; **12**(1):137-151

[12] Zou S, He X, Wang H. Numerical investigation on the crosswind effects on a train running on a bridge. Engineering Applications of Computational Fluid Mechanics. 2020;**14**(1):1458-1471

[13] Xiang H, Li Y, Chen B, Liao H. Protection effect of railway wind barrier on running safety of train under cross winds. Advances in Structural Engineering. 2014;**17**(8):1177-1187

[14] Yang W, Deng E, Lei M, Zhu Z, Zhang P. Transient aerodynamic performance of high-speed trains when passing through two windproof facilities under crosswinds: A comparative study. Engineering Structures. 2019;**188**: 729-744

[15] Catanzaro C, Cheli F, Rocchi D, Schito P, Tomasini G. High-speed train crosswind analysis: CFD study and validation with wind-tunnel tests. In: International Conference on Engineering Conferences International 2010 Sep 12. Champions: Springer; 2010. pp. 99-112

[16] Wang M, Li XZ, Xiao J, Zou QY, Sha HQ. An experimental analysis of the aerodynamic characteristics of a high-speed train on a bridge under crosswinds. Journal of Wind Engineering and Industrial Aerodynamics. 2018;**177**: 92-100

[17] Chen S, Doolen GD. Lattice Boltzmann method for fluid flows. Annual Review of Fluid Mechanics. 1998;**30**(1):329-364

[18] Aidun CK, Clausen JR. Lattice-Boltzmann method for complex flows. Annual Review of Fluid Mechanics. 2010;**42**:439-472

[19] Benzi R, Succi S, Vergassola M. The lattice Boltzmann equation: Theory and applications. Physics Reports. 1992; **222**(3):145-197

[20] Mohebbi M, Rezvani MA. Two-dimensional analysis of the influence of windbreaks on airflow over a high-speed train under crosswind using lattice Boltzmann method. Proceedings of the Institution of Mechanical Engineers, Part F: Journal of Rail and Rapid Transit. 2018;**232**(3):863-872

[21] Wang Y, Wang Y, An Y, Chen Y. Aerodynamic simulation of high-speed trains based on the Lattice Boltzmann Method (LBM). Science in China, Series E Technological Sciences. 2008;**51**(6): 773-783

[22] Mohebbi M, Rezvani MA. Analysis of the effects of lateral wind on a high speed train on a double routed railway track with porous shelters. Journal of Wind Engineering and Industrial Aerodynamics. 2019;**184**:116-127

[23] Qian YH, d'Humières D, Lallemand P. Lattice BGK models for Navier-Stokes equation. EPL (Europhysics Letters). 1992;**17**:479

[24] Bose ST, Park GI. Wall-modeled large-eddy simulation for complex turbulent flows. Annual Review of Fluid Mechanics. 2018;**50**:535-561

[25] Schober M, Weise M, Orellano A, Deeg P, Wetzel W. Wind tunnel investigation of an ICE 3 endcar on three standard ground scenarios. Journal of Wind Engineering and Industrial Aerodynamics. 2010;**98**(6-7):345-352

Section 3

Service Delivery and Operations

Chapter 6

ComplexTrans Total Ground Mobility Solution Based on Mutual Adaptation and Deep Cooperation of Road and Rail

Jiri Hofman, Roman Cermak and Jiri Korinek

Abstract

Current land transport is not optimal. Road transport is congested and rail transport is under-utilised and unprofitable. Land transport is based on the burning of fossil fuels and contributes to climate change. Hence the EU's desire to push for electric propulsion on the road and to make rail the backbone of Europe's transport system. Developments in transport are solving some problems but creating others. The ComplexTrans project addresses private and public transport of people and freight in and between cities and removes current and upcoming transport problems in a natural way (without restrictions and subsidies), based on the mutual adaptation of electric road and rail vehicles and their deep intermodal and multimodal cooperation and using fast mixed passenger/freight trains. The solution for land transport is not competition but cooperation between road and rail.

Keywords: transport, road, rail, intermodal, multimodal, e-mobility, battery exchange, recharging infrastructure, city transport, renewable energy

1. Introduction

1.1 The current state of transport

The main current modes of transport are

- land (road and rail) transport, which provides transport of passengers and goods mainly over short and medium distances (up to about 1000 km), with road transport dominating over rail transport

- air transport, which mainly carries passengers over medium and long distances (from about 500 km)

- and shipping, which mainly transports goods over medium distances (river transport) and long distances (maritime transport).

All these modes of transport have their advantages and disadvantages.

A common problem with all modes of transport (except electrified rail transport) is their dependence on fossil fuels and the production of greenhouse gases that contribute to climate change.

1.2 Current state of transport development

The current state of transport development can be characterised by these main objectives.

1.2.1 In the field of emissions

Switch the majority of continental passenger and freight transport from road to rail in order to reduce the energy intensity of transport and emissions, both as a result of the lower energy intensity of rail and the increasing possibility of powering electric rail with renewable sources of electricity.

Replace continental passenger air transport over medium distances by high-speed rail, which can offer shorter or comparable overall travel times over distances of up to about 1000 km.

Use more river and sea transport for freight. However, this mode of transport is linked to waterways and sufficient water in rivers.

Replace combustion engines in cars with emission-free powertrains. At present, the development of emission-free drives is mainly focused on the following directions

a. development of synthetic fuel for internal combustion engines,

b. battery-powered independent electric motor propulsion,

c. independent electric propulsion powered by a fuel cell drawing energy from hydrogen and

d. dependent electric propulsion powered by trolley or non-contact (inductive).

All these methods of eliminating emissions can only be successful if the electricity is generated by emission-free sources.

1.2.2 In the field of transport

Reduce the number of private cars by (autonomous) car-sharing.

Replace human driving with autonomous driving. And autonomous driving means also need for connectivity and 5G networks.

By creating and extending MaaS.

So, the aim of the development in land transport is to shift the transport of goods and passengers from road to rail, to shift continental air transport up to 1000 km to rail and to reduce air transport, to strengthen water transport and to have full battery or hydrogen electromobility in road transport, probably also associated with a reduction in the number of cars, which will be replaced by autonomous car sharing. The use of public transport should be made as easy as possible by the introduction of the IT platform MaaS. Developments in land transport, in conjunction with renewable energy sources, are heading towards zero emissions, but also to a restriction of freedom of movement, unless much more suitable traction batteries are available.

Could transport development be organised in such a way that transport emissions are reduced to zero, whilst at the same time individual freedom of movement is not restricted, parking problems are eliminated and traffic density is reduced, the switch from road to rail is voluntary, more air transport can be transferred to rail and rail becomes self-financing? Is there such a transport solution?

Yes—it exists, it is based on a deep cooperation between road and rail and is described in next chapters.

2. Methodology and results

There is one transport solution that can cope with the demands placed on land transport much cheaper and better than the above set of transport solutions currently being developed and implemented. It shifts most passenger and freight traffic from road to rail without any restrictions, effectively replaces continental air transport with rail transport up to a distance of about 3000 km, removes the problems of parking and congestion in and between cities, makes rail self-financing, increases speed, comfort and safety, makes land transport independent of fossil fuels and, in cooperation with renewable sources of electricity, makes it carbon-free. And brings many other benefits.

This solution is called ComplexTrans [1–4] and is based on the mutual adaptation of road and rail vehicles and their deep cooperation with each other and with renewable electricity and on exploiting the synergy of all systems working together (**Figure 1**).

ComplexTrans deals with private and public door-to-door transport of people and goods within and between cities, using almost exclusively electricity for propulsion. Whilst local transport is carried out by road, long-distance transport is mainly carried out by rail. A variant of long-distance transport is the use of the road, with electricity supplied to the vehicles by interchangeable range-extenders powered by hydrogen, synthetic fuel or, temporarily, fossil fuels.

Figure 1.
ComplexTrans "takes everything"—people, cars, transport modules, wagons—like an "overland ferry".

3. Results

3.1 ComplexTrans rail vehicles

The ComplexTrans system has basically two types of rail vehicles (**Figure 2**).

3.1.1 Fast train units for mixed passenger and goods transport

Are made up of end and inserted double-deck carriages with Jacobs bogies (**Figure 3**), whereby the lower deck is used for the transport of adapted passenger vehicles (so-called coupemobiles) including their crew and/or freight transport modules; and the upper deck is used for the transport of passengers and/or delivery boxes which are carried in the places not occupied by passengers. Small variable compartments of two sizes for 2(1) or 4(2) seated (lying) persons are used for transportation, providing sufficient privacy and comfort for passengers.

The 400 m train unit consists of either two nine-carriages units with integrated service carriage or four four-car units with two separate service carriages. From the rear the fast freight waggons are attached to the train units (**Figure 4**).

The units are equipped with automatic couplers and allow easy splitting of trains. The front carriages allow passage to the adjacent unit. Propulsion is electric and distributed.

Figure 2.
ComplexTrans train consists from fast units with double deck mixed carriages.

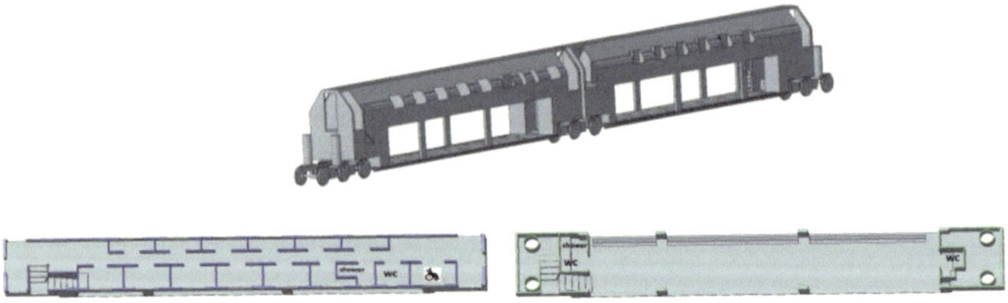

Figure 3.
ComplexTrans double-deck rail carriage for mixed passenger and freight transport. The upper deck carries passengers or delivery boxes in small compartments (left), while the lower deck carries adapted passenger and freight vehicles. Sanitary facilities are available on both floors.

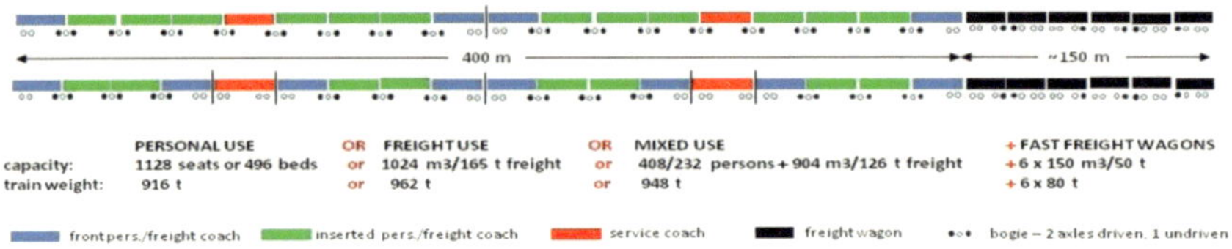

Figure 4.
The mixed train ComplexTrans consists of two or four fast train units for mixed transport of passengers and goods with a length of 400 m (left) and fast freight cars (right).

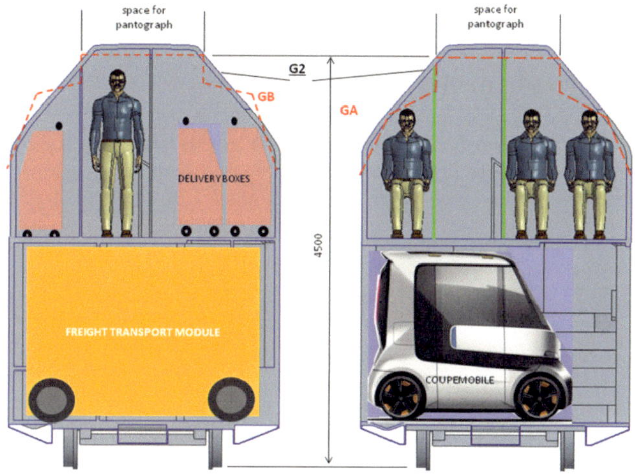

Figure 5.
ComplexTrans double-deck rail carriages can be operatedon all European electrified rail networks with outline GA, GB, G2 and larger.

The carriages are compatible with the European GA, GB outline for electrified lines and can stop at ComplexTrans double-deck platforms as well as at all common European platforms of all heights (550, 750 and 950 mm). The operating speed is 230 km/h and the maximum axle load is 17 t. With a distance of terminals approximately 50 km and a stopping time of 3 minutes, the average speed of ComplexTrans fast trains is 150 km/h.

Thanks to the flexible interior of the trains as well as the passenger and freight transport modules, the entire capacity can be flexibly changed from 100% passenger to 100% freight (**Figure 5**).

3.1.2 Fast rail freight waggons

Are attached individually or in small groups to the ends of ComplexTrans trains and are used to transport larger or heavier loads between sidings. These are covered flat waggons with an integrated pocket/basket for transporting vertically manipulable and non-manipulable freight trailers.

The drive of the car can be controlled by the train bus, remotely or by its own control unit autonomously, with the control unit also carrying all information about the load. An automatic coupling allows quick coupling and uncoupling and an electrically operated brake allows a quick two phase brake test (**Figure 6**).

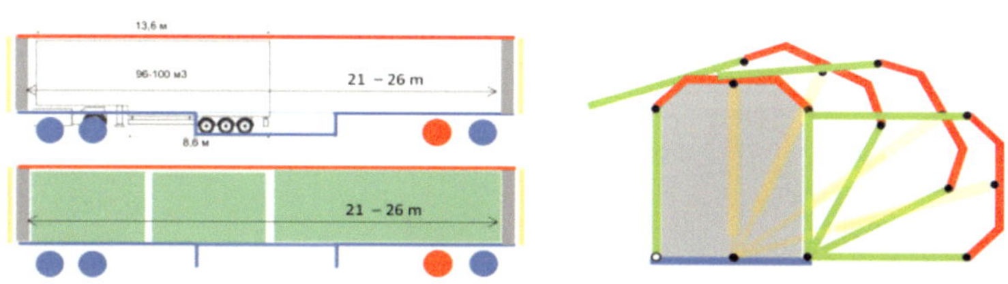

Figure 6.
The ComplexTrans fast rail freight wagon has an automatic coupling, self-propulsion ~500 kW with energy storage and is equipped with a hinged cover. Operating speed 230 km/h and axle load max. 20 t. It can transport sea containers, cargo trailers, transport modules and other large goods.

3.1.3 Fast direct trains

Connect the main stations or terminals of the main and major cities, do not stop at intermediate passenger-cargo terminals and stop only every approximately 250–300 km. At the same operating speed of 230 km/h they reach a cruising speed of over 200 km/h. They may or may not be of the ComplexTrans type.

3.2 ComplexTrans road vehicles

The road transport in the ComplexTrans system is mainly provided by adapted electric cars.

3.2.1 Coupemobiles

Are four-wheeled passenger cars adapted for fast loading, unloading and transport on ComplexTrans trains. Their main features are a spacious cabin, approximately in the shape of a cube with rounded edges, with a variable interior for 5 adults and their luggage, forward and reverse retractable axles with adjustable ground clearance and impact protection function, and an electrically driven rear axle from the back replaceable traction battery. The variable interior can be easily adapted for on-road driving, for work, play or rest for up to five people, or for two people to sleep comfortably (**Figure 7**).

3.2.2 Road freight transport modules

Are enclosed four-wheel "vans without driver's cab" and are used for the transport of piece goods, Euro pallets or air containers. They are propelled by a battery-powered electric motor with a capacity sufficient for a range of 50–100 km (**Figure 8**).

Similar to the coupemobile, the freight transport module has forward and reverse extendable axles with height adjustability and crash protection. Both axles are retractable and equipped with a coupling and at least one wheel of each axle is driven

Figure 7.
Coupemobile, a passenger e-car of the ComplexTrans system for the comfortable transport of five adults and their luggage. Retractable axles, from the back swappable battery, roof grip points and couplers give these cars a whole new range of possibilities, an "unlimited" range and a widely acceptable price.

Figure 8.
Long and short freight transport module for transporting goods in ComplexTrans mixed trains. It has its own electric drive, is transported to the end users by road platoons with a guiding vehicle, and can be remotely controlled for local manipulation.

by an electric motor (cross-placed). All wheels of the long transport module can be swivelled 90° when loading into the train.

The maximum weight of the loaded transport module is 3.5 t, the maximum speed in a platoon is approx. 90 km/h and in a solo journey approx. 25 km/h.

3.3 Terminals: interfaces between road and rail for the transfer of people and goods

In order for road and rail vehicles to work together, road-rail interchange terminals need to be built. In the ComplexTrans system, terminals are shared for passenger and freight transport, but passenger and freight transport are separated for safety and traffic reasons. Two-storey platforms are used for passenger boarding and alighting and loading and unloading of parcel delivery boxes takes place on the upper platform floor, whilst loading and unloading of passenger and freight vehicles takes place on the lower platform floor (**Figure 9**).

The terminals are located on the outskirts of towns so that the road vehicles carried on the trains do not clog up the town centres. In accordance with the different sizes of cities, there are two types of terminals—terminals and semi-terminals (**Figure 10**).

The terminals will also include preparatory and staging car parks for the vehicles being transported.

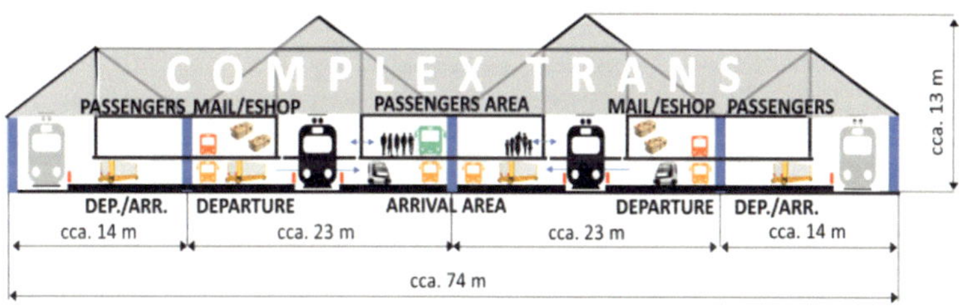

Figure 9.
Two main and two additional double-deck platforms for passenger-freight exchange. The train stopping time in the terminal can be reduced to about 3 min.

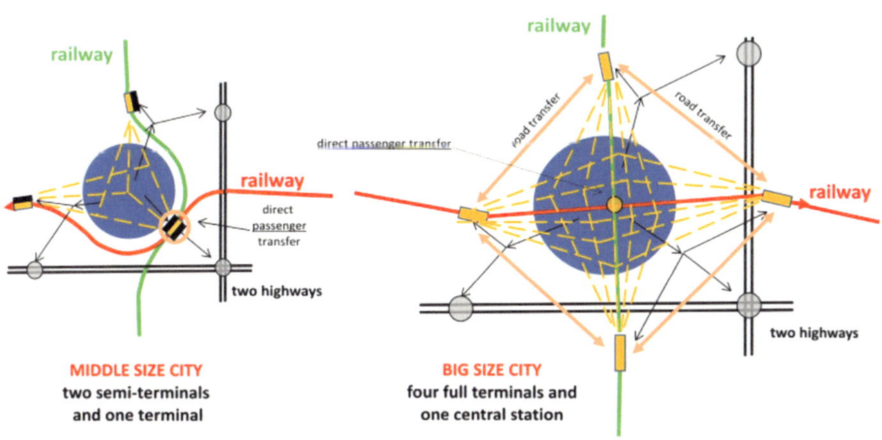

Figure 10.
ComplexTrans (semi-)terminals on the outskirts of cities are used for passenger boarding and alighting as well as cargo loading and unloading; central stations can be used for passengers only.

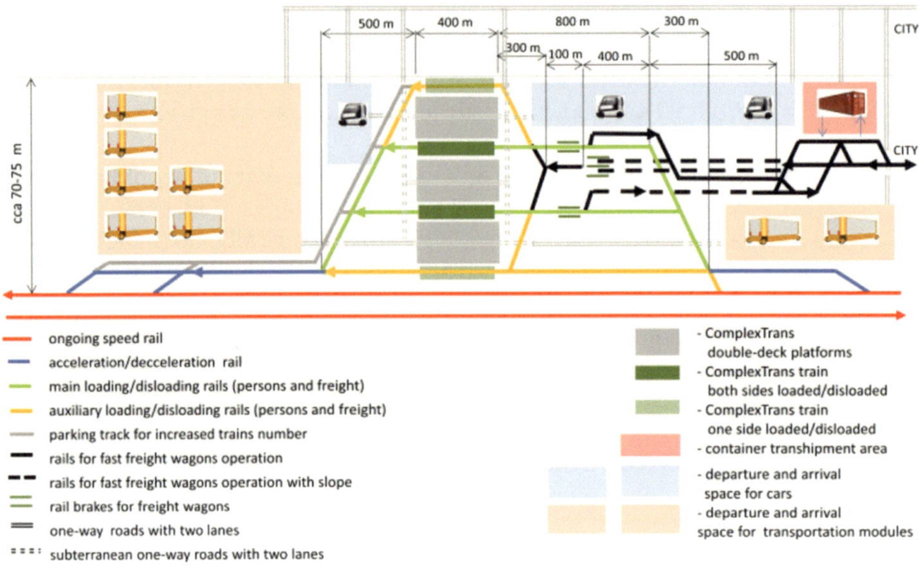

Figure 11.
ComplexTrans road-rail terminal for common passenger-freight transport.

The terminals will also include freight tracks where fast freight waggons will wait to connect to and depart from ComplexTrans trains and will be loaded and unloaded at the terminal if they will be carrying shipping containers or car trailers, or will be transported on regional tracks and sidings to their consignees (**Figure 11**).

4. Main contributions

The main benefit of the ComplexTrans system is the fact that, thanks to the synergistic cooperation of road and rail door-door transport, it solves virtually all the problems of interurban and urban transport of passengers and goods, converts a significant part of continental air transport to rail transport and electrifies the entire land transport. In addition, they are strong supported renewable sources of electricity and reduced the energy consumption.

4.1 Rail: door-to-door mixed transport of ComplexTrans system makes the rail a backbone of ground transport

The double-deck ComplexTrans trains operate at 230 km/h between terminals about 50 km apart at short intervals (3–15 min) and carry passengers, coupemobiles and freight in mobile transport modules, enabling door-to-door transport of passengers and freight (**Figure 12**).

The mixed transport of passengers and freight and the high frequency of trains will create a self-financing [5]. European backbone transport system similar in nature to ferry shipping (**Figure 13**). If about 10% of the European rail network (TEN-T corridors, approx. 20,000 km) is adapted to the ComplexTrans rail transport system, a transport system capable of accommodating all current transport needs in Europe in volume terms will be created. However, in order to bring the transport system sufficiently close to the transport needs, the ComplexTrans system will have to be implemented on about a quarter of the European rail network (approx. 50,000 km, approx.

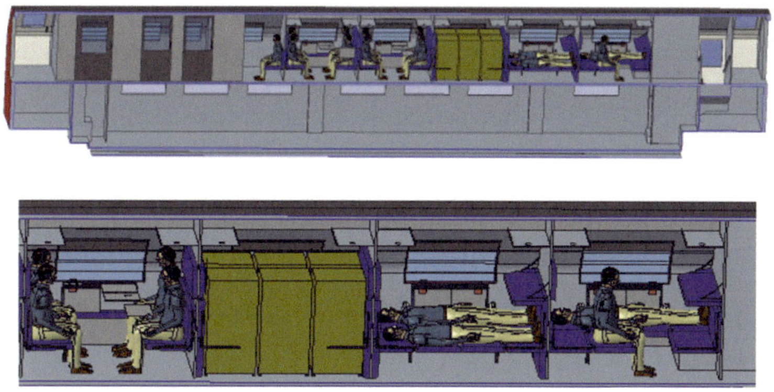

Figure 12.
Small compartments for 2/1 or 4/2 seated/lying persons on the upper floor allow comfortable individual transport day and night. Compartments not occupied by passengers are used for the transport of delivery boxes. On the lower floor, passenger transport in own coupemobiles or passenger transport modules as well as cargo transport in freight transport modules is possible. The flexible combination of passenger/freight transport ensures full capacity utilisation and self-financing of ComplexTrans trains.

Figure 13.
The ComplexTrans network of fast trains resembles "land ferries" that traverse Europe, taking everything—passengers, passengers with their cars, freight transport modules with goods and fast rail wagons with bulky or heavy cargo.

1000 terminals). Even at this scale, ComplexTrans will be fully self-financing, with transport prices comparable to or lower than today's market prices.

It can be said that ComplexTrans more than meets the EU/EC's objective of shifting the majority of freight and passenger transport to rail in order to reduce both energy intensity and CO2 emissions. Importantly, this task can be met by ComplexTrans without any restrictions on road and air transport and without operating subsidies in favour of rail transport. It is sufficient for the EU/EC to support only the initial development and evolution of the system and then the rail system will transform itself into the backbone of European transport. ComplexTrans also helps to protect against the spread of contagious diseases in public transport.

4.2 Replacement of continental air connections by fast night trains

The ambition of the ComplexTrans system is to take over most of the road transport of passengers and goods over medium (from 150 km for passenger transport and from 50 to 100 km for freight transport) and long distances.

In addition, the ComplexTrans system is also capable of taking over a significant part of continental air passenger transport. Thanks to the possibility of using night time for transport in fast ComplexTrans trains (and thus a head start of 8 × 200 = 1600 km), rail transport up to a distance of almost 3000 km is more time-efficient than air transport and up to a distance of about 2300 km. It is possible to make a working round trip in one day and two nights, with more than 6 h available for the actual handling of work matters at the destination (**Figure 14**).

In this way, up to about 90% of continental European air destinations can be replaced by rail, and passenger volumes are even higher, because in the destinations with the highest passenger numbers (Berlin, Brussels, Zurich, Milan, Paris, ...) it will be possible to replace air transport by rail to an even greater share. (**Figure 15**).

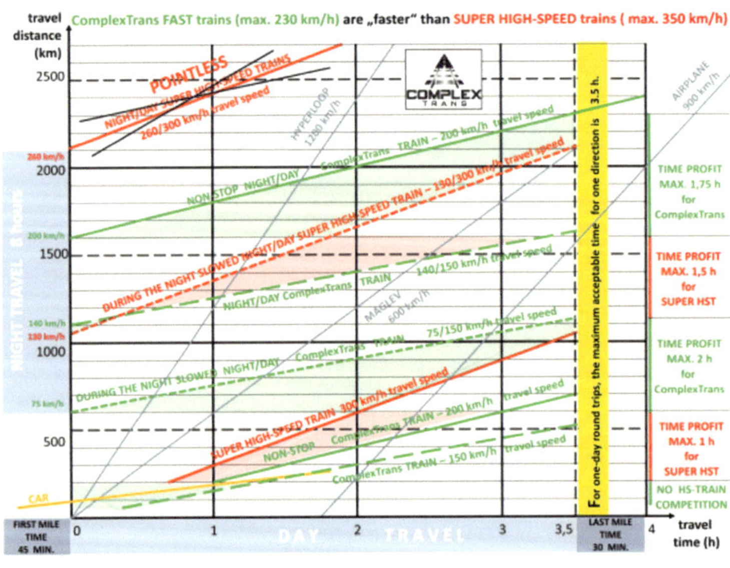

Figure 14.
ComplexTrans day-night trains with an operating speed of 230 km/h and three operating modes (direct, stopping and during night slowed stopping train) will replace very fast trains and planes up to distance of 3000 km and make super fast technologies (Maglev, Hyperloop) unnecessary.

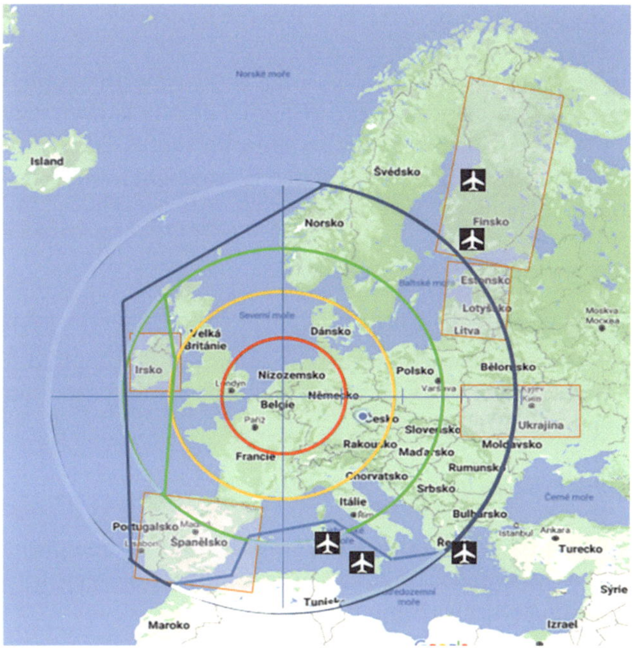

Figure 15.
From Brussels by fast train in one day and two nights to almost every corner of Europe and back. Red—3,5 h, violet—3,5 h + one night by slowed stop train, green—3,5 h + one night by stop train, blue—3,5 h + one night by direct train).

4.3 Urban transport

4.3.1 Urban passenger transport

The road passenger transport in the ComplexTrans system is based on private cars called coupemobiles—i.e. passenger cars with a large cabin for five passengers with extendable axles, which have four roof grip holes for vertical handling and, if necessary, couplers for coupling into road trains (platoons). These features enable the following changes in urban transport.

4.3.2 Parking

Coupemobiles allow to reduce parking requirements in two ways—by retracted axles during parking and by lifted parking (**Figure 16**).

Figure 16.
Parking perpendicular to the sidewalks reduces the need for parking spaces by a factor of two (left). Advanced car overhead parking systems (right); parking over pavements, parking towers or balcony parking—reducing the need for parking areas at least by a half. During advanced parking, the batteries of the coupémobiles are automatically connected to the power grid.

4.3.3 Driving private vehicles in platoons to reduce traffic density

The front and rear axles of coupemobiles can be fitted with couplers. By means of couplings several (2–6) coupémobiles can be joined into a platoon, controlled by the driver of the front vehicle. The platoons also allow for increased traffic light crossings throughput and will be able to use standardised lightweight intersection bridges to further reduce traffic density (**Figure 17**).

4.3.4 Use of private vehicles in public transport

It is likely that drivers registered to ride in the platoons will travel through the city frequently and over a sufficiently long route. Therefore, they could offer free seats in their car (coupemobile) to public transport passengers during their journey. Similar to Uber, for example (**Figure 18**).

4.3.5 Freight transport in the city (door-to-door)

For road freight transport in cities, the ComplexTrans system primarily uses four-wheeled freight transport modules with their own electric battery drive, retractable axles and 90° swivelling wheels (**Figure 19**).

4.3.6 Delivery boxes

For the transport of small consignments (up to 0.5 × 0.5 × 1 m) there are delivery boxes with variable partitions, transported in ComplexTrans trains on the upper deck in places not occupied by passengers (**Figure 20**).

Figure 17.
Private coupemobiles, travelling in a similar direction, can be grouped together in tight platoons during the journey (especially in cities, but also outside them). The platoons reduce traffic density by up to three times and can take advantage of some traffic benefits (e.g. lightweight intersection bridges—see right).

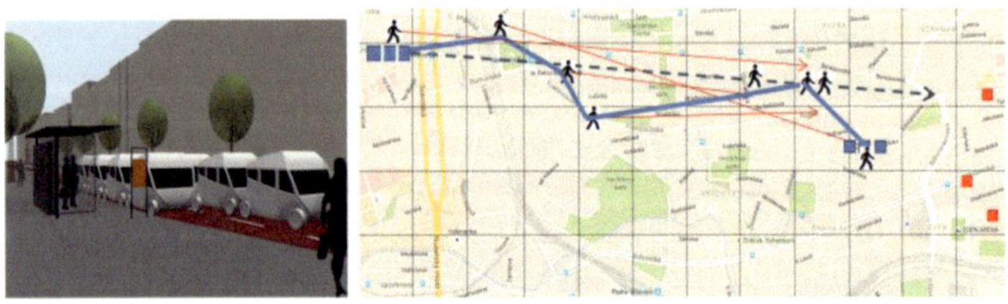

Figure 18.
Platoons of private coupemobiles can also be used for organised public transport on a voluntary basis. The owners of the coupemobiles enter into a contract with a transport operator and the latter organises via an app ad-hoc the use of the free spaces in the coupemobiles for the transport of registered passengers during their journey through the city. This creates a complementary high-capacity, comfortable transport system that is able to reduce traffic density by up to three times and benefits all its users—passengers, coupemobile owners and transport operators.

DOI: http://dx.doi.org/10.5772/intechopen.110184

Figure 19.
After unloading from the lower deck of the ComplexTrans train, the freight transport module is transported in the platoon to the end user. Loading, unloading and local handling (via remote control) is carried out by the user himself.

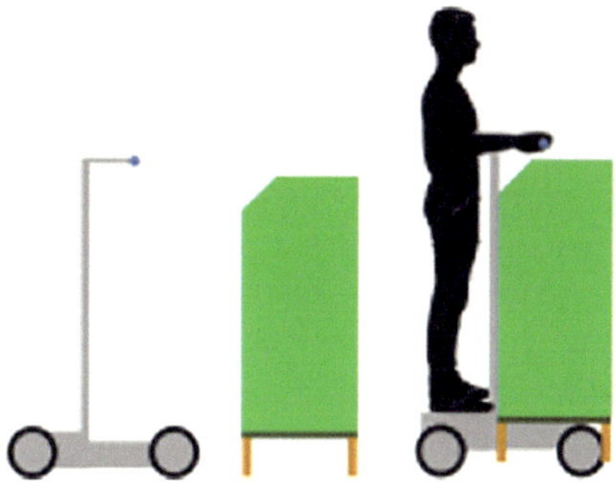

Figure 20.
For the transport of smaller consignments, delivery boxes with a flexible internal layout are used. The first/last mile is covered either on the road in distribution vehicles or autonomously on pavements. With the delivery box can also be carried an (un)accompanying person.

4.4 Intercity road transport

Although all ComplexTrans vehicles are electric, their intercity operation will also be more efficient and at least as convenient as that of internal combustion engine vehicles.

4.4.1 ComplexTrans intercity passenger road transport

The private passenger intercity road transport in the ComplexTrans system is primarily provided by private cars, called coupemobiles (**Figure 21**—left). The maximum operating speed of coupemobiles is approximately 130 km/h and the range up to 200 km. For longer distances, it is advisable to use also train transport (**Figure 21**—right).

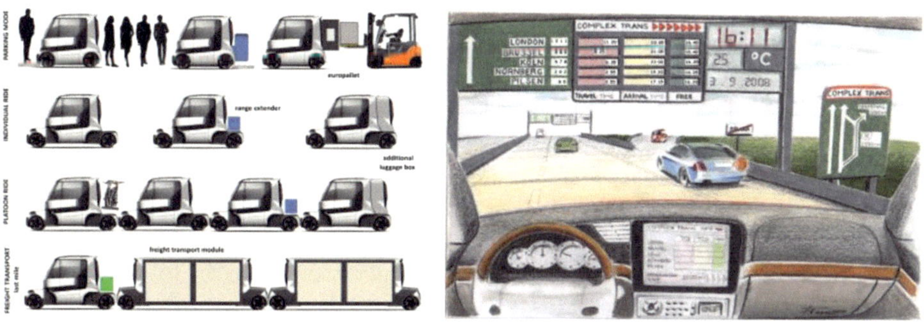

Figure 21.
Left—parking and ride of the coupemobiles—individual or in platoon, electric or with range extender. Bottom left—Freight transport modules are also carried in vehicle platoons over distances up to 50–100 km. Right—when leaving a large city, the driver of the coupemobile has the choice between road and rail.

For bigger range (about 500 km) the traction battery can also be replaced by a so-called range extender (**Figure 21**—left), which is an electric generator powered by a combustion engine (fossil fuel or e-fuel) or a hydrogen cell (H2).

The coupémobiles, equipped with a coupling, will be able to be combined into platoons for longer distance transport (**Figure 21**—left). This will save energy consumption due to the reduction in aerodynamic drag and the crews of some coupémobiles will be able to work, amuse or rest during the journey.

4.4.2 ComplexTrans intercity road freight transport

It is assumed that distances of over 50–100 km will be covered by freight transport modules primarily using the train. For shorter distances, freight transport modules will be transported by road in road platoons with a guide vehicle (**Figure 21**—bottom left).

Trucks will continue to be used for the transport of large or heavy goods, but there will be fewer of them. Truck trailers or sea containers can be transported by ComplexTrans fast freight waggons.

4.5 The contribution of ComplexTrans to electromobility

The ComplexTrans ground transport system uses only electrically powered vehicles with three types of power supply for transporting people and goods.

4.5.1 Types of power supply for electric drives

The first type is to supply the electric drive of double-deck railway units from the overhead line.

The second type of power supply is battery power, which is used for passenger cars (coupemobiles), for road freight and passenger transport modules and local for fast freight rail waggons. The traction battery is unified with dimensions of about 60 × 40 × 35 cm, has a capacity of 30–40 kWh and is easily replaceable from the back/side using standardised handling equipment.

The third type is the supply of power to the electric drive by a range extender, powered primarily by a hydrogen fuel cell, but also by an internal combustion engine burning synthetic or fossil fuel, which is temporarily placed at the rear of the coupemobile.

4.5.2 Attractive e-mobility

E-mobility, as it is practiced today, is not very attractive because a battery of sufficient capacity is very expensive and increases the price of vehicles disproportionately, because charging the battery even with the most powerful chargers (350 kW) takes 15 times longer than refilling the tank with fossil fuel, because building a sufficient charging network will be very difficult, and because searching for refuelling places a disproportionate burden on the owner of the electric vehicle. Today's EVs provide half the utility value for twice the price.

ComplexTrans cars, however, eliminate all these problems in the following way (**Figure** 22):

- Possibility to replace the battery from the back and parking of the coupemobiles perpendicular to the pavement allow discharged batteries to be swapped for charged ones anywhere via distribution vehicles. The traction battery can remain

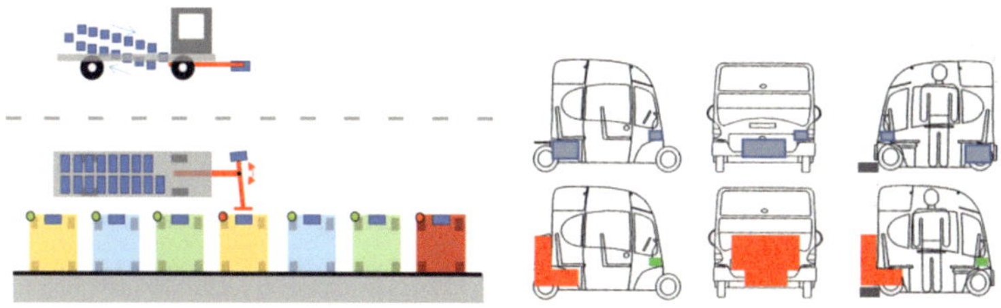

Figure 22.
Left—energy replenishment is done during parking by replacing the discharged battery with a charged one via a distribution vehicle that takes the discharged batteries to a central charger and brings back the charged batteries. The battery exchange is self-scheduled by a car and the driver does not have to be present during the exchange as the batteries remain the property of the distributor. Right—for longer journeys, the traction battery can be replaced by a range extender (red), powered by hydrogen, e-fuel or fossil fuel.

in the ownership of the distributor and the price of a new vehicle is comparable with the price of a conventional internal combustion engine car. This will have a positive effect on the interest in electric vehicles even amongst less well-off people.

- The fact that the battery is not part of the vehicle will also boost the used EV market, as the condition of the battery will not affect the marketability of the vehicle and the used EV will have the same battery as a new EV.

- There is no limit to the range of the car because it takes only tens of seconds to change the battery and the battery can be replaced by a range extender, with which the range increases to at least 500 km and the tank can be refilled in a few minutes. The battery can also be changed anywhere and there is no risk of "range anxiety".

- The owner of the electric car will also be free of the need to worry about charging the battery forever.

4.6 Supporting renewable energy with system ComplexTrans

The ComplexTrans system strongly supports renewable electricity (solar and wind) by creating a giant electricity storage that is connected to the grid much more frequently than other EV systems. The batteries of coupémobiles are automatically connected to the electricity grid not only when they are recharged at central charging stations, but also when they are transported on trains or when they are parked in all non-standard ways. And importantly, the owner of the traction batteries is the electricity distributor, who will find it much easier and more willing to switch to V2G mode and easily decide to switch the batteries to stationary energy storage mode.

4.7 Benefits of ComplexTrans to reduce energy consumption and emissions and protect the climate

The ComplexTrans system can be powered entirely by electricity, partly by hydrogen produced from water using electricity. If all sources of electricity are renewable, ComplexTrans ground transport will become fully renewable and completely emission-free.

The ComplexTrans system is capable of transporting all passengers and most of the goods, carried by the freight transport modules, using only electricity and largely by rail, consuming less energy.

The ComplexTrans rail transport is also capable of completely replacing continental air transport up to a distance of about 3000 km and replaces the fossil fuels by electricity.

Large and heavy goods will continue to be transported by trucks or sea containers. Also semitrailors and containers are transported on the ComplexTrans system by fast electric rail waggons, and first and last mile services may be provided by electric or hydrogen powered trucks.

Thus, the vast majority of passenger and freight transport can be provided emission-free by ComplexTrans.

5. Conclusion and future works

The ComplexTrans system, full combination of rail and road, is able to take over the vast majority of overland transport of people and goods between and also within cities, and a large part of continental air transport, using only electricity supplied to the vehicles via catenary, traction battery or in the form of hydrogen converted into electricity in the fuel cell.

In doing so, the ComplexTrans system retains all the advantages of individual mobility and also mass transit takes on the character of individual transport, making it more attractive.

With the same mobility, the traffic load on cities is reduced (up to 3–10 times), parking requirements are reduced (up to 2 times) and the supply of parking is increased without taking up new space (at least 2 times).

The electromobility of the ComplexTrans system becomes attractive, and in cooperation with renewable sources of electricity, land transport becomes independent of fossil fuels and climate neutral.

The speed and safety of land transport will be increased by transferring a significant part of the transport capacity from the road to day and night fast trains, crossing Europe at speed of over 200 km/h and effectively replacing continental air transport up to a distance of about 3000 km.

The efficient combination of passenger and freight transport makes the ComplexTrans railway self-financing, and through efficient cooperation with road the rail is able to provide door-door transport for people and freight.

The implementation of the system can be phased over decades, but it is advisable to start as soon as possible because climate change will not wait, fossil fuel shortages loom and current European transport is not optimal. However, several steps need to be taken before implementation can begin.

First of all, there needs to be a broad European consensus on the implementation of ComplexTrans, because the ComplexTrans project will not happen without broad European support.

Next, the joint development of all major components needs to be done and a standard has to be developed to ensure the compatibility of all system components. All components will need to be manufactured and tested.

Furthermore, a pilot project will need to be put into operation and the ComplexTrans standard will need to be refined on this basis. It can be estimated that the pilot project can be put into operation within 10 years from the start of the project (kick-off).

As soon as the pilot project starts to show positive results, it will be possible to start the implementation of ComplexTrans on a wide scale.

It would be highly advisable for the development and provision of the basis for the production of ComplexTrans components to be carried out under the auspices of the EU, as the project is too large for the business community and, moreover, its development could be hampered by an overly competitive environment.

Acknowledgements

This article was written with the support of the SGS-2022-009 project.

Author dctails

Jiri Hofman*, Roman Cermak and Jiri Korinek
University of West Bohemia, Pilsen, Czech Republic

*Address all correspondence to: jihofman@kks.zcu.cz

References

[1] Hofman J, Cermak R. ComplexTrans—Global Land Transportation System, TRA2020, the 8th Transport Research Arena, 7/2020. Available from: https://www.traficom.fi/sites/default/files/media/publication/TRA2020-Book-of-Abstract-Traficom-research-publication.pdf (page 49)

[2] Hofman J, Cermak R. ComplexTrans—Global land transportation system. The best way ahead for railways and roads is coherent cooperation, not the competition. South Florida Journal of Development. 2021;**2**(1):313-338. Available from: https://southfloridapublishing.com/ojs/index.php/jdev/article/view/112

[3] Hofman J, Cermak R. ComplexTrans—Global rail-road transportation system: The Best WayAhead for railways and roads is coherent cooperation, not the competition. Journal of Strategic Innovation and Sustainability. 2021;**16**(1):11-32. Available from: https://articlegateway.com/index.php/JSIS/article/view/4180

[4] Hofman J. ComplexTrans RAIL and Road, Not Published. Available from: jihofman@kks.zcu.cz

[5] Hofman J. ComplexTrans RAIL and Road—ECONOMY, Not Published. Available from: jihofman@kks.zcu.cz

Chapter 7

Automation of the Freight Wagon Subsystem

Raphael Pfaff, Bernd D. Schmidt and Manfred Enning

Abstract

The wagon as part of the rail freight system is lagging behind other vehicle classes in terms of automation and monitoring. This stems partly not only from a fierce competition with regard to the vehicle cost but also from a long vehicle lifecycle and long innovation cycles. The drawback of the low amount of automation and monitoring of the vehicle is the additional manual labour required for operation and inspection, as well as the many use cases that cannot be served in an optimal fashion. The Wagon 4.0 concept sets out to mitigate these drawbacks by implementing monitoring and automation technologies into the wagon subsystem in an economically feasible manner. Minor extensions to the rail freight systems already lead to major reductions in labour intensity and use cases that can be served in a novel fashion. This chapter introduces the Wagon 4.0 concept and illustrates the benefits based on use cases.

Keywords: rail freight, automation, industry 4.0. smart vehicle, brake tests, smart logistics

1. Introduction

For decades, freight rail has been vital to growth in countries around the world. This development took place despite the fact that freight takes significant manual labour for handling and checking the wagon subsystem. The manual labour is spent mostly during train formation as well as on the first and last mile.

While in the original applications of rail freight, mostly the efficient transport of mass goods such as coal and ore, these inefficiencies were outweighed by the energy efficiency of rail freight, such goods are set to decrease due to decarbonisation and the circular economy. Instead, shipments are getting increasingly individualized and require more timely and efficient handling. **Figure 1** shows the development of lifted goods in the UK, with the decline in fossil fuels and raw materials clearly visible.

The wagon subsystem differs from most vehicle types in the railway system in that it is in many cases optimised to provide low investments, with wagon prices ranging from 50.000 € to 140.000 € [1]. For this reason, little to no advanced technology is supplied with the wagon, such as monitoring or actuation. Telematics systems are considered feasible and economic, despite providing only limited assistance in the handling of the rail freight system [2].

While low investment in the wagon subsystem leads to very economic freight rail operation for the original mass goods, it leads to higher costs when handling high-value, comparatively light goods to individual sites.

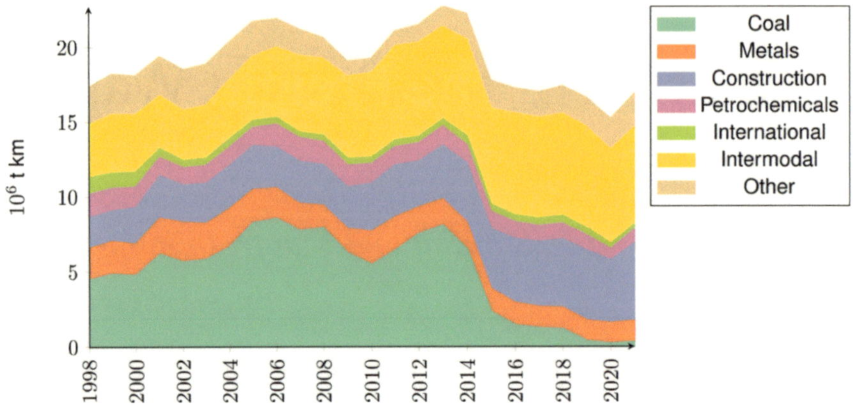

Figure 1.
Freight moved in the UK (financial year) [3].

The costs occur mainly due to

- Manual preparation of the wagons for the current train setup, including coupling and brake settings;
- Manual inspections before train departure, most notably routine inspections and brake tests;
- Unplanned maintenance due to lack of condition monitoring.

Further, this leads to a high requirement of track capacity for serving sidings, mostly due to the fact that the little advanced wagon subsystem cannot be operated safely together with passenger services within their schedule due to these time-consuming activities.

The Wagon 4.0 concept [4, 5] was conceived aiming to provide solutions to these challenges while at the same time yielding a sufficiently high return to justify the inevitable cost increase. This is mainly achieved by adding automation and assistance systems to replace manual activities.

2. Current handling of freight wagons and trains

2.1 Introduction

Wagon handling is required multiple times per journey due to the multi-stage nature of the railway network (**Figure 2**), in particular for single wagon loads. These need to be collected from the shipment points and regrouped to form trains, potentially multiple times, in order to maximise capacity utilization.

A loss in single wagon load cargo moved, as observed in most economies in the past decades, leads to a reduced point-to-point connection between regional distributions and thus further reduces the advantages of the rail freight system due to longer train running times. For this reason, it is vital to increase the overall portion of rail freight in order to keep it vital.

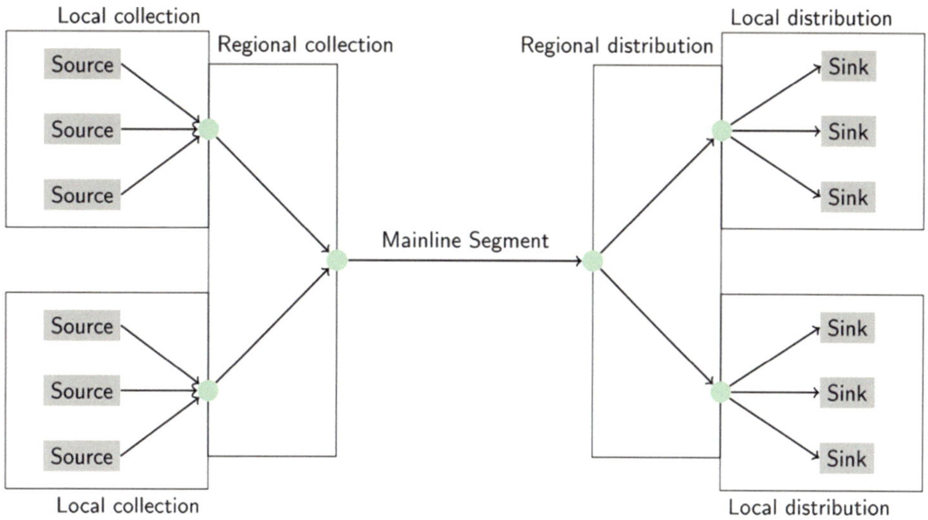

Figure 2.
Multi-stage rail network example.

The local regulations in place typically require a full train inspection at each of the collection or distribution site, which includes a brake test and a routine inspection besides the manual effort of coupling and preparation.

The current routines and practices for train preparation and inspection stem partly from the quasi-universal usage of the pneumatic air brake system, for example, as laid out in Ref. [6]. This system relies on the de-energise-to-activate principle, requiring a distributed system with local energy storage as well as a continuous brake pipe. It is depicted in **Figure 3**.

The distributed nature of the system, together with the lack of sensors and communication equipment, requires manual preparation action as well as pre-departure checks in order to ensure safe operation.

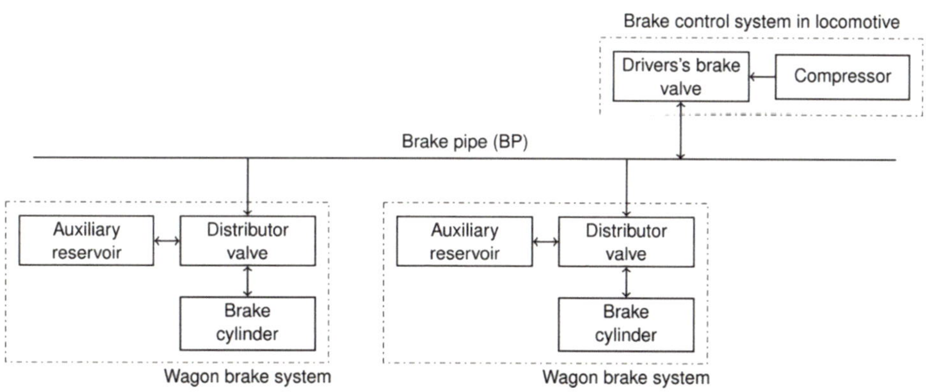

Figure 3.
Pneumatic brake system.

2.2 Preparation

During preparation, the train needs to be automatically or manually coupled, and the brake mode is set manually. This requires knowledge of the train currently handled as well as of the particular wagon since the brake mode is selected according to train mass and length, while for many wagons, the empty-loaded selection is also handled manually.

These steps are executed manually, which requires an operator (typically especially skilled yard personnel or the train driver) to walk along the train. This wasteful motion of the operator consumes time, depending on the length of the train and the number of required actions to be taken.

Further preparation steps include filling of the brake pipe and the associated reservoirs, which may be time-consuming depending on the train's state.

2.3 Inspections

The preparation of the overall consist is followed by inspections of the individual wagons within the consist. The most notable step is the brake test, which aims to avoid critical malfunctions in the brake system.

Hazards of a malfunctioning brake system include:

- Non-continuity of BP: In this case, the portion of the train after the non-continuity (in running order) does not apply the brakes as requested, leading to reduced retardation and potentially the inability to maintain a safe velocity on a downhill gradient.

- Failure to apply brakes upon request: An individual wagon or coach does not apply the brakes when requested by the leading vehicle, resulting in a minor increase in the braking distance of the train consist depending on the total number of wagons.

- Untimely application of the brakes: An individual wagon or coach applies the brake without being requested. Such behaviour may lead to wheel flats or even catastrophic failures such as derailments due to an overheated wheel tread.

- Reduced braking effort due to low pressure in the BP: The reduced BP pressure, often resulting from leakages, leads to reduced pneumatic energy stored in the local auxiliary reservoirs and consequently reduced brake cylinder pressures.

In freight trains, the application to and release from the wheel of the individual brake blocks is required to be checked by visual inspection, requiring additional walking times for the pre-departure checks.

The procedure following German regulations as laid out in Ref. [7] (similar regulations are in place, e.g., in the US [8]) consists of the following steps:

1. Visual inspection of all brakes in train consist, including setting of brake mode and empty-loaded selection.

2. Filling of BP to release pressure.

3. Check the released state by inspection of all wagons for release of brake blocks from wheel tread or released state of visual indicators.

4. Tightness check of BP with the help of a pressure drop of less than 50 kPa per minute.

5. Brake application by a pressure drop of approximately 80 kPa in BP.

6. Inspection of all wagons for application of brake blocks to wheel tread or applied state of visual indicators.

7. Release brakes by filling of BP to release pressure.

8. Check the released state of all wagons by block position or visual indicator.

9. Continuity check of BP by opening of end cock on the last wagon.

10. Check the removal of scotches.

11. Report brake test.

The first step was described under preparation. Stages of the brake test set in bold font require a full walk of the train length; in the case of a single operator executing the test, a double walk of the train length is required due to the necessity to return to the leading vehicle. For the maximum European train length of $l = 750\,\mathrm{m}$. including locomotives, this results in $d = 2 \cdot 4 \cdot l = 6000\,\mathrm{m}$. At a realistic walking velocity in the track area of $v_{\mathrm{walking}} = 1\,\mathrm{m/s}$, this relates to a time consumed just for walking of $t_{\mathrm{walking}} = 100\,\mathrm{min}$

Following German regulations, a brake test has to be executed on each newly assembled train, after adding wagons or changing the direction of the train, and is repeated at least every 24 hours. In certain cases, for example, during a direction change, a so-called simplified brake test is acceptable, which does not check each wagon individually but rather focuses on the continuity of the BP and the braking of the last wagon.

In addition to the brake test, the wagon subsystem is visually inspected for its technical state, for example, the state of the suspension system, as well as for the integrity of the wagon and its load. Since the operator passes by each wagon, this is currently included in the inspection.

2.4 Siding operation

The operation in sidings, as part of the more general shunting operation, is typically executed in order to serve the individual loading points. Wagons arrive in a closed consist to the siding, typically in shunting mode from the last station passed on the way. Shunting operation is executed on-sight, and thus the velocity on the mainline is limited, in the German case, to a maximum of 40 km/h with further restrictions based on infrastructure or a particular operation.

The mainline track between the last station and the siding switch remains blocked during the whole time that the shunter operates on this section of track, as it can only be safely considered free once the shunter returns to the station (**Figure 4**).

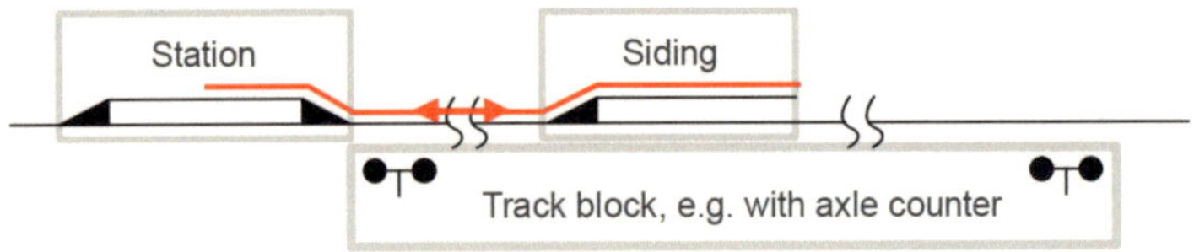

Figure 4.
Line segment with siding and train motions required to serve it.

This mode of operation uses a lot of capacity as the line is blocked for two trains, with a typically long stop and pushing operation into the siding, both operated at low velocities.

In the siding, the consist frequently needs to be split up into wagon groups or single wagons, which are then transported with the help of a shunting locomotive to the loading point for processing. After processing at the loading point, the wagons need to return to a collection point to be picked up in the same fashion as they were brought to the siding.

Different designs of sidings exist, listed either as passing siding or station siding, which yield improvements over the basic design. In the former, the line is only blocked for service to and from the siding point; in the latter, a full train movement is possible up to the siding. Both, however, come with significant investments in command, control and signalling infrastructure. This makes such designs only viable for sidings handling large quantities of wagons.

3. Wagon 4.0 concept

The development and the concept of Wagon 4.0 (W40) are largely driven by the interest to maintain an attractive freight rail system despite adverse trends such as autonomous lorries or reduced amounts of mass goods to be transported. The main aim of the work is to improve the competitiveness of single wagon loads, as this is to be considered a major driver of future logistics demands [9].

The W40 is based on conventional freight wagons, with conventional bogies and couplings according to local standards. It generates much of its added value, thanks to its local control hardware and software, supplied by a wheelset generator with a buffer battery as well as sensors, actuators, communication units and a shunting drive.

The W40 is self-sufficient, self-aware and recognizes other W40 in its vicinity. Thanks to a battery that is charged during mainline operation, it is also smart when stationary and without a locomotive. Due to the operating system and other interfaces to the power supply, it can be optimised for various applications.

Details about the staged introduction scenario, using the idea of classes as well as more information on potential advantages in mainline operation can be found in the tripartite paper [10–12].

The W40 is based on five key structural elements, which are as follows:

1. Power supply: Electrical power plays a vital role in automation as well as in condition monitoring. Freight wagons are typically unpowered assets. In order to have sufficient power for a variety of actuators, a 24 V power network supplied by electricity produced by local generators and stored in local batteries is implemented for W40 with a standardised architecture defined in Ref. [13].

2. Data network: The data network uses standard technologies to establish a connection with wagons as well as within the train consist.

3. Sensors: The sensors of the W40 will mainly consist of position and pressure sensors, aiming to reduce manual inspection and use the sensor feedback to increase safety and efficiency. Also, monitoring of vital components of the wagon is provided in order to replace inspections and enable predictive and condition-based maintenance.

4. Actuators: While sensors and networks provide useful data, the actuators enable operators to operate certain aspects of the wagon subsystem remotely, for example, the brake mode setting.

5. Operating system: The so-called WagonOS, an open-source operating system, unifies the above-mentioned base elements to allow for extending the capabilities of the W40 and to standardise communication protocols, data formats and related standards. A central operating system will furthermore enable currently disjointed efforts to unite under the umbrella of a single industry standard.

The W40 concept follows a class structure aiming to modularise the system, ease introduction into fleets and allow the selection of the most appropriate and economical class for a certain wagon or fleet. **Figure 5** shows a schematic of a W40, indicating the class of the respective functionality.

The classification starts with a W40 of class 1, allowing communication with and over this wagon. A class 2-W40 adds an interface to the brake system, including measurements of brake cylinder and brake pipe pressures as well as the remote activation of brake mode or empty-loaded selection. A further sensible addition to the system is the so-called ep-light brake, an indirect electro-pneumatic brake. This is the only additional functionality of class 3 over the other classes. A class 4 wagon aims to fully automate the wagon subsystem with an additional end-of-train signal, whereas a class 5 wagon adds a drive system to enable automated movements in the siding.

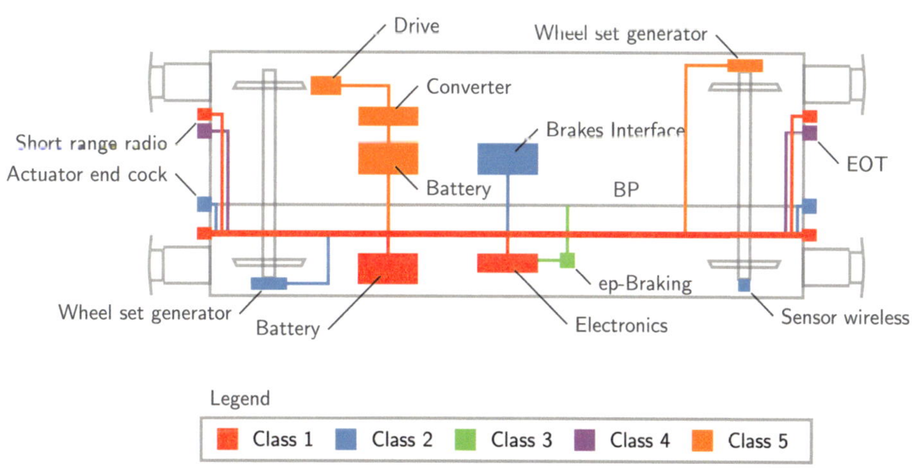

Figure 5.
Wagon 4.0 structure and functional classes.

4. Wagon 4.0 contributions

4.1 Train formation and setup

Assisted brake tests as well as automation in train data handling rely on information on train set-up, especially the order and orientation of wagons is important. Although human operators intuitively capture the order and completeness of the wagon group and convey information, a technical system needs to gather and transfer information explicitly.

In automatic brake test systems currently implemented, the completeness of the wagon group is assured by comparison of the rake to a wagon list transmitted beforehand [14]. This information regarding wagons in the rake and their order may be potentially supplemented by GNSS localisation. The intra-train communication relies on point-to-point radio or mobile communication.

The authors expect that both point-to-point and mobile communication may lead to problems limiting the availability of the systems, for example, in areas with poor cell coverage. Further, the usage of wagon lists for train topology generation may yield disadvantages over the detection of the actual state of the rake, since such lists may contain errors and need to be generated first.

For this reason, the W40 follows a different approach. Each wagon is equipped with controllers at both ends of the wagon. A local area network connects both controllers as well as sensors and actuators installed in the wagons. Adjacent wagon ends are connected by a V2V communication system. This creates a linear network closely replicating the rake structure. Each wagon is able to identify its neighbouring wagons and may share this information on the network. This makes it possible to maintain a digital representation of the wagons in the rake as well as their state on each wagon.

This yields the advantage that communication throughout the train can be implemented comparatively easily and reliably. The physical layer of the wagon-to-wagon interface can be formed by the recently proposed digital automatic coupler [15] or with the help of short-range radio communication. It is reasonable to assume that such a radio connection is at least as reliable as a galvanic connection since a line-of-sight connection is always available. The radio connection between wagon ends is realised with the help of a WifiDirect (trademark of Wifi Alliance) connection. Additionally, a Bluetooth Low Energy connection may be used to measure distances in order to safely connect to the next wagon [16].

Since the W40 concept implements remote-controlled end cocks and BP pressure sensors, as indicated in **Figure 5**, this distance measurement is not strictly necessary since adding wagons to the train consists that are not pneumatically coupled will not lead to a successful brake test, and thus any wrong connections will be contained and will not endanger the mainline operation.

Operating personnel connect to the wagon with the help of Wifi formed *ad hoc* using a smart device such as a tablet. **Figure 6** shows the corresponding user interface.

In the target market of freight wagons, it is reasonable to expect non-equipped wagons. The W40 concept is not intended to pass by any non-equipped wagons. Instead, the basic equipment of the wagon to allow communication and detection of neighbouring wagons (termed W40 class 1) comprises only low-cost, low-maintenance components. The class 1 equipment enables the wagon to identify its neighbouring wagons and to participate in the V2V communication. This equipment uses semiconductors from the consumer range and can be operated for years on battery power; thus no wheel set generator is required.

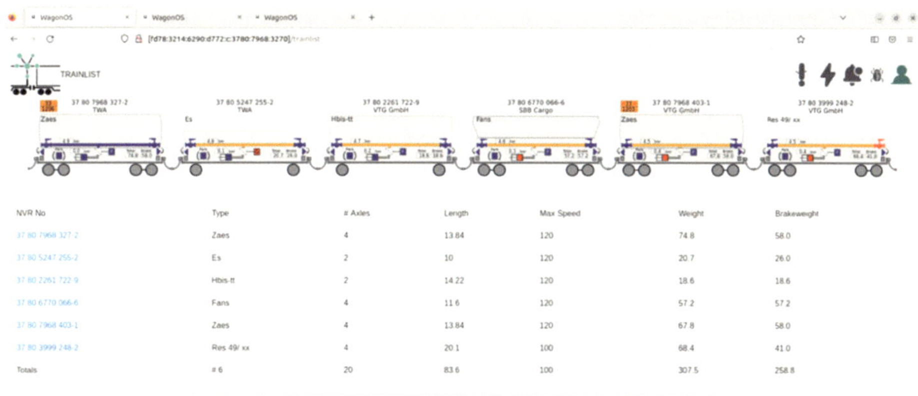

Figure 6.
Web interface of a device connected with WagonOS showing the train list.

4.2 Brake test

An analysis of accidents [17] highlights discontinuities in the brake pipe (BP), untimely brake applications and the inappropriate use of hand brakes and scotches as causes for catastrophic outcomes. Irregularities or damages to the brake rigging or brake calipers reduce the brake effort only for single bogies or wagons, with no significant effect on the train.

The observed failures and errors causing the accidents are

- discontinuity or limited continuity of the BP,
- leakage of the BP,
- failure of the distributor valve (static and dynamic) and
- untimely application of the service or parking brake.

In the investigated cases, all of the above went undetected using the existing brake test procedures conducted prior to the accidents. Most of these are difficult to detect using only visual inspection and the common static brake test procedure. Further, in some cases, the failure developed during the train mission, such as in the Llangennech rail disaster [18].

The W40 approach introduces a novel definition of system borders for brake tests based on the brake-related accidents analysed in Ref. [18]. Based on this analysis, a different split of test steps between brake tests and inspections is proposed.

The effectiveness of visual checks for brake application and release or cylinder stroke needs to be questioned. Such visual checks are costly and do not sufficiently mitigate errors such as discontinuities in the BP.

On the other hand, irregularities in and damages to the brake rigging appear at a very low frequency in accident reports. In contrast to the failures in handling and operation of brake systems reported above, these typically result in the unavailability of brake functionality on single wheelsets, bogies, or wagons. Such singular failures are not likely to endanger the safety of the train as a whole. From a perspective of the overall safety of the railway system, an automated test based on brake cylinder pressures rather than visual checks of brake block travel may be an appropriate alternative.

The required sensors are robust and cost-efficient pressure sensors for brake cylinder and BP pressure, whereas for the detection of an untimely application of the brake, a position sensor on the brake cylinder is required.

A pneumatic scheme with added pressure sensors is depicted in **Figure 7**.

This set-up is able to detect the following states:

1. Brake released (by position sensor attached to the brake cylinder),

2. Brake cylinder pressurised (by cylinder pressure sensor) and

3. Brake command state (by BP pressure sensor).

Thanks to continuous measurements of brake pipe and brake cylinder pressures, it is also possible to observe both the propagation of the brake command in the BP and the filling and release time of the brake cylinder. This enables the development of further diagnostic systems, for example, to detect deterioration of the distributor valve or an incorrect brake mode. The propagation also serves as a second channel beside WIFI for the verification of the consist order.

The information on the ongoing brake test is displayed on the user interface screen, effectively assisting the operator in the brake test by providing information formerly obtained by walking to the brake system in question. The user interface is shown in **Figure 8**.

Further, the sensor equipment can continue to observe these values during mainline operation, which improves safety over the singular observation in classical brake tests. The continuous observation is capable of detecting untimely service brake applications as well as inappropriately applied hand brakes.

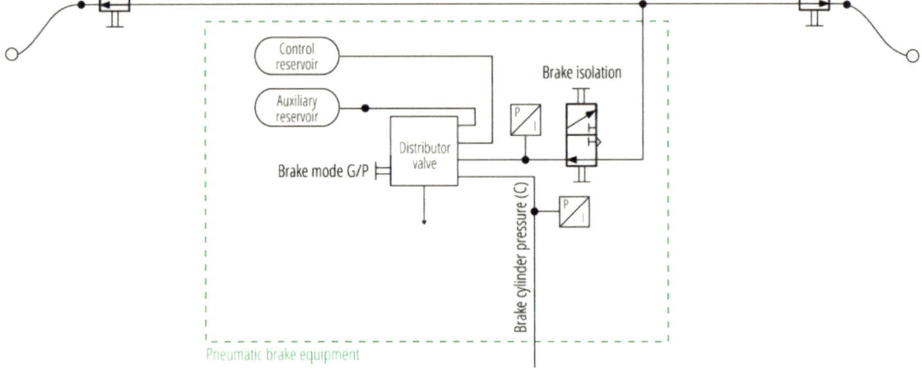

Figure 7.
Pneumatic scheme of a wagon brake system with sensors for BP and brake cylinder pressure, from [13] (CC-BY 4.0).

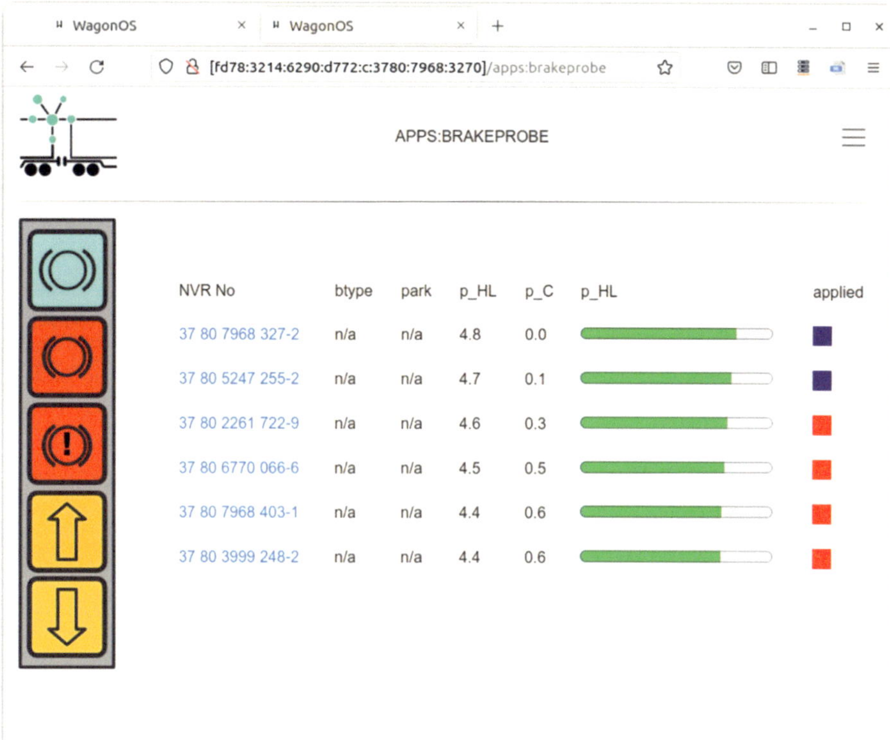

Figure 8.
User interface displayed during the assisted brake test.

4.3 Technical inspection

Under German as well as most other regulations, it is common to execute a visual inspection together with the brake test. When automating the wagon subsystem and assisting in brake tests, it is important for economic viability to automate or reduce these inspection steps as well.

The approach of W40 is not able to detect

- immobility of brake cylinders when pressure is applied,
- uneven distribution of braking force,
- the full release of brake blocks after venting of brake cylinder and,
- wear on the slack adjuster.

Such failure modes typically evolve slowly since they are related to the wear of the equipment. These failure modes, as well as other slowly developing deteriorations, such as the inappropriate state of the brake blocks, should be addressed by additional technical inspection as opposed to a brake test. In this way, ensuring the brake functionality for the next mission is separated from determining the technical condition of the wagons. Wagon condition monitoring may be automated without large investments into the individual wagons by providing a wayside train monitoring system (WTMS).

In the current state of development, WTMS uses visible light imaging to estimate brake block thickness, and this may easily be extended to detect further visible failures, enabling WTMS to replace most visual checks by human operators [19]. The remaining inspection items such as the closing of hatches, which need to be checked prior to departure, can be automated using cheap and reliable sensors on the rolling stock. The overall system combining an extended network of wayside monitoring with sensible and economical equipment on the wagons is capable of yielding improved safety of the freight rail system at a reduced cost.

5. Use cases

5.1 Letterbox style operation (handover tracks)

Rail freight operations in an open access multi-stage wagon load network (shown in **Figure 2**) generate, as for all similar networks (e.g., road traffic and electric grids), the highest specific costs in the last collection and distribution stages. The reason is that the costs of a valuable locomotive and corresponding staff can only be allocated to one or two wagons on the last metres of the journey. These last metres consume many loco/staff hours but produce almost no traffic volume given in tons or kilometres. The problem is well known for more than a century. So historically, railway companies avoided using expensive mainline steam locos for this kind of operation. A typical solution is a group of handover/holding tracks. The first of these tracks holds the wagons delivered by the mainline loco. A second track contains the wagons, which need to be picked up. Further (empty) tracks may be needed for locomotive movements (**Figure 9**).

The customer can then move the wagons based on his internal timing needs to the loading docks using horses (historic solution) or nowadays road-rail vehicles. But we must note the disadvantages: the railway operator streamlines his operations, but the customer now needs an expensive device.

The W40 class 5 features a shunting drive system, which can overcome these disadvantages. Without needing an additional rail-road vehicle, the customer may now move the wagons himself on his premises by using the shunting drive. The user interface as well as the training of the staff will be aligned with current practice used for forklifts, which makes it also a viable solution for customers who only receive a single-digit number of wagons a week.

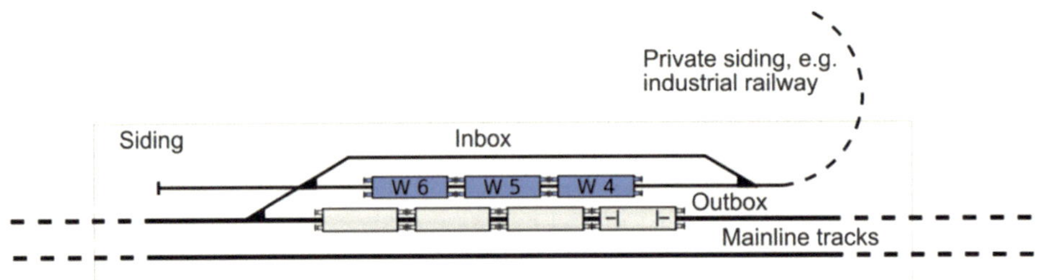

Figure 9.
Letterbox—handover tracks.

5.2 Port interaction

Cooperation with cranes and other facilities in the port area provides plenty of potential for operational optimisation. Since the infrastructure is costly and frequently the landside rail operation forms a bottleneck, this optimisation can vastly improve the overall performance of the port.

In many cases, port rail operation comprises reversing of the freight train, for example, onto a dock track. In this case, the intra-train Wi-Fi mesh network can transfer real-time video and sensor data from a mobile device temporarily attached to the unmanned end of the train. This enables a single person to safely perform reversing operations.

In many ports, a train consist is broken down into small groups of wagons, with distances introduced between these groups to allow for the passing of container carriers. Using the local control and shunting drive, the related uncoupling and motion can be achieved by the wagons autonomously, speeding up preparation for loading.

Using industry 4.0-style self-organisation, container cranes and wagons can communicate on the sequence of containers to be loaded and adjust the trunnions according to the wagons' trunnions.

After loading, the train can be formed again quickly using the assisted brake test, while communicating with the container carriers and other automated vehicles in the harbour. The train can leave almost immediately, thanks to the assisted brake test and train formation. The cycle of operations is illustrated in **Figure 10**.

5.3 Intra-site logistics

Similar to port operations, there may also be other loading sites that require intensive cooperation with the customer and the moving of wagons during loading operations. Typically, this may be the filling and unloading of tank wagons, where the loading site has a limited length and offers only unloading single or small groups of wagons simultaneously, or hopper wagons that need to be filled or emptied at a loading site. If these sites are highly frequented, they are often equipped with electric tractors or steel cables/winches to move wagons during (un-)loading operations. But for smaller sites, this kind of infrastructure is too expensive. So, for small customers, the railway operator keeps his locomotive and staff busy supporting the loading operations, which is not very cost-efficient. The same applies for loading sites inside a plant, for example, a household appliances factory doing metal forming. Normally, these factories receive single wagons with coils of steel sheets, which must be positioned under a crane for unloading. Currently, at such sites, the only option is to ask the railway operator to move the wagons or have their own shunting tractor. This is especially costly for small factories that do not receive high volumes of wagons. The Wagon 4.0 class 5 overcomes these challenges as it features a shunting drive system. Without needing additional assets, the customer may now move the wagons himself on his premises by using the shunting drive.

5.4 Power boost by distributed power

Depending on the velocity range of operation and the local legislation, the shunting drive may remain activated in certain velocity ranges during mainline operation. Using cooperative control, that is, the wagons apply tractive power if their neighbouring wagons apply it, the train may be able to help the locomotives in certain situations, for example, when starting or on steep uphill grades.

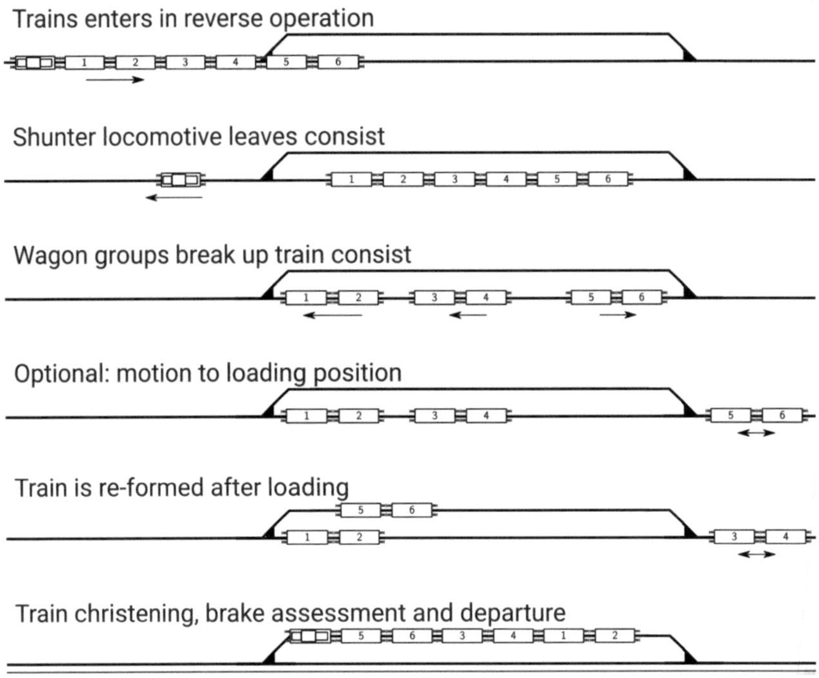

Figure 10.
Port operation stages.

This reduces longitudinal forces in the train consist and makes higher commercial speeds possible.

Figure 11 shows a simulation of a train set up with 30 container cars type Sgjs, each of them loaded with three 20'-containers. The total train mass is in the range of 1100 t. As a loco, a DB class 145 (mass 82 t) is assumed, delivering 250 kN of tractive effort and a power of 4.2 MW.

W40 class 5 is equipped with a traction system. In this way, the wagons may operate like multiple-unit vehicles and support the locomotive during acceleration.

When using the tractive capabilities of the W40 for a short term during the acceleration of the train, the tractive effort of the whole composition will be 880 kN, signifying an increase by a factor of 3.5.

The short-term power of the system will only increase slightly by 10%. That means, as shown in **Figure 11**, that use of the tractive capabilities of the W40 makes only sense for speeds below 20 km/h.

This velocity range, however, is the range that is important for freight traffic. Normally, turnouts in freight yards are designed for branch speeds of 40 km/h due to cost reasons. Consequently, the most important task is accelerating up to 40 km/h as fast as possible to reduce the occupation and locking times of train path elements in the interlocking system. As an initial estimate, the acceleration will take approximately 1 minute less with W40 traction applied.

On heavily used mainlines, headways are normally in the order of four minutes, so saving one minute during the acceleration of a freight train may significantly increase capacity in congested nodes.

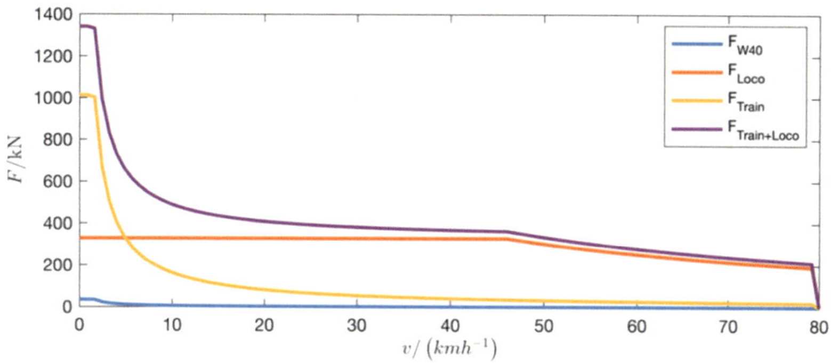

Figure 11.
Traction curves of individual W40 (F_{W40}), a locomotive (F_{Loco}), a train of 30 W40 (F_{Train}) as well as the total tractive effort of the train setup.

5.5 ep-light brake

Thanks to the continuous power supply of the W40 and the intra-train network capability, it is possible to extend the brake system of the wagon by a valve to command an indirect electro-pneumatic brake application locally.

This is achieved by locally venting the brake pipe to the atmosphere. The benefit of locally commanded brake application (ep-apply) is a faster propagation of the brake request through the brake pipe, which leads to three effects:

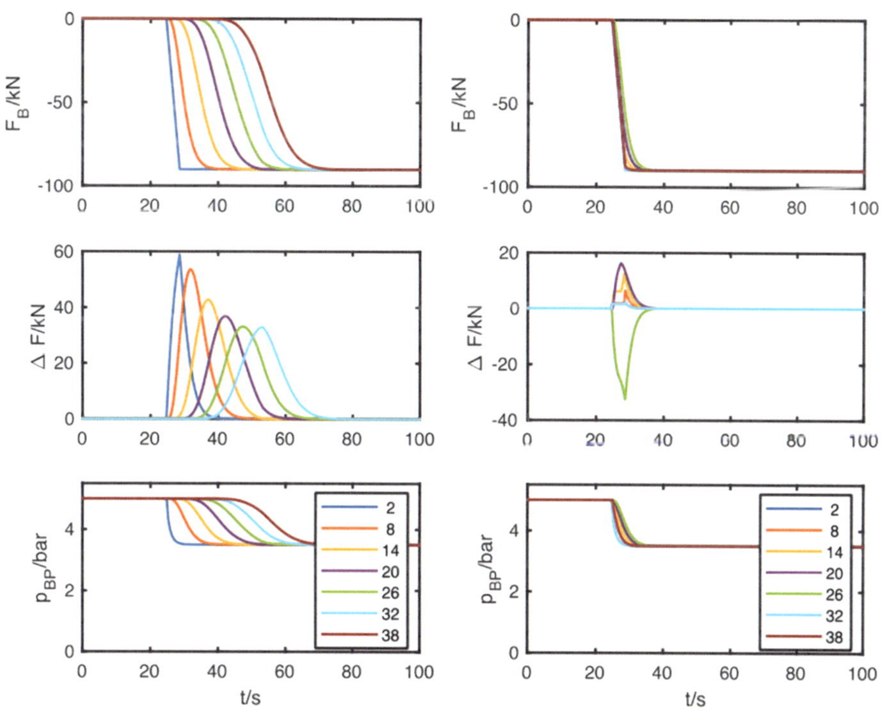

Figure 12.
Traction power boost by distributed power.

- On the operations level, the braking distances and, therefore, the headway are reduced.

- On the train level, the longitudinal buffer forces within the consist are lower than for a brake request from the front of the train (Refer to **Figure 12** left column conventional braking, a right column with ep-assist).

- On the wagon level, a more equalised brake application leads to less wear on certain wagons, such as the first P-braked wagon in an LL-braked train consist.

The reduction of buff forces may lead to increased train masses being braked in the P-regime, which in turn also increases maximum velocities. In terms of industry 4.0, the ep-assisted brake can be considered a collaborative function, as no master is required for this functionality. Instead, the wagons support each other in braking the train, with the most sensible way to react to the neighbouring wagon's braking being to support the process.

Depending on the treatment of the improved functionality with respect to operations (i.e., braked weight, train length and masses), such a function does not require particularly high equipment safety levels, since failures in individual wagons do not impede the overall safety of braking at the whole train level due to the continuous brake pipe.

6. Economic advantages and return-on-invest

The freight Wagon 4.0 as well as the concept of industry 4.0 has economic advantages in many situations but not in all. Often, digitalised concepts are only paying off if, for instance, the cost of internet access is priced based on marginal costs, which is reasonable because a mobile phone base station is nearby anyway for other purposes. In the case of Wagon 4.0, the user base is limited, and economic feasibility must be checked for the individual use case. Normally, wagons that do not move very often are used as storage rooms rather than vehicles and will never give a positive return on investment. In all other cases, the operational setup and the chosen class determine the results. Class 1 wagons allow communication and certainly can also be position tracked via the mobile network. Thus, they may already allow some operational benefits (e.g., knowledge of miles travelled and adapted maintenance schedule), but savings are limited.

Bigger savings of operational costs will be found for class 2–4 wagons. This especially applies in cases when handling times for train inspection and brake checks can be significantly reduced. It is mostly not the costs of staff that count, since in many railway industry applications, staff are relatively cheap compared to the capital costs of assets used. More significant are the savings in turnaround times of trains and track occupancy, for example, in a port setup or a bimodal terminal. When applied together with loading site automation in a seaport environment, the benefits may even increase further. So intelligent freight wagons may be handled like autonomous container movers by the port automation system, removing inefficient manual interfaces and risky or heavy manual labour:

- Control of all vehicles can be centralized in the port operations centre with seamless integration of container handling and train movements.

		Time/minutes		Savings/minutes			
Step	**Description**	**Current**	**Novel**	**Labour**	**Loco**	**Train**	**Track**
1	Arrival and split-up	30	5	60	30	25	25
2	Unloading (4 stackers, manual vs. automatic)	45	30	180	0	15	15
3	Train inspection	30	0	45	0	30	30
4	Trunnion setting and loading (as above)	45	30	225	0	15	15
5	Assembly and brake test	90	10	120	80	80	80
	Total	240	75				

Table 1.
Time savings in port for container trains.

- Increase in worker safety as there is no need for heavy manual labour on tracks anymore.
- Faster train setup and brake check.
- Less tracks needed or higher capacity with existing tracks, higher throughput and efficiency of tracks and vehicles compared to legacy operations.

The most important point is the higher throughput and efficiency. This especially applies to train turnaround times as well as track occupancy. Comparing the time needed for current (legacy) operations compared to a setup with wagons 4.0, significant time savings can be achieved, as given in **Table 1**.

The handling time of a 30-wagon trainset at the terminal facility falls from 4:00 to 1:15 hours. As can be seen, more than 66% of turnaround time may be saved, and therefore, rolling stock and track assets show significantly better utilisation. This especially becomes true if a complete trainset can be saved in the timetable.

Also, the port obtains benefits; without changing track infrastructure, the capacity of the facility increases by 200%. On an average port, this may generate an additional 20% savings on the cost of the track infrastructure.

Taking higher classes of Wagon 4.0 (e.g., class 5) into account, savings may not be as big for a standard use case from the railway operators' side, but they may enable rail services in cases where customers do not have their own shunting devices or in cases where crowded mainlines with lots of commuter passenger traffic will not allow freight services with conventional methods.

7. Conclusions

The W40 concept is able to yield significant operational and logistical advances, yet it relies on proven technologies. The concept can be scaled depending on the intended application, ranging from communication to full automation of wagon movements.

While the concept yields operational and economic benefits, it is at the same time a means of reducing the manual labour required, a further contribution to the future of the rail freight system by providing meaningful and safe jobs to future railway workers.

W40 class 2 was implemented as demonstrators in the funded project.

Acknowledgements

The authors acknowledge funding for the Wagon 4.0 project by the German Ministry of Science and Education under grant number 16ES085 as well as a research grant from FH Aachen University of Applied Sciences.

Conflict of interest

The authors declare no conflict of interest.

Thanks

The authors thank numerous collaborators who contributed to our work. Particularly, we wish to thank Daniela Wilbring, Stefan Sutter, Marcel Ochsendorf, Constantin Zborowska and Rudi Hilgers.

Author details

Raphael Pfaff*, Bernd D. Schmidt and Manfred Enning
FH Aachen, Aachen, Germany

*Address all correspondence to: pfaff@fh-aachen.de

References

[1] Baumgartner JP. Prices and Costs in the Railway Sector. Lausanne: École polytechnique fédérale de Lausanne; 2001

[2] Behrends V, Haunschild M, Galonske N. Smart telematics enabling efficient rail transport–development of the ViWaS research and development project. Transportation Research Procedia. 2016;**14**:4430-4439

[3] Office of Road and Rail. Statistics table 1310 – Freight moved. ORR Data Portal. 2021. https://dataportal.orr.gov.uk/statistics/usage/freight-rail-usage-and-performance/table-1310-freight-moved/. Accessed 10 December 2022

[4] Pfaff R, Schmidt BD, Enning M. Towards inclusion of the freight rail system in the industrial internet of things - Wagon 4.0. In: Proceedings of the Stephenson Conference. London, UK; 2017

[5] Pfaff R, Schmidt BD, Wilbring D, Enning M, Franzen J. Wagon4.0 – the smart wagon for improved integration into Industry 4.0 plants. In: Proceedings International Heavy Haul STS Conference. Narvik, Norway; 2019

[6] Railway applications – Braking – Requirements for the brake system of trains hauled by locomotives. 2018. EN 14198:2016+A1:2018

[7] Verband Deutscher Verkehrsunternehmen. Bremsen im Betrieb bedienen und prüfen – Bremsvorschrift. 2015. VDV 757:2015

[8] Class I brake test-initial terminal inspection .2020. 49 C.F.R § 232.205

[9] Lammgard C. Intermodal train services: A business challenge and a measure for decarbonisation for logistics service providers. Research in Transportation Business and Management. 2012;**5**:48-56

[10] Enning M, Pfaff R. Güterwagen 4.0 - Der Güterwagen für das Internet der Dinge. Teil 1: Gesamtsystembetrachtung und grundlegendes Konzept. ETR - Eisenbahntechnische Rundschau. 2017;**66**(1/2):12-16

[11] Pfaff R, Enning M. Güterwagen 4.0 - Der Güterwagen für das Internet der Dinge. Teil 2: Ausgewählte technische Aspekte und Prozesse. ETR - Eisenbahntechnische Rundschau. 2017;**66**(5):74-77

[12] Schmidt BD, Enning M, Pfaff R. Güterwagen 4.0 - Der Güterwagen für das Internet der Dinge. Teil 3: Einführungsszenarien für aktive, kommunikative Güterwagen. ETR - Eisenbahntechnische Rundschau. 2018;**67**(5):60-64

[13] Verband deutscher Ingenieure. VDI 5905 Blatt 1 E. Schnittstellen aktiver kooperierender Güterwagen – Stromversorgung. 2021

[14] PJ Monitoring [Internet]. Automatische Bremsprobe. 2021. https://pjm.co.at/wp-content/uploads/2019/05/success-storie-automatische-bremsprobe.pdf. [Accessed: 29 December 2020]

[15] Hecht M, Leiste M, Discher S. Erstellung eines Konzeptes für die EU-weite Migration eines Digitalen Automatischen Kupplungssystems (DAK) für den Schienengüterverkehr. 2020. Report to Federal Ministry for Digital and Transport

[16] Bluetooth SIG. Bluetooth core specification version 5.0 Specification of the Bluetooth System. 2020

[17] Blukii GmbH. Data Sheet Smart Beacon Outdoor Extreme 4000. 2020. https://www.blukii.com/wp-content/uploads/DS_SmartBeaconOutdoorExtreme 4000_EN.pdf [Accessed: 2 February 2022]

[18] Pfaff R, Enning M, Sutter S. A risk-based approach to automatic brake tests for rail freight service: Incident analysis and realisation concept. SN Applied Science. 2022;**4**:115. DOI: 10.1007/s42452-022-05007-x

[19] Railway Accident Investigation Branch. Derailment and fire involving a tanker train at Llangennech, Carmarthenshire. 2020. Report 01/2022. 2022.

Section 4

Performance Analysis and Assessment

Chapter 8

Accident Prediction Modeling Approaches for European Railway Level Crossing Safety

Ci Liang and Mohamed Ghazel

Abstract

Safety is a core concern in the railway operation. Particularly, in Europe, level crossing (LX) safety is one of the most critical issues for railways. LX accidents often lead to fatalities and weighted injuries and seriously hamper railway safety reputation. Moreover, according to statistics, collisions between trains and motorized vehicles contribute most to LX accidents. With this in mind, we will elaborate on accident prediction modeling for train-vehicle collisions at LXs in this chapter. The methods and findings discussed in this chapter will offer an in-depth insight for interpreting significant aspects underlying collision occurrence and facilitate identifying technical countermeasures to improve LX safety.

Keywords: level crossing safety, train-vehicle collisions, accident prediction modeling, nonlinear least-squares method, negative binomial regression method, Poisson regression method, zero-inflated Poisson regression method, zero-inflated negative binomial regression method, model performance evaluation

1. Introduction

The level crossing (LX) is railway property upon which road users are given permission to cross [1]. Accidents at LXs give rise to serious material and human damage, and the majority of accidents are caused by vehicle driver violations. As demonstrated by accident statistics, LX safety is one of the most critical issues that railway stakeholders need to deal with [2, 3]. In 2012, there were more than 118,000 LXs in the 28 countries of the European Union (E.U.) [4]. In some E.U. countries, LX accidents account for up to 50% of railway accidents [5]. In the UK, LXs account for 11.8 fatalities and weighted injuries on average per year, comprising 8.4% of the total system risk for the railway network [6]. There were 49 collisions between road vehicles and trains at LXs in Australia in 2011 [7]. In France, the railway network incorporates more than 18,000 LXs for 30,000 km of railway lines and around 13,000 LXs show heavy road and railway traffic [8]. In 2016, 111 train-vehicle collisions at French LXs led to 31 deaths [9]. This number was half the total number of collisions per year at LXs a decade ago, but still too large [10]. Due to nondeterministic causes, complex operation background, and the lack of thorough statistical analysis based on detailed accident/incident data, the risk assessment of LXs remains a challenging task. Therefore, there is a pressing need for a series of thorough analyses to understand the potential reasons for these accidents and to identify practical countermeasures to prevent accidents at LXs, thus significantly reducing the LX accidents.

In recent years, the Poisson regression model, negative binomial (NB) regression model, and other variants of the Poisson regression model [11, 12] have gained popularity to deal with risk/accident statistics. Ref. [13] adopted the expressions of the estimated expectation value $\hat{\lambda}$ as shown in Eq. (1) corresponding to the Poisson regression and NB regression models, respectively. Ref. [14] employed the variants of Poisson regression model, namely, the zero-inflated Poisson (ZIP) model and the hurdle Poisson model, to deal with LX accident prediction involving the data in North Dakota. Ref. [15] compared the zero-inflated negative binomial (ZINB) model with the USDOT model [16] by using the LX accident data from Illinois, in terms of accident prediction accuracy. The results of this study show that the ZINB model has higher accuracy of prediction. It is worth noticing that the expressions of estimated $\hat{\lambda}$ as shown in Eq. (1) are not appropriate in our current study, since they are limited to handling zero observations and some impacting variables should not be in the exponential form. Ref. [17] developed another model of $\hat{\lambda}$ as shown in Eq. (2). In this model, the product of the average daily road traffic V and the average daily railway traffic T (known as the conventional traffic moment) is adopted. However, using the conventional traffic moment hinders improving the accuracy of the prediction model:

$$\begin{aligned} \hat{\lambda}_{Poi} &= \exp\left(\sum_{j=1}^{m} \beta_0 + \beta_j x_j\right), \\ \hat{\lambda}_{NB} &= \exp\left(\sum_{j=1}^{m} \beta_0 + \beta_j x_j + \varepsilon\right), \end{aligned} \tag{1}$$

where β is the estimated regression coefficient, x is the impacting variable, and ε is the gamma-distributed error in NB regression model:

$$\hat{\lambda} = (V \times T)^{\beta_1} \exp\left(\sum_{j=1}^{m} \beta_j x_j + \sigma\right), \tag{2}$$

where $\sigma = \beta_0$ in Poisson regression model or $\sigma = \beta_0 + \varepsilon$ in NB regression model.

Based on these investigations, it is clear that there is a pressing need for an appropriate accident prediction model that should comprehensively consider contributing factors toward LX safety. Moreover, such a model should have high predictive accuracy. Therefore, in the present study, a new accident prediction model is developed to predict the accident frequency at LXs. Specifically, we focus on the SAL2 type of LX (i.e., an automated LX system with two half barriers and flashing lights), which is the most widely used type of LX in France and contributed most to the total number of accidents at French LXs from 1974 to 2014.

2. Method

In this section, an advanced accident prediction model is developed, which enables to rank risky LXs accurately and identify the significant impacting parameters efficiently. The model considers the average daily road traffic, the average daily railway traffic, the annual road accidents, the vertical road profile, the horizontal road alignment, the road width, the crossing length, the railway speed limit, and the geographic region. The nonlinear least-squares (NLS) method, Poisson regression method, NB regression method, ZIP regression method, and ZINB regression method are employed to estimate the respective coefficients of parameters in the prediction model.

2.1 Data sources and coding

The dataset used in our study, which cover SAL2 LXs in 21 administrative regions in mainland France from 2004 to 2013, has been provided by SNCF Réseau (the French national railway infrastructure manager). Moreover, the dataset includes 10 years of information about annual LX accident frequency, annual roadway accident statistics and railway, roadway, and LX characteristics. In total, there are 8332 public SAL2 LXs involved in our investigation. The impacting parameters relevant to LX accidents considered in our investigation can fulfill the following characteristics: (1) important in determining accident frequency, (2) more permanent in nature (e.g., sight obstruction noted as a problematic factor due to involved alterable construction topography, vegetation, and other environmental elements), and (3) not accident-dependent [18]. The statistical characterization of parameters considered in this investigation are shown in **Table 1**. It is worth noticing that the road accident factor is reflected by the ratio of the annual number of road accidents in a given year to the average number of road accidents per year over the period of 10 years considered, while the region risk factor is reflected by the general accident frequency per SAL2 in the corresponding region. Overall, the data coding is shown in **Table 2**.

2.2 Advanced accident prediction model

Here, we define that the formula of the conventional traffic moment is given as: Traffic moment = Road traffic frequency × Railway traffic frequency [19]. However, based on some previous analyses [20], we adopt a variant called "corrected moment," or CM for short. $CM = V^a \times T^b$, where $a + b = 1$ and the optimal value of a in terms of fitting is calculated to be $a = 0.354$ according to the statistical analysis performed by SNCF Réseau [21]. Therefore, we consider $\left(V^{0.354} \times T^{0.646}\right)$ as an integrated parameter that reflects the combined exposure frequency of both railway and road traffic.

The developed advanced model takes into account various variables as interpreted in **Table 2**. The general form of the model is shown as follows:

$$\lambda_{10Y} = K \times F_{RAcc} \times \left(V^a \times T^b\right) \times \exp\left(C_{Profile} \times I_{Profile} + C_{Align} \times I_{Align} + C_{Wid} \times Wid + C_{Leng} \times Leng + C_{RSL} \times RSL + C_{Reg} \times F_{Reg}\right), \quad (3)$$

Parameter	Description	Mean	Std. dev.
Railway traffic characteristics			
Average daily railway traffic	The average number of trains crossing the LX daily;	26.1	30.2
Railway speed limit	The maximum permission speed of train within the LX section;	92.5	42.4
Roadway traffic characteristics			
Average daily road traffic	The average number of road vehicles crossing the LX daily;	826.8	1.8e+03
Annual road accidents	The number of road accidents in a given year;	7.1e+04	9.7e+03
LX characteristics			
Alignment	Horizontal road alignment shape: "straight", "curve," or "S";	N/A	N/A
Profile	Vertical road profile shape: "normal", "hump," or cavity";	N/A	N/A
Length	The entering road width;	9.7	3.9
Width	The distance that road vehicles need to cross through the LX;	5.5	1.4
Region	The region of the LX considered;	N/A	N/A

Table 1.
Statistical characterization of parameters considered.

where λ_{10Y} represents the annual accident frequency at a given SAL2 for a period of 10 years; F_{RAcc} is the road accident factor, which is a time-dependent variable and reflects the variation of annual road accidents as time advances; K is the coefficient of F_{RAcc}; V denotes the average daily road traffic; T denotes the average daily railway traffic; $I_{Profile}$ is the profile indicator and $C_{Profile}$ is the coefficient of $I_{Profile}$; I_{Align} is the alignment indicator and C_{Align} is the coefficient of I_{Align}; Wid is the LX width and C_{Wid} is the coefficient of Wid; $Leng$ is the crossing length and C_{Leng} is the coefficient of $Leng$; RSL is the railway speed limit and C_{RSL} is the coefficient of RSL; F_{Reg} is the region factor and C_{Reg} is the coefficient of F_{Reg}. Note that this model does not only rank risky LXs accurately but also allow for identifying significant parameters efficiently.

2.2.1 Regression approaches

In this section, several regression approaches are adopted to estimate the coefficients associated with the parameters of our model. The nonlinear least-squares (NLS) technique and Gauss-Newton algorithm [22] are firstly considered to estimate the variable coefficients in our model. Considering a fitting model function $y = f(x, \boldsymbol{\beta})$, where variable x depends on a vector of l parameters: $\boldsymbol{\beta} = (\beta_1, \ \beta_2, \ \dots, \ \beta_l)$. The goal is to find the vector $\boldsymbol{\beta}$ which can let the model function fit best the actual observed data in the least-squares sense. In other words, minimize the sum of residual squares S expressed as follows:

Parameter	Data coding
Railway traffic characteristics	
Average daily railway traffic	Numerical, used directly;
Railway speed limit	Numerical, used directly;
Roadway traffic characteristics	
Average daily road traffic	Numerical, used directly;
Annual road accidents	Road accident factor: *Annual road accidents in a given year/Average road accidents per year over the period observed*;
LX characteristics	
Alignment	Alignment indicator: 0, 1, and 2 represent "straight", "curve," and "S," respectively;
Profile	Profile indicator: 0 and 1 represent "normal" and "hump or cavity," respectively;
LX width	Numerical, used directly;
Crossing length	Numerical, used directly;
Region	Region risk factor, highlighting the general LX-accident-prone region: *The number of SAL2 accidents over the observation period in the region considered/ The number of SAL2 LXs in the region considered*;

Table 2.
Parameters considered and data coding.

$$S=\sum_{i=1}^{m} r_i^2,\quad m\geq l, \tag{4}$$

where r_i is the residual between the fitting model estimation and the actual observation, $r_i=y_i-f(x_i,\boldsymbol{\beta})$.

The minimum value of S is obtained by solving the gradient function $\partial S/\partial\beta_j=0$, i.e.,

$$\begin{aligned} &\partial S/\partial\beta_j=2\sum_i r_i\partial r_i/\partial\beta_j=0,\\ &\beta_j\approx\beta_j^{k+1}=\beta_j^k+\Delta\beta_j, \end{aligned} \tag{5}$$

where k is the iteration number and $\Delta\beta_j$ is the shift parameter.

At each iteration step, the model is linearized by approximation to the first-order Taylor series expansion about $\boldsymbol{\beta}^k$:

$$f(x_i,\boldsymbol{\beta})\approx f\left(x_i,\boldsymbol{\beta}^k\right)+\sum_{j=1}^{l}\left(\beta_j-\beta_j^k\right)\partial f\left(x_i,\boldsymbol{\beta}^k\right)/\partial\beta_j\approx f\left(x_i,\boldsymbol{\beta}^k\right)+\sum_{j=1}^{l}J_{ij}\Delta\beta_j, \tag{6}$$

where J_{ij} is the element of Jacobian matrix $\mathbf{J}$ and $\partial r_i/\partial\beta_j=-J_{ij}$.

Therefore, r_i can be rewritten as:

$$r_i = \Delta y_i - \sum_{s=1}^{l} J_{is} \Delta \beta_s,$$
$$\Delta y_i = y_i - f\left(x_i, \beta^k\right). \tag{7}$$

By substituting the above expressions into the gradient equation in Eq. (5), we obtain the normal equation and its matrix notation:

$$\sum_{i=1}^{m} \sum_{s=1}^{l} J_{ij} J_{is} \Delta \beta_s = \sum_{i=1}^{m} J_{ij} \Delta y_i,$$
$$(\boldsymbol{J}^T \boldsymbol{J}) \Delta \boldsymbol{\beta} = \boldsymbol{J}^T \Delta \boldsymbol{y}. \tag{8}$$

For an NLS model, S should be modified as follows:

$$S = \sum_{i=1}^{m} W_{ii} r_i^2, \quad m \geq l. \tag{9}$$

Therefore, the matrix notation of normal equation for an NLS model is expressed as follows:

$$(\boldsymbol{J}^T \boldsymbol{W} \boldsymbol{J}) \Delta \boldsymbol{\beta} = \boldsymbol{J}^T \boldsymbol{W} \Delta \boldsymbol{y}. \tag{10}$$

These aforementioned equations form the basis of the Gauss-Newton algorithm for solving an NLS problem.

In fact, the Poisson regression model shown as Eq. (11) is a natural choice for modeling accident occurrence:

$$Poi(X = k) = \frac{\lambda^k e^{-\lambda}}{k!}, \quad k = 0,1,2, \dots, \tag{11}$$

where $Poi(X = k)$ is the probability of k accidents occurring, $k \in \mathbb{N}$, and λ is the expectation value of the number of accidents.

However, [23] indicates that accident frequency is likely to be over-dispersed (see Eq. (12)) and suggests using the negative binomial (NB) regression model as an alternative to the Poisson model:

$$VAR(X) \begin{cases} = E(X) \\ > E(X), & \text{for over-dispersed} \\ < E(X), & \text{for under-dispersed} \end{cases}. \tag{12}$$

The NB model as a special case of Poisson-Gamma mixture model is a variant of the Poisson model designed to deal with over-dispersed data [11, 24, 25]. The over-dispersion could come from several possible sources, e.g., omitted variables, uncertainty in exposure data, covariates, or nonhomogeneous LX environment [26]. The NB model considered in this study has the following expression:

$$P_{NB}(X = k) = \frac{\Gamma\left(k + \frac{1}{\alpha}\right)}{\Gamma(k+1)\Gamma\left(\frac{1}{\alpha}\right)} \left(\frac{1}{1 + \alpha\lambda}\right)^{1/\alpha} \left(\frac{\alpha\lambda}{1 + \alpha\lambda}\right)^k, \quad k = 0,1,2, \dots, \tag{13}$$

where $P_{NB}(X)$ is the probability of k accidents occurring, $k \in \mathbb{N}$, α is the dispersion parameter, and λ is the expectation of the number of accidents.

The relationship between the mean value and the variance in the NB model is given as follows:

$$VAR(X) = \alpha E(X)^2 + E(X), \tag{14}$$

if $\alpha < 0$, there is an under-dispersion; if $\alpha > 0$, there is an over-dispersion; in the case where $\alpha = 0$, the NB model reduces to the Poisson model.

In practice, the count data may contain extra zeros relative to the Poisson or NB distribution. In this case, the ZIP or ZINB regression model is useful for analyzing such data [27]. The ZIP model is expressed as follows:

$$P_{ZIP}(X = k) = \begin{cases} \omega + (1-\omega)\exp(-\lambda), & \text{for } k = 0 \\ (1-\omega)\exp(-\lambda)\lambda^k/k!, & \text{for } k > 0 \end{cases}, \tag{15}$$

where $P_{ZIP}(X = k)$ is the probability of k accidents occurring, $k \in \mathbb{N}$, λ is the expectation value of the number of accidents, and $log\left(\frac{\omega}{1-\omega}\right) = z'\gamma$ is the ZI link function that z' is the ZI covariate and γ is the corresponding ZI coefficient. The mean value and variance of ZIP model are $E(X) = (1-\omega)\lambda$ and $VAR(X) = (1-\omega)\lambda(1+\omega\lambda)$.

The ZINB model is expressed as follows:

$$P_{ZINB}(X = k) = \begin{cases} \omega + (1-\omega)(1+\alpha\lambda)^{-1/\alpha}, & \text{for } k = 0 \\ (1-\omega)\dfrac{\Gamma\left(k+\dfrac{1}{\alpha}\right)}{\Gamma(k+1)\Gamma\left(\dfrac{1}{\alpha}\right)}\left(\dfrac{1}{1+\alpha\lambda}\right)^{1/\alpha}\left(\dfrac{\alpha\lambda}{1+\alpha\lambda}\right)^{k}, & \text{for } k > 0 \end{cases}, \tag{16}$$

where $P_{ZINB}(X = k)$ is the probability of k accidents occurring, $k \in \mathbb{N}$ and λ is the expectation value of the number of accidents. The mean value and variance of ZINB model are $E(X) = (1-\omega)\lambda$ and $VAR(X) = (1-\omega)\lambda(1+\omega\lambda+\alpha\lambda)$. The ZINB reduces to the ZIP in the limit $\alpha \to 0$.

However, the NB and ZINB models are limited to handling under-dispersed data ($\alpha < 0$) [11]. That is why [13] proposed the Gamma model to handle under-dispersed samples. The Gamma model is given as follows:

$$P_G(X = k) = Gamma(\beta k, \lambda) - Gamma(\beta(k+1), \lambda), \tag{17}$$

where $P_G(X)$ is the probability of k accidents occurring, $k \in \mathbb{N}$, λ is the expectation of the number of accidents, and β is the dispersion parameter. If $\beta > 1$, there is an under-dispersion; while $\beta < 1$, there is an over-dispersion and if $\beta = 1$, the Gamma model reduces to the Poisson model. However, the Gamma model shown in Eq. (18) is limited to the time-dependent observation assumption and zero observations, since general $\Gamma(x)$ restricts discrete responses to positive values:

$$Gamma(\beta k, \lambda) = \begin{cases} 1, & \text{for } k = 0 \\ \dfrac{1}{\Gamma(\beta k)\displaystyle\int_0^{\lambda} u^{\beta k-1}e^{-u}du}, & \text{for } k > 0 \end{cases}. \tag{18}$$

According to the above discussion, the restriction between mean value and variance can be used to identify an appropriate regression model. Therefore, we firstly make preliminary variance analysis by means of group classification. Namely, the annual accidents at a given SAL2 during the 10 years were divided into 100 groups with the same number of samples in each group. Then, the variance and mean value of accidents in each group were calculated, respectively, to analyze the relationship between the group variance and the group mean value. The variance analysis shows that the variance and mean value are very close to each other. Hence, we performed meticulous analyses to assess the NLS regression, the Poisson regression, the ZIP regression, the NB regression, and the ZINB regression methods with regard to SAL2 LXs in our accident dataset so as to identify which model is more effective.

2.2.2 Regression modeling results

NLS regression:

When applying the NLS regression, the form of λ_{10Y} is given by Eq. (3). The estimated coefficients computed by NLS regression are provided in **Table 3.** $|t - statistic| > 1.96$ is introduced to identify the significant parameters corresponding to a 95% confidence level. As a result, the railway speed limit, the average daily railway traffic, the average daily road traffic, the annual road accidents, the LX-accident-prone region, the road alignment, the LX width, and the crossing length have been shown to have significant and positive influence on SAL2 accident frequency. However, the test shows that the road profile is not a significant factor ($|t - statistic| = 0.635 < < 1.96$); thus, the impact of road profile could be neglected. Moreover, the coefficients of the considered variables with the exponential form can reflect the sensitive degrees of the SAL2 accident frequency to these variables, respectively. According to these sensitive degrees (rank indicated in brackets), the LX-accident-prone region factor is the most sensitive contributor among these variables.

In order to assess the predictive accuracy of accident occurrence estimated by the NLS regression model λ_{10Y} combined with the NB and ZINB distributions (see Section 3.1), we adopt the maximum likelihood estimation (MLE) method to estimate the dispersion parameter α of the dataset [28]. As expressed by Eq. (19) and Eq. (20), the values of α in NB and ZINB distributions are estimated, respectively, using R language to solve $\partial l / \partial \alpha = 0$:

Parameter	Coefficient	Estimated value	Standard error	t-statistic	Significant
	K	2.703e-05	5.078e-06	5.322	×
$I_{Profile}$	$C_{Profile}$	3.626e-02	5.706e-02	0.635	
I_{Align}	C_{Align}	3.427e-01 (2)	2.942e-02	11.648	×
Wid	C_{Wid}	9.847e-02 (3)	1.494e-02	6.589	×
$Leng$	C_{Leng}	2.084e-02 (4)	4.284e-03	4.865	×
RSL	C_{RSL}	3.089e-03 (5)	7.586e-04	4.072	×
F_{Reg}	C_{Reg}	4.962e-01 (1)	1.722e-01	2.882	×

Table 3.
Results of the λ_{10Y} NLS regression model.

$$l(\alpha)_{NB} = ln\left(\prod_{i}^{n} P_{NB}(X_i = y_i)\right) = \sum\left(y_i\, ln\,(\lambda_i) - \left(y_i + \alpha^{-1}\right) ln\,(1+\alpha\lambda_i) + \sum_{v=0}^{y_i-1} ln\,(1+\alpha v)\right), \tag{19}$$

$$l(\alpha)_{ZINB} = ln\left(\prod_{i}^{n} P_{ZINB}(X_i = y_i)\right) = \begin{cases} \sum ln\,(\omega_i) + (1-\omega_i)\left(\frac{1}{1+\alpha\lambda_i}\right)^{1/\alpha}, \text{if} y_i = 0 \\ \sum ln\,(\omega_i) + ln\,\Gamma\left(\frac{1}{\alpha}+y_i\right) - ln\,\Gamma(1+y_i) - ln\,\Gamma\left(\frac{1}{\alpha}\right). \\ +\frac{1}{\alpha}\, ln\left(\frac{1}{1+\alpha\lambda_i}\right) + y_i\, ln\left(1-\frac{1}{1+\alpha\lambda_i}\right), \text{if} y_i > 0 \end{cases} \tag{20}$$

Poisson regression:

When applying the Poisson regression, the general form of λ_{10Poi} is given by $e^{\sum_{j=1}^{m}\beta_0+\beta_j x_j}$. Therefore, we need to transform Eq. (3) into the following expression:

$$\lambda_{10Poi} = \begin{cases} 0, \text{if } F_{RAcc} = 0,\ V = 0 \text{ or } T = 0 \\ \exp(K_1 + C_F \times F_{RAcc} + C_{CM} \times CM + C_{Profile} \times I_{Profile} + C_{Align} \times I_{Align} + \\ C_{Wid} \times Wid + C_{Leng} \times Leng + C_{RSL} \times RSL + C_{Reg} \times F_{Reg}), \text{if} F_{RAcc} \neq 0, \\ V \neq 0, \text{and } T \neq 0 \end{cases} \tag{21}$$

The results estimated through the Poisson regression approach are shown in **Table 4**. According to these results, being similar to the NLS case, one can notice that the road profile is not significant ($|t-statistic| = 0.621 < < 1.96$). On the other hand, with an exponential form, the impact of road accident factor F_{RAcc} is weakened, namely the impact of F_{RAcc} with an exponential form is not significant when using Poisson regression approach ($|t-statistic| = 1.913 < 1.96$). Furthermore, according to

Parameter	Coefficient	Estimated value	Standard error	t-statistic	Significant
	K_1	−9.562	0.440	−21.714	×
F_{RAcc}	C_F	0.636	0.332	1.913	
CM	C_{CM}	0.005 (6)	2.949e-04	17.144	×
$I_{Profile}$	$C_{Profile}$	−0.076	0.122	−0.621	
I_{Align}	C_{Align}	0.326 (2)	0.069	4.756	×
Wid	C_{Wid}	0.206 (3)	0.026	8.051	×
$Leng$	C_{Leng}	0.030 (4)	0.009	3.232	×
RSL	C_{RSL}	0.011 (5)	0.001	7.895	×
F_{Reg}	C_{Reg}	1.725 (1)	0.334	5.165	×

Table 4.
Regression results of λ_{10Poi}.

the sensitive degrees of these parameters with the exponential form (rank indicated in brackets), once again the LX-accident-prone region factor is the most sensitive contributor among these parameters.

NB regression:

When applying the NB regression, the general form of λ_{10NB} is given by $e^{\sum_{j=1}^{m}\beta_0+\beta_j x_j+\varepsilon}$, and it still requires to be expressed by Eq. (21). The dispersion parameter α is estimated at 3.2394 in our study through the iterative estimation algorithm automatically. The estimated results of the NB regression are shown in **Table 5**. According to the results associated with the NB regression approach, it is worth noticing that the road profile is still not significant ($|t-statistic| = 0.850 < < 1.96$). One can also notice that the impact of F_{RAcc} with an exponential form is not significant as well, when using the NB regression approach ($|t-statistic| = 1.793 < 1.96$). Moreover, according to the sensitive degrees of these parameters with the exponential form (rank indicated in brackets), the LX-accident-prone region factor is still the most sensitive contributor among these parameters.

ZIP regression:

When applying the ZIP regression, the general form of λ_{10ZIP} is given by $e^{\sum_{j=1}^{m}\beta_0+\beta_j x_j}$, and it still requires to be expressed by Eq. (21). The estimated results of the ZIP regression are shown in **Table 6** and (for nonzero observations) and **Table 7** (for zero-inflation observations).

According to the results associated with the ZIP regression approach, it is worth noticing that, as for the nonzero related model, F_{RAcc}, $I_{Profile}$, I_{Align}, and $Leng$ are not significant (<1.96). Moreover, according to the sensitive degrees of other significant parameters with the exponential form (rank indicated in brackets), the LX-accident-prone region factor is still the most sensitive contributor among these parameters. While as for the zero-inflation model, only the Wid, RSL, and F_{Reg} are significant (>1.96).

ZINB regression:

When applying the ZINB regression, the general form of λ_{10ZINB} is given by $e^{\sum_{j=1}^{m}\beta_0+\beta_j x_j+\varepsilon}$, and it still requires to be expressed by Eq. (21). The values of dispersion parameter α for nonzero observations and zero-inflation observations are estimated at

Parameter	Coefficient	Estimated value	Standard error	t-statistic	Significant
	K_1	−9.424	0.457	−20.615	×
F_{RAcc}	C_F	0.616	0.343	1.793	
CM	C_{CM}	0.006 (6)	3.762e-04	16.493	×
$I_{Profile}$	$C_{Profile}$	−0.107	0.126	−0.850	
I_{Align}	C_{Align}	0.298 (2)	0.072	4.159	×
Wid	C_{Wid}	0.199 (3)	0.028	7.173	×
$Leng$	C_{Leng}	0.031 (4)	0.010	3.201	×
RSL	C_{RSL}	0.010 (5)	0.001	7.034	×
F_{Reg}	C_{Reg}	1.508 (1)	0.351	4.294	×

Table 5.
Regression results of λ_{10NB}.

Parameter	Coefficient	Estimated value	Standard error	t-statistic	Significant
	K_1	−1.128e+01	7.586e-01	−14.867	×
F_{RAcc}	C_F	3.717e-01	4.202e-01	0.885	
CM	C_{CM}	6.221e-03 (4)	4.336e-04	14.347	×
$I_{Profile}$	$C_{Profile}$	−1.855e-01	1.513e-01	−1.226	
I_{Align}	C_{Align}	1.483e-01	8.786e-02	1.688	
Wid	C_{Wid}	4.397e-01 (2)	6.625e-02	6.636	×
$Leng$	C_{Leng}	3.971e-02	1.725e-02	1.904	
RSL	C_{RSL}	1.432e-02 (3)	2.069e-03	6.921	×
F_{Reg}	C_{Reg}	2.319 (1)	6.655e-01	3.484	×

Table 6.
Count model regression results of λ_{10ZIP}.

Parameter	Coefficient	Estimated value	Standard error	t-statistic	Significant
	K_1	−1.574e+01	4.276	−3.680	×
F_{RAcc}	C_F	−1.104	1.646	−0.671	
CM	C_{CM}	1.584e-03	1.450e-03	1.093	
$I_{Profile}$	$C_{Profile}$	−4.355e-01	6.531e-01	0.505	
I_{Align}	C_{Align}	−1.185	6.141e-01	−1.931	
Wid	C_{Wid}	1.024 (2)	2.241e-01	4.571	×
$Leng$	C_{Leng}	8.231e-02	4.190e-02	1.964	
RSL	C_{RSL}	4.117e-02 (3)	1.449e-02	2.840	×
F_{Reg}	C_{Reg}	5.861 (1)	1.748	3.353	×

Table 7.
Zero-inflation model regression results of λ_{10ZIP}.

3.8102 and 1.4069, respectively, in our study through the iterative estimation algorithm automatically. The estimated results of the ZINB regression are shown in **Table 8** (for nonzero observations) and **Table 9** (for zero-inflation observations). According to the results associated with the ZINB regression approach, it is worth noticing that, as for the nonzero related model, CM, I_{Align}, and Wid are significant (>1.96). One can also notice that according to the sensitive degrees of the three parameters (rank indicated in brackets), the LX width is the most sensitive contributor among them. While as for the zero-inflation model, only the F_{RAcc} and CM are significant (>1.96).

3. Model performance evaluation and discussion

In this section, we will assess the performance of our prediction models while determining an appropriate statistical distribution to be combined with the models, in such a way as to ensure the most accurate estimation of the probability of accidents

Parameter	Coefficient	Estimated value	Standard error	t-statistic	Significant
	K_1	−7.128	0.734	−9.709	×
F_{RAcc}	C_F	0.671	0.413	1.624	
CM	C_{CM}	4.486e-03 (3)	4.991e-04	8.990	×
$I_{Profile}$	$C_{Profile}$	−5.886e-02	0.144	−0.406	
I_{Align}	C_{Align}	0.371 (1)	8.274e-02	4.495	×
Wid	C_{Wid}	0.145 (2)	4.558e-02	3.175	×
$Leng$	C_{Leng}	3.219e-03	1.203e-02	0.268	
RSL	C_{RSL}	2.558e-03	1.954e-03	1.309	
F_{Reg}	C_{Reg}	0.795	0.446	1.783	

Table 8.
Count model regression results of λ_{10ZINB}.

Parameter	Coefficient	Estimated value	Standard error	t-statistic	Significant
	K_1	−4.036	2.190	−6.709	×
F_{RAcc}	C_F	0.260 (1)	1.456	2.179	×
CM	C_{CM}	6.685e-02 (2)	1.838e-02	3.636	×
$I_{Profile}$	$C_{Profile}$	0.705	0.544	1.296	
I_{Align}	C_{Align}	0.535	0.328	1.632	
Wid	C_{Wid}	8.873e-02	0.180	0.491	
$Leng$	C_{Leng}	0.114	6.639e-02	1.725	
RSL	C_{RSL}	5456e-03	6.629e-03	0.823	
F_{Reg}	C_{Reg}	1.632	1.679	0.972	

Table 9.
Zero-inflation model regression results of λ_{10ZINB}.

occurring at a given SAL2 in a given year. The Bayesian information criterion (BIC) [29], Akaike's information criterion (AIC) [30], the Pearson chi-square statistic (PCS) test [31], and the degree of freedom (DF) are used to evaluate the goodness of fit (GOF) of the model. They can be respectively expressed as follows:

$$\text{BIC} = n + n \times ln\,(2\pi) + n \times ln\,(\text{RSS}/n) + (l+1)\,ln\,(n), \tag{22}$$

$$\text{AIC} = n + n \times ln\,(2\pi) + n \times ln\,(\text{RSS}/n) + 2(l+1), \tag{23}$$

$$\text{PCS} = \sum_{i=1}^{n} \frac{(O_i - \lambda_i)^2}{\lambda_i}, \tag{24}$$

$$\text{DF} = n - (l+1), \tag{25}$$

where RSS is the sum of the squares of residuals between the annual accident frequencies observed and the annual accident frequencies estimated, n is the sample size, l is the number of independent exponential parameters, λ_i is the annual accident frequency expected, and O_i is the annual accident frequency observed.

The BIC and AIC are used to test the relative quality of models for a given dataset. Smaller BIC and AIC values indicate a better model fitting. The PCS test is used to determine if there is a significant difference between the values expected and the values observed. The PCS is roughly equal to DF if the model fits the data perfectly without any dispersion. Namely, the closer the PCS is to the DF, the better the model fits the data [14].

The log-likelihood statistic test (LL) is adopted to assess the GOF of the accident frequency prediction model combined with a statistical distribution. The larger the LL, the more preferred the model [14]. The mathematical expression of the LL is given as follows:

$$\mathrm{LL} = \sum_{i=1}^{n} ln\left(\hat{P}_i\right), \tag{26}$$

where n is the sample size and $\hat{P}_i$ is the estimated probability of accident frequency observed. $\hat{P}_i$ is computed respectively according to the accident frequency prediction model combined with the Poisson or the NB distribution.

3.1 Model performance comparison among variants of λ_{10Y}

The results of AIC, BIC, and PCS statistical tests are shown in **Table 10** with the goodness ranked in brackets. The following findings are obtained: 1) considering AIC and BIC, the λ_{10Y} model gives better results, since the AIC and BIC values corresponding to the λ_{10Y} model are much smaller than those for the λ_{10Poi}, λ_{10NB}, λ_{10ZIP}, and λ_{10ZINB} models; 2) in terms of PCS test, the λ_{10Y} model is also the most effective one, since the PCS of λ_{10Y} model is closer to DF (DFs of λ_{10Y}, λ_{10Poi}, λ_{10NB}, λ_{10ZIP}, and λ_{10ZINB} are considerably approximative).

LL test results are shown in **Table 10**. One can notice that, for the λ_{10Y} model combined with either the Poisson or NB distribution, its GOFs are significantly better than λ_{10Poi} and λ_{10NB} models' GOFs according to the LL test. Furthermore, the GOF of λ_{10Y} combined with the NB distribution (NB-λ_{10Y}) is better than when combined with the Poisson distribution (POI-λ_{10Y}).

Test	POI-λ_{10Y}	NB-λ_{10Y}	λ_{10Poi}	λ_{10NB}	λ_{10ZIP}	λ_{10ZINB}
AIC	−190,744 (1)	−190,744 (1)	−187,804 (5)	−189,942 (2)	−188,312 (4)	−189,826 (3)
BIC	−190,670 (1)	−190,670 (1)	−187,720 (5)	−189,858 (3)	−188,176 (4)	−189,935 (2)
PCS	65,796 (1)	65,796 (1)	125,495 (5)	123,715 (4)	118,185 (3)	110,496 (2)
DF	83,313	83,313	83,311	83,311	83,311	83,311
LL	−2599 (2)	−2596 (1)	−2732 (6)	−2711 (5)	−2701 (4)	−2631 (3)
Goodness score						
(the lower, the better)	5	4	21	14	15	10

Table 10.
Model GOF comparison among variants of λ_{10Y}.

3.2 A comparison between λ_{10Y} and two existing reference models

In this section, we compare the present model λ_{10Y} with other two models which are widely used in existing related works. As mentioned in Section 1, the first widely used model is given in Eq. (1) [13, 14, 18]. In our study, this model can be specified as follows:

$$\lambda_{TV} = \begin{array}{l} \exp\Big(K_2 + C_V \times V + C_T \times T + C_F \times F_{RAcc} + C_{Profile} \times I_{Profile} + C_{Align} \\ \times I_{Align} + C_{Wid} \times Wid + C_{Leng} \times Leng + C_{RSL} \times RSL + C_{Reg} \times F_{Reg}\Big), \end{array} \tag{27}$$

where the average daily road traffic V and the average daily railway traffic T are applied separately in exponential form.

The second model as shown in Eq. (2) (e.g., [17, 32]) is specified as Eq. (28) in our study:

$$\lambda_{Mon} = \begin{array}{l} \exp\Big(K_3 + C_M \times ln\,(V \times T) + C_F \times F_{RAcc} + C_{Profile} \times I_{Profile} + C_{Align} \\ \times I_{Align} + C_{Wid} \times Wid + C_{Leng} \times Leng + C_{RSL} \times RSL + C_{Reg} \times F_{Reg}\Big), \end{array} \tag{28}$$

where the conventional traffic moment $V \times T$ is applied.

It should be noted that the ZIP and ZINB models were also investigated for λ_{TV} and λ_{Mon} but resulted in no higher goodness-of-fit values and a quite small number of significant parameters compared with the Poisson and NB models and, hence, were not reported in this section. The Poisson and NB regression results of the λ_{TV} and λ_{Mon} are shown in **Tables 11–14**, respectively. One can notice that the impacts of road profile and road accident are still not significant in the λ_{TV} and λ_{Mon}. The AIC, BIC, PCS, and LL tests and observed/estimated accident frequency comparison are given in **Table 15**. According to the quality test results discussed in Section 3.1, the λ_{10Y} combined with the NB distribution (NB-λ_{10Y}) shows the best prediction performance among the four investigated combinations. Therefore, we will only compare the NB-

Parameter	Coefficient	Estimated value	Standard error	t-statistic	Significant
	K_2	−9.807	0.413	−22.223	×
V	C_V	1.098e-04 (7)	1.613e-05	6.811	×
T	C_T	8.777e-03 (6)	1.115e-03	7.869	×
F_{RAcc}	C_F	0.636	0.333	1.913	
$I_{Profile}$	$C_{Profile}$	−1.445e-01	1.209e-01	−1.195	
I_{Align}	C_{Align}	3.319e-01 (2)	6.747e-02	4.919	×
Wid	C_{Wid}	2.059e-01 (3)	2.483e-02	8.292	×
$Leng$	C_{Leng}	3.952e-02 (4)	7.868e-03	5.024	×
RSL	C_{RSL}	1.154e-02 (5)	1.487e-03	7.759	×
F_{Reg}	C_{Reg}	1.750 (1)	3.463e-01	5.053	×

Table 11.
Poisson regression results of λ_{TV}.

Parameter	Coefficient	Estimated value	Standard error	t-statistic	Significant
	K_2	−9.882	4.531e-01	−21.810	×
V	C_V	1.155e-04 (7)	1.683e-05	6.861	×
T	C_T	9.152e-03 (6)	1.234e-03	7.416	×
F_{RAcc}	C_F	0.607	3.402e-01	1.784	
$I_{Profile}$	$C_{Profile}$	−1.532e-01	1.243e-01	−1.232	
I_{Align}	C_{Align}	3.240e-01 (2)	6.988e-02	4.636	×
Wid	C_{Wid}	2.212e-01 (3)	2.579e-02	8.575	×
$Leng$	C_{Leng}	3.895e-02 (4)	8.415e-03	4.629	×
RSL	C_{RSL}	1.160e-02 (5)	1.529e-03	7.589	×
F_{Reg}	C_{Reg}	1.739 (1)	3.575e-01	4.864	×

Table 12.
NB regression results of λ_{TV}.

Parameter	Coefficient	Estimated value	Standard error	t-statistic	Significant
	K_2	−11.816	4.540e-01	−26.023	×
$ln\,(V \times T)$	C_M	4.036e-01 (2)	2.776e-02	14.538	×
F_{RAcc}	C_F	6.359e-01	3.325e-01	1.913	
$I_{Profile}$	$C_{Profile}$	−6.279e-02	1.205e-01	−0.521	
I_{Align}	C_{Align}	2.875e-01 (3)	6.799e-02	4.228	×
Wid	C_{Wid}	1.185e-01 (4)	3.296e-02	3.596	×
$Leng$	C_{Leng}	2.213e-02 (5)	9.530e-03	2.322	×
RSL	C_{RSL}	8.811e-03 (6)	1.350e-03	6.527	×
F_{Reg}	C_{Reg}	1.446 (1)	3.358e-01	4.307	×

Table 13.
Poisson regression results of λ_{Mon}.

Parameter	Coefficient	Estimated value	Standard error	t-statistic	Significant
	K_2	−11.850	4.628e-01	−26.603	×
$ln\,(V \times T)$	C_M	4.034e-01 (2)	2.822e-02	14.297	×
F_{RAcc}	C_F	6.368e-01	3.382e-01	1.883	
$I_{Profile}$	$C_{Profile}$	−7.103e-02	1.230e-01	−0.578	
I_{Align}	C_{Align}	2.848e-01 (3)	6.960e-02	4.092	×
Wid	C_{Wid}	1.214e-01 (4)	3.361e-02	3.612	×
$Leng$	C_{Leng}	2.204e-02 (5)	9.752e-03	2.260	×
RSL	C_{RSL}	8.892e-03 (6)	1.368e-03	6.500	×
F_{Reg}	C_{Reg}	1.480 (1)	3.428e-01	4.316	×

Table 14.
NB regression results of λ_{Mon}.

Test	NB-λ_{10Y}	POI-λ_{TV}	NB-λ_{TV}	POI-λ_{Mon}	NB-λ_{Mon}
AIC	−190,744 (1)	−177,914 (5)	−179,842 (4)	−183,714 (3)	−186,532 (2)
BIC	−190,670 (1)	−177,610 (5)	−179,738 (4)	−183,587 (3)	−186,191 (2)
PCS	65,796 (1)	121,715 (5)	119,133 (4)	118,511 (3)	115,634 (2)
DF	83,313	83,310	83,310	83,311	83,311
LL	−2596 (1)	−2722 (5)	−2703 (3)	−2705 (4)	−2683 (2)
Goodness score					
(the lower, the better)	4	20	15	13	8

Table 15.
Model GOF comparison among λ_{10Y}, λ_{TV}, and λ_{Mon}.

λ_{10Y} with the λ_{TV} and λ_{Mon} combined with the Poisson and NB distributions, respectively, in the following content.

As shown in **Table 15**, the AIC, BIC, and PCS results related to the λ_{10Y} model are better than those for the λ_{TV} and λ_{Mon} models. Moreover, in terms of the LL test, the NB-λ_{10Y} is still the most preferred one.

4. Conclusions

Based on our study, some remarks need to be highlighted as follows:

1. The corrected traffic moment proposed is more effective in estimating automobile-involved LX accidents frequency compared with the conventional traffic moment, single average daily railway traffic or single average daily road traffic. It is worth mentioning that the average daily railway traffic with a power of 0.646 has a more decisive impact on the LX accident frequency than the average daily road traffic with a power of 0.354. Moreover, the higher the combined exposure of railway and roadway traffic, the higher the likelihood of an accident occurring.

2. According to the analyses above, the form of λ_{10Y} highlights the impact of road accident factor F_{RAcc}, while the impact of F_{RAcc} is neglected in λ_{10Poi}, λ_{10NB}, λ_{TV}, and λ_{Mon} models (see **Tables 4, 5, 11–14**). The impact of road accidents on the risk level was likely to be ignored in the previous studies related to LX safety analysis.

3. We originally introduce the region LX-accident-prone factor (see **Table 2**) in this study to interpret the variation of LX accident statistics with regard to various regions. According to the sensitive degrees of variables ranked in **Table 3**, among the LX characteristics, the risk of LX accidents is most sensitive to the region LX-accident-prone factor. However, in many past studies, the impact of LX local region is neglected. In fact, the regional accident history varies from one region to another, which correspondingly has varying degrees of impact on the LX accident frequency in different regions.

To sum up, the develop model λ_{10Y} has trustworthy goodness of fit. Moreover, it shows relatively high prediction accuracy for LX accident frequency prediction when combined with the NB distribution.

Author details

Ci Liang[1,2]* and Mohamed Ghazel[2]

1 Harbin Institute of Technology, Harbin, China

2 Université Gustave Eiffel (ex-IFSTTAR) - COSYS/ESTAS, Villeneuve-d'Ascq, France

*Address all correspondence to: ciliang.lc@gmail.com

References

[1] Read GJM, Salmon PM, Lenné MG, Stanton NA. Walking the line: Understanding pedestrian behaviour and risk at rail level crossings with cognitive work analysis. Applied Ergonomics. 2016;**53**:209-227

[2] Ghazel M. Using stochastic petri nets for level-crossing collision risk assessment. IEEE Transactions on Intelligent Transportation Systems. 2009;**10**(4):668-677

[3] Liu B, Ghazel M, Toguyéni A. Model-based diagnosis of multi-track level crossing plants. IEEE Transactions on Intelligent Transportation Systems. 2016;**17**(2):546-556

[4] ERA. Railway safety performance in the European Union. 9(2) Agency Regulation 881/2004/EC. 2014

[5] Ghazel M, El-Koursi EM. Two-half-barrier level crossings versus four-half-barrier level crossings: A comparative risk analysis study. IEEE Transactions on Intelligent Transportation Systems. 2014;**15**(3):1123-1133

[6] Silmon J, Roberts C. Using functional analysis to determine the requirements for changes to critical systems: Railway level crossing case study. Reliability Engineering and System Safety. 2010; **95**(3):216-225

[7] Australian Transport Safety Bureau. Australian Rail Safety Occurrence Data: 1 July 2002 to 30 June 2012 (ATSB Transport Safety Report RR-2012-010). Canberra, Australia: ATSB; 2012

[8] SNCF Réseau. World Conference of Road Safety at Level Crossings. 2011. Available from: http://www.planetoscope.com/automobile/1271-nombre-de-collisions-aux- passages-a-niveau-en-france.html

[9] Plesse G. Des détecteurs d'obstacles déployés aux passages à niveau. 2017. Available from: http://www.leparisien.fr/info-paris-ile-de-france-oise/transports/des-detecteurs-d-obstacles-deployes-aux-passages-a-niveau-02-06-2017-7011714.php

[10] Liang C, Ghazel M, Cazier O, El-Koursi EM. Analyzing risky behavior of motorists during the closure cycle of railway level crossings. Safety Science. 2018;**110**:115-126

[11] Lord D, Mannering F. The statistical analysis of crash-frequency data: A review and assessment of methodological alternatives. Transportation Research Part A: Policy and Practice. 2010;**44**(5):291-305

[12] Guikema SD, Quiring SM. Hybrid data mining-regression for infrastructure risk assessment based on zero-inflated data. Reliability Engineering and System Safety. 2012;**99**: 178-182

[13] Oh J, Washington SP, Nam D. Accident prediction model for railway-highway interfaces. Accident; Analysis and Prevention. 2006;**38**(2): 346-356

[14] Lu P, Tolliver D. Accident Analysis & Prevention. Accident prediction model for public highway-rail grade crossings. 2016;**90**:3-81

[15] Medina JC, Benekohal RF. Macroscopic models for accident prediction at railroad grade crossings: Comparisons with US Department of Transportation accident prediction formula. Transportation Research

Record: Journal of the Transportation Research Board. 2015;**2476**:85-93

[16] Chadwick SG, Zhou N, Saat MR. Highway-rail grade crossing safety challenges for shared operations of high-speed passenger and heavy freight rail in the US. Safety Science. 2014;**68**:128-137

[17] Miranda-Moreno L, Fu L, Saccomanno FF, Labbe A. Alternative risk models for ranking locations for safety improvement. Transportation Research Record: Journal of the Transportation Research Board. 2005; **1908**:1-8

[18] Austin RD, Carson JL. An alternative accident prediction model for highway-rail interfaces. Accident; Analysis and Prevention. 2002;**34**(1):31-42

[19] Liang C, Ghazel M, Cazier O, El-Koursi EM. Risk analysis on level crossings using a causal Bayesian network based approach. Transportation Research Procedia. 2017; **25**:2172-2186

[20] Liang C, Ghazel M, Cazier O, El-Koursi EM. Developing accident prediction model for railway level crossings. Safety Science. 2018;**101**:48-59

[21] SNCF Réseau. SNCF. Statistical Analysis of Accidents at LXs. France: SNCF Réseau; 2010

[22] Madsen K, Nielsen HB, Tingleff O. Methods for non-linear least squares problems. In: Informatics and Mathematical Modelling. 2nd ed. Denmark: Technical University of Denmark; 2004

[23] Chang LY. Analysis of freeway accident frequencies: Negative binomial regression versus artificial neural network. Safety Science. 2005;**43**(8): 541-557

[24] Buddhavarapu P, Scott JG, Prozzi JA. Modeling unobserved heterogeneity using finite mixture random parameters for spatially correlated discrete count data. Transportation Research Part B: Methodological. 2016;**91**:492-510

[25] Utkin LV, Coolen FPA, Gurov SV. Imprecise inference for warranty contract analysis. Reliability Engineering and System Safety. 2015;**138**:31-39

[26] Miaou SP. The relationship between truck accidents and geometric design of road sections: Poisson versus negative binomial regressions. Accident; Analysis and Prevention. 1994;**26**(4):471-482

[27] Ridout M, Hinde J, DeméAtrio C. A score test for testing a zero-inflated Poisson regression model against zero-inflated negative binomial alternatives. Biometrics. 2001;**57**(1):219-223

[28] Dai H, Bao Y, Bao M. Maximum likelihood estimate for the dispersion parameter of the negative binomial distribution. Statistics & Probability Letters. 2013;**83**(1):21-27

[29] Weakliem DL. A critique of the Bayesian information criterion for model selection. Sociological Methods & Research. 1999;**27**(3):359-397

[30] Bozdogan H. Model selection and Akaike's information criterion (AIC): The general theory and its analytical extensions. Psychometrika. 1987;**52**(3): 345-370

[31] Pearson KX. On the criterion that a given system of deviations from the probable in the case of a correlated system of variables is such that it can be reasonably supposed to have arisen from random sampling. The London, Edinburgh, and Dublin Philosophical Magazine and Journal of Science. 1900; **50**(302):157-175

[32] Saccomanno FF, Fu L, Miranda-Moreno L. Risk-based model for identifying highway-rail grade crossing blackspots. Transportation Research Record: Journal of the Transportation Research Board. 2004;**1862**:127-135

Chapter 9

Railways Passenger Comfort/ Discomfort: Objective Evaluation

Patrícia Silva, Joaquim Mendes, Eurico Seabra and Pedro Pratas

Abstract

Railways are one of the most used public transportation modes. Due to the low environmental impact, governments are promoting their use as a solution for mass transportation for middle-distance. Critical parameters to satisfy passengers and attract new ones are related to comfort, safety, and user conditions. Comfort is classified by dynamic and static methods, ideally combining both. The former is based on the comfort evaluation in the presence of vibration, whilst the latter is applied to standstill situations. Vibration, derived from the train motion, will be presented throughout the trip; thus, it is a major concern. For the same reason, assessing vibration levels associated with comfort gives important information about the train's condition and its major maintenance or repair needs. Vibration transmissibility allows the study of the vibration frequency that is being transmitted to the user. In this chapter, the complete seat structure and the passenger's comfort are evaluated combined with the interface pressure. This chapter presents the commonly used methodologies and results from the tests conducted on railway environment whilst running passenger service. These tests report the seat comfort evaluation and a new methodology to identify train and rail maintenance needs based on comfort analysis.

Keywords: railways comfort, dynamic analysis, static analysis, vibration, interface pressure

1. Introduction

World population is increasing, and therefore the needs for mass transportation. Nowadays, railways are one of the most widely used public transportation systems, mainly because of their safety, superior transportation capacity, reduced boarding time, and the possibility of making better use of the journey time to work or enjoy train facilities. Additionally, trains have been revealed to be the most sustainable transportation system. In Europe, the predominant proportion of rail services are operated by electrified trains; thus, its CO_2 emissions are residual. The emissions per passenger-kilometer are much lower for the train than for air or car transportation [1–3].

Due to its low environmental impact, multiple governments are promoting rail use as the major mass transportation system, especially for middle-range distances. Based on this proposal, train passengers have continuously increased from 2013 until the

2020 COVID pandemic. Between 2015 and 2019, a historic 4000 billion passenger-kilometers were recorded worldwide. However, during the pandemic years, due to multiple lockdowns, train passenger numbers drastically decreased [4–6]. Currently, those numbers are even higher than the pre-pandemic ones [7].

The quality of public transportation services influences travelers' choices. Passengers with previous good travel experiences will probably use the same travel transportation mode again. On the other hand, customers that experienced problems with the journey service may change to a different transportation mode for their next trip. Therefore, to keep increasing continuously the number of passengers, it is crucial to raise trains' attractiveness and provide comfortable journeys [8, 9]. Those are defined based on safety, comfort, and user conditions [10, 11]. In particular, the seat and vibration are strongly linked with discomfort. Passengers spend most of their travel time seated; thus, vibration, derived from train motion and train-rail interaction, is transmitted to the user through this surface, being classified as whole-body vibration (WBV). A specific feature of a seat is its dimensions which must be able to accommodate people with different anthropometric characteristics, and simultaneously provide low fatigue levels and a general feeling of comfort [12]. Although affecting passengers' comfort, vibration is also linked to safety. High safety levels are ensured through adequate maintenance. That will guarantee the reliability and longevity of the rail, which needs to provide a stable and safe platform for train operation. Corrective and preventive interventions are accomplished during track maintenance. Its main goal is to preserve the system's functions and prevent its breakdown or failure [13–16].

Comfort and discomfort are two concepts not consensually defined in the literature. Hence multiple interpretations can be found. However, it is well established that comfort should be evaluated based on objective and subjective methods, ideally a combination of both. Objective methods are derived from mechanical approaches; thus, these methods quantitatively define the physical variables of comfort [17]. In opposition, subjective evaluation methods rate users' feelings and, this way, they quantify the psychological impact of comfort on the passengers based on questionnaires and rating scales [18].

Comfort parameters are divided into "Motion" and "Non-Motion" factors. The latter is characterized by issues such as noise, smell, illumination, humidity, and temperature. To properly classify "Motion" parameters, it is essential to conduct both dynamic (in the presence of vibration) and static (absence of vibration) tests. The former is evaluated by ride comfort, seat effective transmissibility (SEAT) and transmissibility tests. The latter is mainly defined by interface pressure measurements [10]. Seats may present good dynamic behavior but poor static performance. The ideal seat has a combination of optimum dynamic properties to minimize unwanted vibration and the best static behavior to equally distribute pressure at the seat surface and, this way, reduce the interface pressure.

The present chapter intends to introduce a new methodology to identify train and rail track infrastructure sections' maintenance requirements based on ride comfort analysis. Moreover, transmissibility and interface pressure experiments are conducted as complementary comfort analysis methods. Vibration transmissibility allows the study of the vibration frequency that is being transmitted to the user. If associated in combination with the interface pressure (static analysis), it is possible to define the complete seat structure and conclude about passengers' comfort. The chapter is organized into seven sections to report on the aforementioned goals. Multiple comfort and discomfort definitions will be presented, and the one used in this research will be detailed. Based on vibration analysis, the ride quality evaluation methods are defined, leading to the rail vehicle and track infrastructure maintenance needs identification. Finally, transmissibility and interface pressure experiments are described leading to the final passengers' comfort evaluation and seat structure analysis.

2. Comfort and discomfort definition

Whether for pleasure or work, people want to be comfortable. Thus, when designing a seat, it is vital to start by analyzing the comfort that it will provide to the users. Therefore, understanding how comfort and discomfort are defined and evaluating them, is a major assignment. These two parameters do not have a consensus; hence multiple definitions and interpretations for comfort and discomfort can be found in the literature [19].

In 1958, Hertzberg defined comfort as the absence of discomfort. Hertzberg also postulated that comfort and discomfort could not coincide, so discomfort is not present when someone feels comfort [20]. Based on that statement, Shackel et al. and, later, Richards developed a concept where comfort and discomfort were defined as two states placed on opposite extremities of a linear scale [19, 21, 22].

However, in 1992 and 1996, Zhang and Helander and Zhang et al. developed a work that ostracized the previous linear concept and introduced the non-linear model of comfort and discomfort. Zhang et al. defined comfort and discomfort as independent, individual concepts associated with different underlying factors. Comfort is related to feelings of well-being and relaxation, and the esthetic impression of a product or environment influences it. In opposition, discomfort is associated with pain, soreness, numbness, and stiffness and is influenced by the product's physical constraints [23, 24]. In 2003, as a complement to his previous work, Herlander stated that comfort is an emotional state, whilst discomfort is a physical state of being [25].

De Looze et al. built a theoretical model that illustrates the non-linear relationship between comfort and discomfort and its relationship with the physical product. This model also distinguishes the three categories influencing comfort/discomfort assessment: the human, the product, and the environment [26–28].

In this model, it is possible to observe the differences between discomfort (left side of the model) and comfort (right side of the model). Also, it concludes that discomfort has a dominant effect on comfort. De Looze indicated that discomfort levels are influenced by the interaction of the human's physical capacity, like weight, physiological processes, muscle activation, body temperature, intradiscal pressure or nerve conduction. Relatively to the comfort side, it is highly influenced by the user's emotions, expectations, and esthetic design. Regarding the rail context, the emotions and expectations of the user (human level), as well as the train temperature (environment level), can influence the passenger's perception of comfort/discomfort (product level) [19, 26, 27].

In 2005, Moes introduced a different model concerning sitting discomfort. This model defines four initial parameters: the person, the seat, the purpose, and the usage. Then, five steps are taken before discomfort, interaction, effect on the internal body, perceived effects, appreciation of the effects and, finally, discomfort. The author also states that if a person is seated for a specific purpose, the interaction occurs.

For a train journey, the interaction can result in higher-pressure distribution in the seat interface. This pressure will lead to internal body effects, like nerve compression, which can change the user perceived effects and its appreciation and, consequently, lead to discomfort [26, 29].

The previous methods have different backgrounds and approaches but are unanimous in framing comfort as a combination of physical and psychological factors. Whilst the physical state can be influenced by external stimuli such as noise, temperature, or vibration, the psychological state is affected by the previous user experience and expectations [29–31]. Therefore, to be able to evaluate both, physical and psychological impact on the users' comfort, objective and subjective methods should be considered. The present research is developed following the De Looze model. Therefore, the interaction between the product, the user, and the environment is considered and evaluated.

3. Ride comfort evaluation

Ride comfort, also called long-term comfort, of rail vehicles is affected by multiple parameters such as vibration, noise, temperature, smell, and visual stimuli. Derived from wheel-track interaction and rail motion, vibration is the critical factor affecting users' comfort and health, limiting their performance. Therefore, evaluating the vibration transmission in a rail environment is fundamental to quantify passengers' comfort levels and assess the harmful consequences of vibration [31–35].

Three main standards are fully dedicated and employed to evaluating passengers' comfort based on vibration analysis, ISO 2631, EN 12299, and Sperling ride index. The human body has its natural vibration mode, which affects the human vibration feeling. When this mode matches an externally induced vibration, resonance may occur, which, if absorbed, can lead to the physical stress of tissues and organs [10, 33, 35–39]. Furthermore, depending on the human tissue's characteristics, vibration with similar intensities but different spectral content will induce different dynamic responses in the human body. Thus, acceleration needs to be weighted based on human body sensitivity to obtain an index that can reflect the vibration feeling. Although different methodologies exist, the three methods mentioned above share the application of frequency weighting curves. Those capable of producing the highest effect are ranked with the highest weight, and other frequencies are attenuated based on their relative importance [10, 40]. All methods state acceleration measurements in three directions: vertical, fore-and-aft, and lateral.

It should be highlighted that, besides sharing the same assumptions and goals, the methods use different calculation techniques to evaluate comfort. Therefore, one method cannot be transformed into another just by analyzing its results. Instead, a complete analysis and correlation amongst indexes need to be performed [10]. ISO 2631, EN 12299, and Sperling ride index methods and their vibration evaluation techniques will be given. The present work followed the ISO 2631 standard approach. Thus, it is used as a reference methodology to identify rail track infrastructure sections and train maintenance needs will be demonstrated.

3.1 ISO 2631 comfort evaluation

ISO 2631 standard quantifies WBV regarding comfort, human health, and motion sickness. Comfort and health evaluations are related in many ways; thus, frequencies between 0.5 and 80 Hz are defined as the most relevant ones since, at this range, vibration affects the body as a whole, which can lead to discomfort and fatigue.

Acceleration measurements should occur on the vibration transmission interfaces: floor, seat surface, and seatback. Then, the root-mean-square (rms) acceleration is calculated for each axis and the corresponding weighting curve is applied [36–38, 40]. The weighting process is calculated according to Eq. (1):

$$a_w = \left[\sum (W_i a_i)^2\right]^{½} \quad (1)$$

where W_i represents the weighting frequencies and a_i the rms accelerations. Weighting curves application depends on the measurement location and purpose. The total vibration (a_v) is achieved following Eq. (2):

$$a_v = \left(k_x^2 a_{wx}^2 + k_y^2 a_{wy}^2 + k_z^2 a_{wz}^2\right)^{½} \quad (2)$$

where aw are the rms accelerations for each axis, and k represents the multiplying factor dependent on the measuring position, presented in **Table 1**.

Finally, based on a_v, the discomfort is evaluated by a defined scale, **Table 2**, where accelerations higher than 0.315 m/s^2 are ranked as uncomfortable.

3.2 EN 12299—mean comfort method

The EN 12299 standard is a statistical method based on the rms method. The mean comfort is divided into two methods, the *standard method* and the *complete method*. The standard method only considers the floor vibration, whilst the complete method

	***X*-axis**	***Y*-axis**	***Z*-axis**
Floor	W_k and k_x = 0.25	W_k and k_y = 0.25	W_k and k_z = 0.40
Seat surface	W_d and k_x = 1.0	W_d and k_y = 1.0	W_k and k_z = 1.0
Seatback	W_c and k_x = 0.80	W_d and k_y = 0.50	W_d and k_z = 0.40

Table 1.
Frequency weighting curves and multiplying factors defined by ISO 2631 for comfort analysis of a seated passenger.

a_v (m/s^2)	**Ride comfort**
≤0.315	Not uncomfortable
0.5–0.63	Little uncomfortable
0.63–0.8	Little uncomfortable to fairly uncomfortable
0.8–1.0	Fairly uncomfortable to uncomfortable
1.0–1.25	Uncomfortable
1.25–1.6	Uncomfortable to very uncomfortable
1.6–2.0	Very uncomfortable
2.0–2.5	Very uncomfortable to extremely uncomfortable
≥2.5	Extremely uncomfortable

Table 2.
ISO 2631 comfort evaluation scale.

uses floor and seat locations. Thus, the standard method is a simplification of the complete method.

The two method variations quantify the passenger mean comfort during a continuous 5 minutes run. Therefore, the measurement duration shall be a multiple of five, and a minimum of four zones traveled at constant speed must be accomplished to apply the method [11, 41, 42].

In opposition to the ISO 2631 method, the frequencies are initially weighted, and then the rms acceleration over 5 seconds is calculated for each axis. Finally, the 95th and 50th percentiles are determined for periods of 5 minutes, and the mean comfort index is obtained.

The mean comfort (N_{MV}) is calculated following Eq. (3):

$$N_{MV} = 6\sqrt{\left(a_{xP95}^{w}\right)^2 + \left(a_{yP95}^{w}\right)^2 + \left(a_{zP95}^{w}\right)^2} \tag{3}$$

where, a_{P95}^{w} represents the 95th percentile of the weighted accelerations in the three directions, x, y, and z. The evaluation of N_{MV} is defined based on a scale, **Table 3**. The scale considers values between 1 and 5, where a ride comfort index under 1 is considered a "very comfortable ride", and above 5 is considered a "very uncomfortable ride" [42].

This method presents some significant limitations; the use of the 95th percentile leads to data exclusion and the lack of possibility to correspond the track irregularity's location with the N_{MV} values (the highest N_{MV} values can occur during different 5 seconds time intervals). Moreover, measurements must occur at a constant speed for 5 continuous minutes, which is difficult to achieve during passenger service [42].

3.3 Sperling's method

The special characteristic of Sperling's method is that the ride comfort index (W_z) is evaluated individually for vertical and lateral directions. The calculation goes following Eq. (4).

$$W_{Zi} = \left[\int_{0.5}^{30} G_i(f)B_i^2(f)\,df\right]^{1/6.67} \tag{4}$$

N_{MV}	Ride comfort
≤1	Very comfortable
1–2	Comfortable
2–4	Medium
4–5	Uncomfortable
≥5	Very uncomfortable

Table 3.
EN 12299 evaluation scale.

W_z	Ride comfort
1	Just noticeable
2	Clearly noticeable
2.5	More pronounced but not unpleasant.
3	Strong, irregular but still tolerable.
3.25	Very irregular
3.5	Extremely irregular, unpleasant, annoying; prolonged exposure intolerable.
4	Extremely unpleasant; prolonged exposure harmful.

Table 4.
Sperling's method ride comfort evaluation scale.

Where G_i corresponds to the double-side square acceleration $[(cm/s^2)^2]$ and B_i represents the frequency weighting curve. As with the previous methods, the WBV level is evaluated based on a scale, **Table 4**. The passengers will not feel discomfort for values under 3 and will feel extreme discomfort for results above 3.5 [10].

This method is mainly applied to evaluate the vibration level of the vehicle rather than the users. Therefore, Sperling's method is specially used when comparing two or more train comfort rides. This method's major limitation is that vibration influences in different frequency bands and directions relating to sitting comfort are ignored [39].

4. Ride comfort index and rail and train maintenance identification

Besides affecting passengers' comfort, vibration is also associated with safety. High safety levels, guaranteed by maintenance, are crucial to secure the reliability and longevity of rails and trains, furthermore nothing drives passengers away more than safety failures. Maintenance is performed by corrective and preventive interventions. The reactive action after failure recognition characterizes the corrective approach [13–16]. In opposition, preventive interventions are defined as proactive measures to prevent and minimize failures at reduced costs.

Condition-based maintenance (CBM) is the traditional applied preventive track monitoring system. This is executed based on a defined schedule, allowing action to be taken when there is failure evidence. Inspection vehicles, such as EM 120 are usually applied to detect track failures [43–45]. These are expensive vehicles, and their passage introduces traffic disruptions, affecting the regular service operation [16, 44].

Railway track abnormalities influence rail vibration. Higher acceleration peaks are noticed on a defective rail compared to a healthy one. Thus, besides affecting safety, track abnormalities lead to increased passenger discomfort [15]. Due to the connection between vibration, track infrastructure abnormalities, and ride comfort, a new CMB methodology capable of identifying track abnormalities and rail vehicles maintenance needs was proposed. Based on the limitations of EN 12299 and Sperling's method, the ISO 2631 standard was defined as a reference methodology concerning ride comfort analysis. The main goal was to overcome the limitations of the current CBM methods, providing a complementary low-cost solution without disrupting the railway service. Thus, it was hypothesized that if multiple trains with different suspension systems present floor discomfort at the same location, then the track infrastructure requires maintenance. Moreover, it was also considered that the train vehicle needs maintenance if a specific train presents low ride quality at the seat surface.

4.1 Experimental procedure

Different train series have different suspension mechanisms. Those are the critical mechanical structure affecting ride comfort. The suspension mechanism's function is to attenuate the vibration resulting from the train motion and wheel-track interaction. The primary suspension contains wheel-track interaction vibrations, whereas the secondary suspension suppresses the vibration transmission from the bogie to the carbody. This way, vibration is attenuated from the bogie to the seat, so if passengers feel discomfort, a track abnormality is identified [46–48].

Based on the mentioned assumption, it was assumed that the track infrastructure needs maintenance if trains with different suspension mechanisms reported floor discomfort at the same geographic location. Pendolino, Intercity and Urban trains vibrations were monitored for 18 journeys whilst running a passenger service. For the long-distance trains, nine measurements were taken for the Pendolino and six for the Intercity train. Measurements occurred at different places inside the train, namely at the lead, middle, and end cars. Regarding Urban rail, three records happened at the first seat in the motion direction. The experiments run on the Northern Line, downward direction, between Porto–Campanhã and Aveiro stations.

Following ISO 2631 recommendations, three-axial accelerometers were placed on the floor and seat surface and aided by a GPS, the train's geographic location was obtained. This way, track infrastructure maintenance sections were identified by matching the floor discomfort with the train geographic location. Vibration measurements were performed by three-axial seat pad accelerometers (PCE-VDL-24I ± 16 g) at a sample rate of 200 Hz, preventing aliasing and respecting Nyquist's theorem. Moreover, in addition to following ISO 2631 recommendations, the accelerometers also allow data recording on a μSD card [49].

The geographic location was obtained using a RedBoard Qwiic [50] connected with a GPS Logger Shield [51]. The equipment was programmed to retrieve the location, train velocity, and record the data on the μSD card at 1 Hz. Vibration and geographic measurements occurred synchronously.

4.2 Track maintenance needs identification

For this purpose, an algorithm was developed using MATLAB 2020a (from MATHWORKS) [52]. Following ISO 2631 approach, the analysis starts by calculating the instantaneous floor discomfort, corresponding to the total acceleration calculation at each second. Discomfort levels were divided into two categories: equal or under 0.315 m/s^2, rated as "Not uncomfortable", and above the same threshold, ranked as "Uncomfortable". Once identified, the "Uncomfortable" locations were mapped.

The GPS device was installed in the train driver's cabin to get the best GPS signal. Therefore, a maximum distance of 158.9 m (Pendolino length) between acceleration and GPS measurements was observed for measurements taken at the end of the train. Moreover, considering the worst-case scenario where the Pendolino travels at 220 km/h, around 61.1 m are achieved in 1 second. Therefore, a maximum 220 m offset was applied to obtain the matching segment.

Considering all mentioned parameters, the matching locations and maintenance needs segments are those with less than 220 m distance between discomfort places of three different trains, **Figure 1**.

According to **Figure 1**, three maintenance segments were identified. The start and end geographic locations and the respective associated kilometer are presented in **Table 5** for each zone.

The listed locations match those defined by the Portuguese train company (IP) by passing the EM 120 inspection vehicle as requiring maintenance [53]. This way, the methodology reliability is proven, and ride quality as maintenance needs identification may be applied, resulting in a low-cost and non-disruptive CBM method.

Figure 1.
Illustrates the track maintenance segments identified by the MATLAB algorithm.

Segment number	Start coordinates		End coordinates		Line kilometer
	Latitude	Longitude	Latitude	Longitude	
1	40.9771	−8.6372	40.9747	−8.6366	315
2	40.9005	−8.6221	40.8941	−8.6211	306
3	40.8817	−8.6191	40.8797	−8.6188	304

Table 5.
Maintenance segments geographic locations and line kilometer.

4.3 Train maintenance needs identification

Following ISO 2631 standard methodology, comfort levels were calculated for all journeys regarding the floor and seat surface locations. It was hypothesized that if most trains present "Not Uncomfortable" journeys, those ranked with discomfort levels for floor or seat surface needed maintenance. The previous acceleration records calculated the total acceleration for each trip; results for the Pendolino are shown in **Table 6**. The train identification number was replaced to keep the train's ID anonymous.

Pendolino results presented low acceleration levels associated with the "Not Uncomfortable" ranking. The same tendency was not observed for the InterCity trains. Comfort levels, see **Table 7**, noticed high discomfort levels for the seat surface location. This way, three trains were identified as needing maintenance.

In opposition to the Pendolino train, operated as a single unit, the Intercity train is composed of one locomotive hauling five coaches. Although five hauled coaches perform under the same journey conditions, each car's comfort does not depend on the other vehicles' comfort. Trains D, B and E, presented "Little Uncomfortable" comfort levels; thus, it can be concluded that these trains needed maintenance.

Regarding the Urban train, as on Pendolino, all journeys were ranked as "Not Uncomfortable" independently of the measurement location. **Table 8**, shows Urban train results.

Train ID	Measurement place	Measurement location	$a_v\,(m/s^2)$	Ride comfort
A	Beginning	Floor	0.06	Not uncomfortable
		Seat surface	0.27	Not uncomfortable
B	Beginning	Floor	0.06	Not uncomfortable
		Seat surface	0.24	Not uncomfortable
B	Beginning	Floor	0.06	Not uncomfortable
		Seat surface	0.22	Not uncomfortable
C	Middle	Floor	0.05	Not uncomfortable
		Seat surface	0.23	Not uncomfortable
D	Middle	Floor	0.05	Not uncomfortable
		Seat surface	0.24	Not uncomfortable
A	Middle	Floor	0.05	Not uncomfortable
		Seat surface	0.22	Not uncomfortable
C	End	Floor	0.05	Not uncomfortable
		Seat surface	0.24	Not uncomfortable
C	End	Floor	0.06	Not uncomfortable
		Seat surface	0.27	Not uncomfortable
E	End	Floor	0.06	Not uncomfortable
		Seat surface	0.26	Not uncomfortable

Table 6.
Pendolino ride comfort evaluation.

Train ID	Measurement place	Measurement location	$a_v\,(m/s^2)$	Ride comfort
A	Beginning	Floor	0.06	Not uncomfortable
		Seat surface	0.28	Not uncomfortable
B	Beginning	Floor	0.06	Not uncomfortable
		Seat surface	0.23	Not uncomfortable
C	Middle	Floor	0.08	Not uncomfortable
		Seat surface	0.27	Not uncomfortable
D	Middle	Floor	0.06	Not uncomfortable
		Seat surface	0.38	Little uncomfortable
B	End	Floor	0.06	Not uncomfortable
		Seat surface	0.36	Little uncomfortable
E	End	Floor	0.06	Not uncomfortable
		Seat surface	0.39	Little uncomfortable

Table 7.
Ride comfort evaluation for Intercity trains.

Train ID	Measurement place	Measurement location	$a_v\,(m/s^2)$	Ride comfort
A	Beginning	Floor	0.06	Not uncomfortable
		Seat surface	0.20	Not uncomfortable
B	Beginning	Floor	0.07	Not uncomfortable
		Seat surface	0.26	Not uncomfortable
C	Beginning	Floor	0.06	Not uncomfortable
		Seat surface	0.22	Not uncomfortable

Table 8.
Urban trains ride comfort evaluation.

Train seats have the potential to modify the vibration transmission from the floor to the user. Vibration tends to be amplified by the seat. Thus, ride comfort needs to be evaluated on both floor and seat surface to draw conclusions. Based on this assumption, ride quality assessment was conducted for both locations on multiple Pendolino, Intercity and Urban trains.

Pendolino started its operation in 1999 and was renovated in 2017. This renovation changed the seat design but kept the seat structure and carbody. The Intercity service was introduced earlier than the Pendolino in 1980 and renovated in 2002. In the same year, the 3400 series Urban train was introduced. Thus, besides having different characteristics, it was expected that the Intercity and Urban trains presented similar ride comfort values due to the coincident renovation and service introduction years [54, 55].

However, whilst Pendolino and Urban journeys were ranked as "Not Uncomfortable", the same tendency was not observed on the Intercity journeys. Three of those vehicles reveal high discomfort levels. Thus, once the journeys were run on the same track, it is possible to conclude that Intercity coaches present maintenance needs.

5. Transmissibility

Passengers spend most of their time seated; thus, vibration is transmitted to the user due to the contact with the seat and floor. This way, the seat is essential to reduce vibration transmission and increase passengers' comfort. The seat and human body's dynamic responses affect seat vibration transmission as they constitute a coupled dynamic system [10, 11, 33, 35, 37, 56]. Therefore, seat dynamics are quantified regarding transmissibility, which verifies seat efficiency in vibration discomfort and is an indicator of ride comfort. The seat transmissibility presents the ratio of vibration at the user-seat interface and the floor, according to Eq. (5):

$$H(f) = \frac{G_o(f)}{G_i(f)} \tag{5}$$

where $H(f)$ represents the transmissibility, $G_o(f)$ is the output acceleration at the seat-user interface, and $G_i(f)$ represents the input acceleration at the floor [57].

Transmissibility differs in direction (vertical, fore-and-aft, and lateral) and location (seat surface and seatback). Laboratory experiments demonstrated a vertical transmissibility peak between 4 and 6 Hz when sitting upright with backrest support [57–60]. However, these studies did not implement transmissibility tests on train seats. Instead, they considered a rigid seat frame with distinct characteristics of the standard train seat, such as frame dimensions and support points. Moreover, in opposition to the natural rail environment, the experimented seats are individual instead of double, and the foam is placed on top of the surface without a restricting cover [57–60]. These are essential parameters because seat transmissibility is affected by the physical properties of the foam, such as thickness, density, or yield strength [59]. Doubling the foam thickness roughly halves its stiffness; thus, transmissibility and discomfort also increase. Patelli and Griffin [57] conducted a study where foam thickness's effect on transmissibility was observed. When the thickness increased from 40 to 80 mm, the transmissibility reduced from 4 to 7 Hz to 3–5 Hz, respectively, with maximum frequencies of 3–4 Hz for the 40 mm foam and 2–3 Hz for the 80 mm foam. In this study, the individuals did not have contact with the backrest [58]. Zhang et al. [61] also reported a similar tendency and observed higher transmissibility when the seat pan foam was increased from 60, 80, and 100 mm.

The stomach has a resonance frequency between 2 and 20 Hz; thus, passengers may feel sick if the seat transmits vibration within that range.

Ribeiro [62] performed transmissibility tests on Alfa Pendular trains in 2012 before the train renovation using the same method as the one used in this study. Transmissibility peaks around 4.3 Hz were found. At that time, seats were covered by tissue without seams instead of the leather covers with seams now used [54]. Moreover, Ribeiro numerically identified the rigid body frequencies of the carbody and bogie. The former reported frequencies under 1.42 Hz, whilst frequencies between 4 and 12 Hz characterize the latter [62].

To properly quantify the dynamic performance of the Pendolino seats, transmissibility tests were conducted on its two-seat classes, namely, comfort and touristic. **Figure 2** illustrates both seat types. The main difference regarding those seats is their dimensions, especially the seat surface thickness. The comfort class seat has a thickness of 190 mm, whereas the touristic class has a thickness of 130 mm. Moreover, the

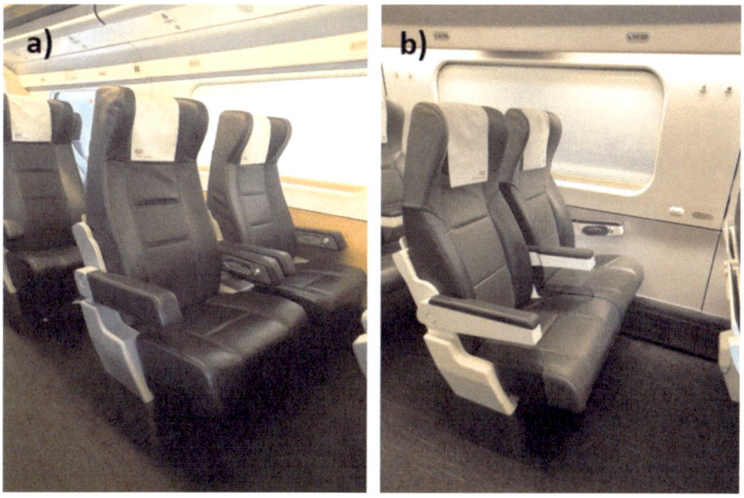

Figure 2.
Pendolino seats: (a) comfort seat, (b) touristic seat.

Subject	Gender	Age (years)	Weight (kg)	Stature (m)
M1	Female	9	33	1.33
M2	Female	25	58	1.70
M3	Male	26	80	1.87
M4	Male	39	115	1.85

Table 9.
Subjects' characteristics.

2017 renovation introduced significant changes to seat covers and foams. All foams were replaced with new ones, and a leather cover with seams was introduced [54].

A set of dedicated experimental tests were accomplished. According to **Table 9**, four volunteers (two males and two females), aged between 9 and 39, weighing 33–115 kg and 1.33–1.87 m in height, participated in the study. The subjects sat in a normal posture, placed both hands on their thighs, and made complete contact with the seatback.

Two seats, one of each class, were instrumented. Seats had the same exact location in the vehicle, particularly near the bogie. Acceleration at the seat surface was measured using a three-axial seat pad accelerometer (PCB 356B41). Uniaxial accelerometers (PCB 393A03) were placed on the floor (1 unit) and the metallic support frame (4 units) for measuring the acceleration at the seat frame. Accelerometers were positioned as shown in **Figure 3**.

The data acquisition system was composed of a NI cDAQ-9172 with NI 9234 IEPE modules connected to a PC to acquire and record data measurements. The vibration was induced by a group of people randomly walking and jumping nearby the seat. A time series of 3 minutes with a sampling frequency of 2048 Hz, posteriorly decimated at 100 Hz, was saved. Data processing was performed using MATLAB scripts previously validated.

Figure 3.
Accelerometer positioning: (a) seat frame and floor accelerometers, (b) seat surface pad.

5.1 Transmissibility curves

Transmissibility peaks were identified for all individuals within the two seats. **Figure 4** shows the transmissibility curves for the comfort and touristic seats for individuals M1 to M4. Three transmissibility peaks were identified, T1 (yellow zone), T2 (gray area), and T3 (green zone).

The comfort seat shows T1 transmissibility equal to 0.2 Hz, the T2 transmissibility presents two main peaks, at 1.4 Hz and 2.5 Hz resulting from the same type of movement but with different amplitudes. T3 appears at frequencies equal to 4.9 Hz for M2 and M3, 4.7 Hz for M4, and 4.1 Hz for M1.

The touristic seat presented a peak of 0.2 Hz concerning T1. On this seat, T2 presented two prominent peaks, a common peak of 1.8 Hz for all individuals and a second peak between 2.3 and 3.3 Hz, depending on the subject. The T3 transmissibility area starts at 4.3 Hz for M3, and increases for M1 (4.5 Hz), M2 (4.7 Hz), and M4 (4.9 Hz).

Comparing comfort and touristic transmissibility results, T1 characterizes both seat types at lower frequencies with values of 0.2 Hz. Increased frequencies define the T2 area. The frequency ranges for the comfort and touristic seats are observed to be between 1.4–2.5 Hz and 1.8–3.3 Hz, respectively. Concerning the last transmissibility zone, T3, this presents frequencies between 4.1 and 4.9 Hz for the comfort seat and a range of 4.3–4.9 Hz for the touristic one.

The frequencies reported in this experiment are lower than those reported in the study performed by Ribeiro (4.3 Hz). Introducing new foams and covers with limited

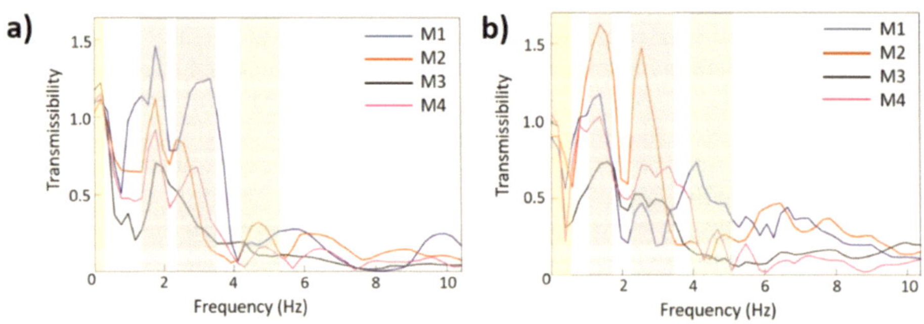

Figure 4.
Transmissibility curves: (a) comfort seat, (b) touristic seat.

seams significantly modifies vibration transmission. These factors restrict the foam cellular morphology movements, and the vibration absorption varies compared to the free foam of the pre-renovation seats. Moreover, the new foams may present different mechanical properties from the previous ones, affecting their vibration-transmission capability. Transmissibility values reported by Ribeiro [62] are similar to those of T3.

The frequency values presented in this study are also lower than those presented by several authors [57–60], who found vertical transmissibility around 4.3 Hz, similar to Ribeiro's findings. However, these tests were conducted in a laboratory environment instead of a real rail environment. Moreover, a single seat was considered, and the seat frame was replaced by a simplified rigid metallic structure (approximately 1000 kg) with free foam on top of the surface without covering or other types of movement restrictions.

Matching the present results with the carbody and bogie modal identification conducted by Ribeiro, T1 zone can be associated with foam movements induced by carbody vibrations. T2 and T3 present frequencies within the bogie rigid body movements. Therefore, these transmissibility zones can be associated with this range, which may lead to seat structural movements such as rigid body, torsion, or bending [62].

In agreement with the research of Patelli and Griffin [57] and Zhang et al. [61], transmissibility peaks decreased for higher foam thicknesses; that is, T1 and T2 transmissibility frequencies were higher for the touristic seat than those for the comfort seat. Moreover, the Pendolino seat foam thickness is higher than those of the previous studies; therefore, following the reported trend of those experiments, lower transmissibility frequencies were expected in the present results.

6. Interface pressure

The overall seat discomfort is influenced by both dynamic and static seat characteristics. These two factors are firmly connected, the importance of one depends on the influence of the other. In low or absent vibration, discomfort evaluation depends mostly on the static seat characteristics. On the other hand, dynamic features dominated discomfort evaluation when vibration magnitude increases. Therefore, it is essential to consider both static and dynamic seat characteristics when evaluating passengers' comfort. Interface pressure is commonly used as a static comfort evaluation technique [63, 64].

When a passenger sits, the pressure at the interface between the seat and the user's buttocks varies over the seat surface area. This pressure variation defines the "average pressure", representing the mean pressure induced on the seat surface. On the other hand, some pressure is concentrated around the ischial tuberosities leading to pressure peaks [64, 65]. A maximum 32 mmHg threshold should not be exceeded. This value corresponds to the capillary pressure value, and above that the pressure is considered harmful as it can obstruct the capillaries, restricting blood circulation and, consequently, result in a deprivation of oxygen to the tissues, causing discomfort [63, 66–68]. The ideal foam equally distributes pressure on the seat surface.

After evaluating the dynamic seat characteristics of the Pendolino train, the static conditions were assessed by measuring its interface pressure. Experiments run with the same conditions (seats and subjects) as the transmissibility tests. The interface pressure was recorded using a CONFORMat sensor (from TekScan) for 10 minutes for each individual, and maximum pressure was recorded. **Figure 5** illustrates the

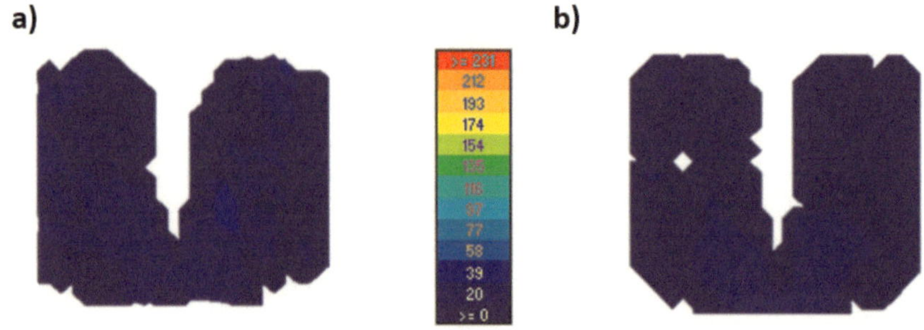

Figure 5.
Pressure distribution: (a) comfort seat, b) touristic seat.

pressure distribution for Pendolino seats regarding M2 subjects. Similar tendencies were observed for all individuals.

Maximum pressure of 32 and 37 mmHg were found for the comfort and touristic seats, respectively. Those pressures correspond to the discomfort threshold; thus, passengers may feel discomfort. Apart from the ischial tuberosities, the pressure was equally distributed on the seat surface.

7. Conclusions

Nowadays, railways are one of the most used mass transportation systems worldwide. Compared to air travel or car, their high transportation capacity and low environmental impact are significant advantages. Higher passenger numbers will be achieved by increasing those vehicles' attractiveness, which depends mainly on comfortable journeys. Passengers define a comfortable journey based on comfort, safety, and user conditions. Seat comfort is characterized by the interaction between seat dynamic and static characteristics. The former is associated with seat behavior under vibration conditions, whilst the latter concerns comfort evaluation in the absence of vibration. Therefore, both evaluation types should be performed to evaluate users' comfort properly.

The rail-wheel interaction associated with motion leads to a complex vibration environment. Besides affecting comfort, vibration is strongly connected to safety. Based on that assumption, a new CBM methodology capable of identifying train and rail maintenance needs was proposed. Following ISO 2631 approach, ride comfort was calculated for 18 journeys conducted on Pendolino, Intercity and Urban trains connecting Porto–Aveiro stations. Results matched those of IP obtained by the EM 120 inspection vehicle. The low-cost system was then validated, and its high precision was proved.

Improving the seating capability to absorb vibration is crucial to increasing passenger comfort. The extent to which vibration is amplified or attenuated depends primarily on the dynamic properties of the seat. Transmissibility quantifies these properties. Therefore, transmissibility tests were performed on comfort and tourist seats on the Pendolino train. The prominent transmissibility peaks were lower than those obtained for the simplified seat version in the laboratory experiments.

After classifying seat dynamic characteristics, the seat static parameters were evaluated by pressure tests. Maximum pressures were obtained at the discomfort threshold near the ischial tuberosities. Apart from this location, the pressure was equally distributed on the seat surface.

To the author's knowledge, this is a pioneering study considering a complete passenger comfort evaluation. Moreover, new low-cost and non-disruptive maintenance identification methodologies were proposed based on the passengers' comfort evaluation. The results of this research can be applied to improve users' comfort levels and increase the attractiveness of the rail mode of transportation.

Acknowledgements

The first author expresses gratitude to the Fundação para a Ciência e Tecnologia (FCT) for a PhD scholarship under the project iRail (PD/BD/143161/2019). The authors would like to acknowledge the support of the projects FCT LAETA–UIDB/50022/2020, UIDP/50022/2020, and UIDB/04077/2020.

Conflict of interest

The authors declare no conflict of interest.

Author details

Patrícia Silva[1,2]*, Joaquim Mendes[3], Eurico Seabra[4] and Pedro Pratas[5]

1 CONSTRUCT-LESE, Faculdade de Engenharia, Universidade do Porto, Porto, Portugal

2 Escola de Engenharia, Universidade do Minho, Guimarães, Portugal

3 Faculdade de Engenharia, Universidade do Porto, Porto, Portugal

4 Departamento de Engenharia Mecânica, Escola de Engenharia, Universidade do Minho, Guimarães, Portugal

5 INEGI, Faculdade de Engenharia, Universidade do Porto, Porto, Portugal

*Address all correspondence to: ppsilva@fe.up.pt

References

[1] Consulting M. Tractebel Engineering. European High Speed Rail – An easy way to connect. 2009:1-213

[2] Román C, Espino R, Martín JC. Competition of high-speed train with air transport: the case of Madrid-Barcelona. Journal of Air Transport Management. 2007;**13**:277-284

[3] European Federation for Transport and Environment. Air2Rail Reducing CO2 from intra-European aviation by a modal shift from air to rail. Delft, www.ariebleijenberg.nl/en. March 2020

[4] Railway passenger transport statistics - quarterly and annual datadata. Eurostat - Statistics Explained. [Accessed on November 2020]

[5] Statistics Explained. Railway passenger transport statistics-quarterly and annual data. Eurostat. Available from: https://ec.europa.eu/eurostat/statisticsexplained/. 2020

[6] Office of Rail and Road. Passenger Rail Usage 2020-21 Quarter 3. 11 March 2021. Available from: https://dataportal.orr.gov.uk/media/1936/passenger-rail-usage-2020-21-q3.pdf

[7] Comboios de Portugal. CP transportou mais 3% de passageiros do que em. 2019. Available from: https://www.cp.pt/institucional/pt/comunicacao/notas-imprensa/terceiro-trimestre

[8] Eboli L, Mazzulla G. A new customer satisfaction index for evaluating transit service quality. Journal of Public Transportation. 2009;**12**:21-37

[9] Lin C-Y, Chen L-J, Chen Y-Y, et al. A comfort measuring system for public transportation systems using participatory phone sensing. Proceedings of PhoneSense. Available from: http://sensorlab.cs.dartmouth.edu/phonesense/papers/Lin-Comfort.pdf. 2010

[10] Kim YG, Kwon HB, Kim SW, et al. Correlation of ride comfort evaluation methods for railway vehicles. Proceedings of the Institution of Mechanical Engineers, Part F: Journal of Rail and Rapid Transit. 2002;**217**:73-88

[11] Jiang Y, Chen BK, Thompson C. A comparison study of ride comfort indices between Sperling's method and EN 12299. International Journal of Rail Transportation. 2019;7:279–296. DOI: 10.1080/23248378.2019.1616329

[12] Gokila P, Suresh P. Parameters influenced in car seat design from ergonomic perspectives - a review. Applied Mechanics and Materials. 2014; **592–594**:2534-2538

[13] Fontul S, Fortunato E, De Chiara F, Burrinha R, Baldeiras M. Railways track characterization using ground penetrating radar. In: Procedia Engineering. Elsevier Ltd. Volume 143. 2016. pp. 1193–1200. DOI: 10.1016/j.proeng.2016.06.120

[14] Al-Douri YK, Tretten P, Karim R. Improvement of railway performance: a study of Swedish railway infrastructure. Journal of Modern Transportation. 2016; **24**:22-37

[15] Falamarzi A, Moridpour S, Nazem M. A review on existing sensors and devices for inspecting railway infrastructure. Jurnal Kejuruteraan. 2019;**31**:1-10

[16] Bressi S, Santos J, Losa M. Optimization of maintenance strategies for railway track-bed considering probabilistic degradation models and different reliability levels. Journal of

Physics: Conference Series. Volume 207. Epub ahead of print 1. March 2021. pp. 107-359. DOI: 10.1016/j.ress.2020.107359

[17] Yang Z, Sun S, Chen G. Evaluating sitting comfort with questionnaire and body pressure distribution: overview and design. In: Proceeding 2009 IEEE 10th International Conference on Computer-Aided Industrial Design and Conceptual Design: E-Business, Creative Design, Manufacturing - CAID and CD'2009 2009. pp. 1443–1447

[18] Dama KK, Suresh Babu V, Rao RN, et al. A review on automotive seat comfort design. International Journal of Engineering Research and Technology. 2015;**4**:678-683

[19] Shen W, Vértiz A. Redefining Seat Comfort. SAE Technical Paper 970597. Epub ahead of print 1997. SAE International. Volume 106. pp. 1066-1073. DOI: 10.4271/970597

[20] H.T.E. (Ed.) H. Appendix 1, Seat Comfort. Lee and Ferraiuolo: SAE Technical Paper. 930105. DOI: 10.4271/930105

[21] Shackel B, Chidsey KD, Shipley P. The assessment of chair comfort. Ergonomics. 1969;**12**:269-306

[22] Richards L. On the psychology of passenger comfort. New York. 1980;**2**:15-23

[23] L. Z, M.G. H. Identifying factors of comfort and discomfort: a multidisciplinary approach. 1992; 395-402. DOI: 10.1518/001872096778701962

[24] Zhang L, Helander MG, Drury CG. Identifying factors of comfort and discomfort in sitting. Human Factors. 1996;**38**:377-389

[25] Helander MG. Forget about ergonomics in chair design? Focus on aesthetics and comfort! Ergonomics. 2003;**46**:1306-1319

[26] Vink P, Hallbeck S. Editorial: comfort and discomfort studies demonstrate the need for a new model. Applied Ergonomics. 2012;**43**:271-276

[27] Hiemstra-van Mastrigt S, Groenesteijn L, Vink P, et al. Predicting passenger seat comfort and discomfort on the basis of human, context and seat characteristics: a literature review. Ergonomics. 2016;**60**:889-911

[28] de Looze MP, Kuijt-Evers LFM, van Dieën J. Sitting comfort and discomfort and the relationships with objective measures. Ergonomics. 2003;**46**:985-997

[29] Zemp R, Taylor WR, Lorenzetti S. Are pressure measurements effective in the assessment of office chair comfort/discomfort? A review. Applied Ergonomics. 2015;**48**:273-282

[30] Huang W, Shuai B. A methodology for calculating the passenger comfort benefits of railway travel. Journal of Modern Transportation. 2018;**26**:107-118

[31] Wibowo RKK, Soekarno S, Puspitasari I. Analysis of train passenger seat using ergonomic function deployment method. International Journal of Transport and Vehicle Engineering. 2017;**11**:1747-1751

[32] Bosso N, Gugliotta A, Zampieri N. Design and simulation of a railway vehicle for the transport of people with reduced mobility. Shock and Vibration. Epub ahead of print 2018. DOI: 10.1155/2018/9207639

[33] Hostens I. Analysis of Seating during Low Frequency Vibration Exposure.

Belgium: Katholieke Universiteit Leuven; 2004

[34] Picu L, Picu M. An analysis of whole-body vibration and hand-arm vibration exposure on the Danube ship crew. Journal of Physics: Conference Series. 2019;**1297**: 012011. DOI: 10.1088/1742-6596/1297/1/012011

[35] Park SJ, Subramaniyam M. Evaluating methods of vibration exposure and ride comfort in car. Journal of the Ergonomics Society of Korea. 2013;**32**:381-387

[36] Bovenzi M. Health effects of mechanical vibration. Giornale Italiano di Medicina del Lavoro ed Ergonomia. 2005;**27**:58-64

[37] Năstac S, Picu M. Evaluating methods of whole-body-vibration exposure in trains. The Annals of 'Dunarea de Jos' 2010;**2**:55-60

[38] Griffin MJ. Handbook of Human Vibration. London, UK: Academic Press, 1990. Epub ahead of print 1990; DOI: 10.1016/C2009-0-02730-5

[39] Peng Y, Wu Z, Fan C, et al. Assessment of passenger long-term vibration discomfort: a field study in high-speed train environments. Ergonomics. 2021;**65**:659-671

[40] International Standard Organization. ISO 2631 - Mechanical vibration and shock - Evaluation of human exposure to whole-body vibration

[41] European Committee for Standardization. EN 12299 - Railway applications — Ride comfort for passengers — Measurement and evaluation

[42] Kufver B, Persson R, Wingren J. Certain aspects of the CEN standard for the evaluation of ride comfort for rail passengers. WIT Transactions on the Built Environment. 2010;**114**:605-614

[43] Kim SS, Park C, Kim YG, et al. Parameter characteristics of rail inspection measurement system of HSR-350x. Journal of Mechanical Science and Technology. 2009;**23**:1019-1022

[44] Nijland F, Gkiotsalitis K, van Berkum EC. Improving railway maintenance schedules by considering hindrance and capacity constraints. Transportation Research, Part C: Emerging Technologies; Volume 126. Epub ahead of print 1 May 2021. pp. 103-108. DOI: 10.1016/j.trc.2021.103108

[45] Sharma S, Cui Y, He Q, et al. Data-driven optimization of railway maintenance for track geometry. Transportation Research, Part C: Emerging Technologies. 2018;**90**:34-58

[46] Fu B, Giossi RL, Persson R, et al. Active suspension in railway vehicles: a literature survey. Railway Engineering Science. 2020;**28**:3-35

[47] Sugahara Y, Kazato A, Koganei R, et al. Suppression of vertical bending and rigid-body-mode vibration in railway vehicle car body by primary and secondary suspension control: results of simulations and running tests using Shinkansen vehicle. The Proceedings of the Institution of Mechanical Engineers, Part F: Journal of Rail and Rapid Transit. 2009;**223**:517-531

[48] Kargarnovin MH, Younesian D, Thompson D, et al. Ride comfort of high-speed trains travelling over railway bridges. Vehicle System Dynamics. 2005;**43**:173-197

[49] Silva P, Seabra E, Mendes J. Design, development, and validation of a whole-body vibration measurement device. ASME Open Journal of Engineering; 1. Epub ahead of print 1 January 2022. pp. 011035. DOI: 10.1115/1.4055191

[50] Sparkfun. SparkFun RedBoard Qwiic. https://www.sparkfun.com/products/15123

[51] SparkFun. SparkFun GPS Logger Shield. https://www.sparkfun.com/products/13750

[52] The Mathworks I. MATLAB R2020a; Mathworks: Natick, MA, USA, 2020

[53] Infraestruturas de Portugal. Network Statement 2022; Infraestruturas de Portugal: Lisbon, Portugal. 2021

[54] Comboios de Portugal. CP - Comboio Elétrico Pendular, Séries 4000. Comboios de Portugal: Lisbon, Portugal. 2017

[55] Comboios de Portugal. CP—Carruagem Corail; Comboios de Portugal: Lisbon, Portugal, 2002

[56] Maier J, Zierke O, Hoermann HJ, et al. Effects of personal control for thermal comfort in long-distance trains. Energy and Buildings; Volume 247. Epub ahead of print 15 September 2021. DOI: 10.1016/j.enbuild.2021.111125

[57] Patelli G, Griffin MJ. Effects of seating on the discomfort caused by mechanical shocks: measurement and prediction of SEAT values. Applied Ergonomics. 2019;**74**:134-144

[58] Gong W, Griffin MJ. Measuring, evaluating and assessing the transmission of vibration through the seats of railway vehicles. The Proceedings of the Institution of Mechanical Engineers, Part F: Journal of Rail and Rapid Transit. 2018;**232**: 384-395

[59] Basri B, Griffin MJ. The application of SEAT values for predicting how compliant seats with backrests influence vibration discomfort. Applied Ergonomics. 2014;**45**:1461-1474

[60] Toward MGR, Griffin MJ. The transmission of vertical vibration through seats: Influence of the characteristics of the human body. Journal of Sound and Vibration. 2011; **330**:6526-6543

[61] Zhang X, Qiu Y, Griffin M. Transmission of vertical vibration through a seat: efect of thickness of foam cushions at the seat pan and the backrest. International Journal of Industrial Ergonomics. 2015;**48**:36-45

[62] Ribeiro D, Calçada R, Delgado R, et al. Finite-element model calibration of a railway vehicle based on experimental modal parameters. Vehicle System Dynamics. 2013;**51**:821-856

[63] Park SJ, Min SN, Lee H, et al. Express train seat discomfort evaluation using body pressure and anthropometric data. Journal of the Ergonomics Society of Korea. 2014;**33**:215-227

[64] Ebe K, Griffin M. Quantitative prediction of overall seat discomfort. Ergonomics. 2000;**43**:791-806

[65] Kyung G, Nussbaum MA. Driver sitting comfort and discomfort (part II): Relationships with and prediction from interface pressure. International Journal of Industrial Ergonomics. 2008;**38**: 526-538

[66] Ciaccia FRDAS, Sznelwar LI. An approach to aircraft seat comfort using interface pressure mapping. Work. 2012; **41**:240-245

[67] Stinson MD, Porter-Armstrong A, Eakin P. Seat-interface pressure: apilot study of the relationship to gender, body mass index, and seating position. Archives of Physical Medicine and Rehabilitation. 2003;**84**:405-409

[68] Kärki S, Lekkala J. Pressure mapping system for physiological measurements. 18th IMEKO World Congress 2006: Metrology for a Sustainable Development. 2006;**3**:2247-2251

Section 5

Maintenance and Upgrades

Chapter 10

Assessing Average Maintenance Frequencies and Service Lives of Railway Tracks: The Standard Element Approach

Stefan Marschnig and Peter Veit

Abstract

As track behavior varies in a wide range of service lives and maintenance demands must be specified. The Standard Element Approach provides a specification based on the most important boundary conditions influencing track behavior such as transport load, alignment, superstructure components as well as substructure qualities, and last but not least the functionality of the dewatering system. These parameters show several possible values. The mix of all of these parameter values describes tracks in the entire network. This clustering helps for decision making for strategic asset management: superstructure component use for different parameter sets, calculating average maintenance and renewal demands and thus the respective budgets for a sustainable track strategy avoiding or reducing backlogs. The Standard Element Approach works for all railway infrastructure assets, this chapter focuses on track in detail.

Keywords: track maintenance, track components, superstructure, track asset management, sustainability

1. Introduction

Asset Management aims for identifying and implementing a sustainable and cost-efficient track structure. Furthermore, this track structure needs proper maintenance in order to achieve a sustainable permanent way. However, the one "best track" does not exist as boundary conditions change. These conditions are described by different sets of parameters such as traffic load, subsoil condition, dewatering system, and many more. Thus every "best track" must be a sustainable solution for the given specific situation—sustainable from the technical perspective as well as from the economic one. This requires not just evaluating innovative technical solutions but also proving its economic efficiency, over the entire life cycle.

The evaluation needs a sound technical description of maintenance being then transposed into maintenance costs. The economic evaluation needs to give answer to the three main questions of asset management:

1. Which type of track is to be selected under certain conditions? (Investment strategy)

2. How to maintain this structure? (Maintenance strategy)

3. When should the entire track structure be reinvested? (Economic service life)

These questions cannot be answered separately as an investment strategy cannot be formulated without considering the maintenance demand, a maintenance strategy must be based on the specific investment, and the economic service life is a consequence of both investment and maintenance.

This results in a two-stage model: first, we need technical descriptions of different options describing investment, maintenance, and service life and thus track deterioration. Life cycle management is therefore a prerequisite for asset management. The second step is to transform the technical information into costs in order to identify the economically most sustainable solution from the different options. The presented Standard Element approach provides a strategic view and acts on average track behavior. To answer question 3 on project level, so for a specific track section somewhere in a network, the Standard Element approach is too general. In this case, we need to use on-site measurement data and deterioration function [1, 2]. The economic appraisement is similar.

Looking at the literature, we find a lot about the proper organization of maintenance within an asset management scheme [3–5]. The different options for applying maintenance as reactive, preventive, condition-based, or even predictive maintenance are well discussed in many papers. Besides these theoretical concepts, also maintenance scheduling for railway tracks is being analyzed and published [6–8].

Most scientific papers discuss track or component behavior, the wear- and damage processes, and the deterioration of components or the entire track. This is especially true for the ballast-related issues, mainly track geometry [9–11] and rail behavior [12–16]. While these papers study quality loss, maintenance is often not addressed. And if maintenance actions are covered, mainly the quality improvement is focused on [17–19] in order to identify if the maintenance action was successful or not. We also see papers dealing with proper intervention levels for different track maintenance actions [20–22].

However, maintenance frequencies for tracks facing different boundary conditions are rare and provide very rough and general figures. These maintenance frequencies are by far not sufficient to answer the questions of track asset management. Therefore, this paper delivers maintenance frequencies for varying boundary conditions, for different transport volumes, track radii, track components, and subsoil conditions.

2. Methodology

A Standard Element describes track behavior in its specific maintenance demands and service lives—based on different sets of parameters, different investment options, and its associated maintenance demand. The Standard Element approach was developed for cyclic track investment and maintenance but also can be used for permanent component exchange. Among the different options the most sustainable needs to be identified, finally in its economic efficiency.

One Standard Element describes the track behavior of one specific set of parameters in its investment, maintenance demand, and service life. All parameters influencing track behavior in a relevant way, either the maintenance demand or the service life—or both, needs to be looked at. The relevant parameters must be identified. The entire network is clustered according to these parameters. The aim is to describe common situations precisely (Standard Element Approach). Individual cases are not pursued further within the framework of this method.

Every Standard Element is linked to a working cycle (**Figure 1**) consisting of the maintenance frequencies over the service life. Evaluating general track strategies demands analyzing different options of maintenance like minimized/reactive maintenance or preventive maintenance and their consequences on the service life. In the end, the working cycle linked to the Standard Element depicts the economically most sustainable option, a sufficient, mostly preventive maintenance, and the optimal service life.

The year zero shows the investment, followed by all planned maintenance actions including one line for small maintenance, which sums up all maintenance actions which are just reactive (e.g., rail breakages).

The core part of the Standard Element is the Working Cycle, describing maintenance demand and service life and thus track behavior for the given parameter set. This should be based on data and experience. It is generally carried out in working groups. Their members are decision-makers from the headquarter, technicians, and economists. Their knowledge is mainly based on a big amount of data. However, there is a second source of knowledge, the experiences of track engineers out at the track sections. Their knowledge is based on observation, does not fundamentally contradict data knowledge, but complements it in essential details. These two different sources lead to important discussions necessary for a proper definition of the working cycles.

The Standard Elements should cover the main part of the network. For verification, the number of kilometers of the various Standard Elements in the network must be calculated. Furthermore, the age of the sections is relevant. The verification of the working cycles is possible by comparing real investment and maintenance demand of the network with the demand described within the Standard Elements.

For identifying the most sustainable track solution a cost evaluation of all options is executed. Therefore, Standard Elements are transposed into time sequences of costs, starting in the year zero (year of investment) lasting until the next re-investment. The various options in general show different investment costs, different maintenance demand, and different service lives. Due to the different service lives the net present value cannot be used for ranking the options. The life cycle cost evaluation and thus a ranking of different options must be based on their average annual costs, including costs of fixing money. Within such an evaluation the cheapest option is the most sustainable one, as “cheap” is always taking all life cycle costs into account, from investment to next re-investment. Thus, the sustainable option requires an optimal balance of investment and maintenance and an optimal balance of maintenance and service life.

Based on verified Standard Elements, the most sustainable investment and maintenance options are identified for all relevant track situations in a network. This is also true for innovations, of course after a relatively short testing period.

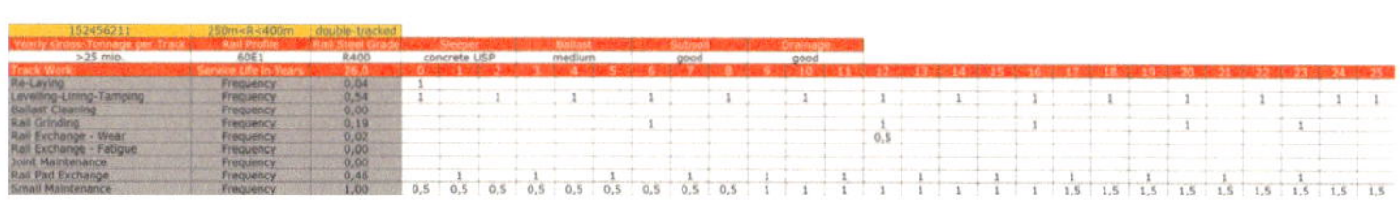

152456211	250m<R<400m	double-tracked																										
Yearly Gross-Tonnage per Track	Rail Profile	Rail Steel Grade	Sleeper			Ballast			Subsoil			Drainage																
>25 mio.	60E1	R400	concrete USP			medium			good			good																
Track Work	Service Life in Years	26,0	0	1	2	3	4	5	6	7	8	9	10	11	12	13	14	15	16	17	18	19	20	21	22	23	24	25
Re-Laying	Frequency	0,04	1																									
Levelling-Lining-Tamping	Frequency	0,54	1		1		1		1		1		1		1		1		1		1		1		1		1	1
Ballast Cleaning	Frequency	0,00																										
Rail Grinding	Frequency	0,19							1						1				1				1			1		
Rail Exchange - Wear	Frequency	0,02													0,5													
Rail Exchange - Fatigue	Frequency	0,00																										
Joint Maintenance	Frequency	0,00																										
Rail Pad Exchange	Frequency	0,46		1		1		1		1		1		1		1		1		1		1		1		1		
Small Maintenance	Frequency	1,00	0,5	0,5	0,5	0,5	0,5	0,5	0,5	0,5	0,5	1	1	1	1	1	1	1	1	1,5	1,5	1,5	1,5	1,5	1,5	1,5	1,5	1,5

Figure 1.
Standard element including working cycle.

2.1 Clustering the network

2.1.1 Parameters

Within the Standard Element Approach "Parameters" are the boundary conditions that lead to certain track behavior and thus trigger the maintenance demands and service life. While some of those parameters are given, some are topics of track strategies.

The alignment of track is a major aspect when it comes to track maintenance demand. It is obvious that a curved track needs another maintenance regime or simply more maintenance than a straight track section. Consequently, service life is shorter on curved tracks compared to straight tracks.

The transport load is obviously triggering maintenance: the more trains operated, the higher the maintenance demand and the shorter service life.

Next to those two parameters hardly being influenced, the construction of the track itself defines another set of parameters.

The subsoil forms the foundation of a track and thus has a major impact on the amount of maintenance needed. Subsoil with sufficient bearing capability and a properly working water drainage system needs the least maintenance and delivers maximum service life.

The used track components are important parameters. Every single component goes along with either specific wear or damage phenomena or at least with different severances of those. The rail profile defines the durability of the rails. Smaller profiles lead to the necessity of through-going rail exchange in case of high transport volumes within the track's service life. Rail steel grade drives both (side) wear in sharp curves and rolling contact fatigue phenomena in larger curves. The sleeper type, mainly concrete and wooden sleepers, but also steel sleepers in some cases and newly concrete sleepers with under sleeper pads (USP), plays an important role in ballast maintenance: the different types lead to changes in the sleeper-ballast interface, decreasing or increasing the contact pressure and thus ballast bed deterioration and consequently impact tamping needs. Ballast and subsoil quality directly influence not only track service life but also tamping demand. The same is true for the quality of the dewatering system.

2.1.2 Parameter values

For the parameters concerning superstructure, the values are simply the components to be used, for example, wooden sleepers or concrete sleepers. To keep the number of combinations low, parameter values might be restricted to some combinations. An example would be "49E1 rails on low loaded lines only".

Values for the parameters "Subsoil" and "Drainage" are an issue as values are not available, especially for existing substructures. In case of executed subsoil rehabilitation, E_{vd} values would be possible, but can also not cover the linked aspect of dewatering. Most infrastructure managers go for a "smart" characterization: "good" is derived from a situation in which superstructure quality and its maintenance are not influenced negatively by subsoil or dewatering topics, while "poor" would mean the other extreme.

Traffic load and alignment need clustering of discrete values. For higher traffic loads, a range of some 10 mio. Gross-tons per year (and track) is feasible looking at international experience with the Standard Element Approach. For lower traffic loads, smaller ranges are necessary though.

The topic of alignment of track is dealt with between the two extrema "straight" (no side wear of rails, rail surface failures, tamping due to restoration of cant, or similar occurs) and the minimum radius for continuous welding of rails (jointed rails lead to different track behavior and maintenance actions anyhow).

The "number of tracks" parameter with its values "single-tracked" and "double-tracked" is not decisive for track behavior in the overwhelming part of effects, but is in case of track work costs due to different logistics and track closure times, respectively operational consequences of such closures.

We summarized the most typical parameter values for mixed traffic networks in **Table 1**.

2.1.3 Parameter sets: The standard element

A combination of exactly one value for every parameter delivers the description of one technical situation within the network, one Standard Element. On the lift top corner in **Figure 2**, we see the coding according to **Table 1**.

Theoretically, there is an enormous amount of possible Standard Elements combining all parameter values. Practical use shows that some 100 Standard Elements help to cover 90 percent of a network being sufficient for achieving valid and robust values for maintenance and renewal.

Transport Volume in Gross-tons per Year	[1]	**Alignment**	[2]	**No. of Tracks**	[3]	**Rail Profile**	[4]	**Rail Steel Grade**	[5]
>25 mio.	1	R > 3000 m	1	single-tracked	1	Xa	1	R200	1
15–25 mio.	2	1000 m < R < 3000 m	2	double-tracked	2	49E1	2	R260	2
10–15 mio.	3	600 m < R < 1000 m	3	others	2	54E1	3	R320Cr	3
5–10 mio.	4	400 m < R < 600 m	4			54E2	3	R350HT	4
3–5 mio.	5	250 m < R < 400 m	5			60E1	4	R400	5
1–3 mio.	6	R < 250 m	6			60E2	4		
<1 mio.	7					46E1	5		
Sleepers	[6]	**Ballast**	[7]	**Subsoil**	[8]	**Drainage**	[9]		
wooden	1	good	1	good	1	good	1		
concrete	2	medium	2	poor	2	poor	2		
steel	3	poor	3	new line	3				
bridge sleepers	4	—	7						
others	5								
concrete USP	6								
slab track	7								
concrete HDS USP	8								

Table 1.
Parameters and parameter values.

152456211	250m<R<400m	double-tracked				
Yearly Gross-Tonnage per Track	Rail Profile	Rail Steel Grade	Sleeper	Ballast	Subsoil	Drainage
>25 mio.	60E1	R400	concrete USP	medium	good	good

Figure 2.
Characteristic of a standard element.

2.2 Working cycles: Input data

In the next step, it is necessary to attach the average maintenance regime to the Standard Elements. This is done by depicting the frequency of different maintenance works between one re-investment and the next one. As this should be done based on existing data, some aspects need to be considered, which are discussed below.

It might be that

- the existing maintenance is not the optimal one guaranteeing the economic service life.
- different maintenance regimes are executed in different regions of the network.
- maintenance regimes use slow orders to stretch maintenance cycles.
- financial or resource restrictions lead to the execution of different maintenance works.

Besides different aspects evaluations of the set of Standard Elements may cover, it can be stated that one Standard Element (set of parameter values) can have different working cycles (maintenance regimes).

The process followed in the latter is based on the idea to depict a sustainable maintenance regime, so neither includes speed restrictions nor insufficient maintenance.

The most challenging task is forming the working cycles for different parameter sets. This is true for depicting the necessary maintenance frequency as well as for the service life. The latter can be approached by using track records. In most cases, infrastructure managers store the "year of track relaying" as status data in their data warehouses. In assessing track age parameter-set-wise, we come up with a survival analysis showing reached service lives so far. This works in case of evenly distributed track ages and/or parameter sets already used longer than one service life (e.g., wooden sleepers). It fails, if parameter sets are relatively new (e.g., new track components or higher transport volumes) as we only see the first part of the service life in the survival analysis. Alternatively (or additionally), we can analyze the reached service lives of tracks at the point in time of relaying. Doing statistics on these data also leads to robust values while delivering the deviations from the mean in addition. **Figure 3** shows such a frequency distribution analysis as an ex post evaluation for wooden sleepers. In this figure, 0 is no deviation from the depicted average service life, all values to the right stand for longer service lives, and all values to the left for shorter ones. **Figure 3** shows additionally the La Place distribution in blue that fits sufficiently good for wooden sleepers.

Generally, maintenance demand increases with increasing track age. When using maintenance records, we thus need to consider track age as an important variable. Taking sections with similar parameters from all over the network, we have executed maintenance frequencies for different time frames in the service live. Averaging these frequencies leads to a first approach of the working cycles. As maintenance is

DOI: http://dx.doi.org/10.5772/intechopen.110488

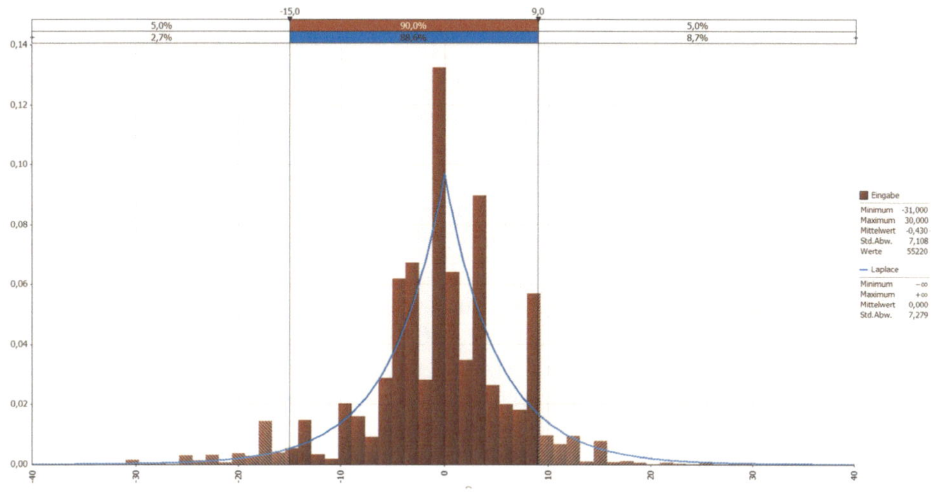

Figure 3.
Distribution of deviations from the average service life—Wooden sleepers.

executed based on on-site track quality, these frequencies of course vary from section to section. Depending on the number of sections available, the averages of those frequencies might be more or less robust. Therefore, a consolidation process is necessary. This process is based on the experience of the track engineers. There are some rough plausibility checks that can be executed quite straightforwardly: Does maintenance increase over the service life? Does higher transport volume lead to tighter maintenance frequencies? Does poorer subsoil condition show intensified maintenance? With the knowledge and experience of the trained maintenance staff, we can derive consolidated maintenance frequencies.

In the end, the depicted maintenance frequencies in the working cycles need to deliver the executed maintenance amount of the analyzed network. The same is true for service lives: in calculating the average service life for all existing parameter sets and applying this to the total lengths of tracks, we can easily calculate the average amount of track renewal. If calculated values deviate from the existing maintenance and renewal works, the working cycles need to be adopted in an iterative way. Adding the unit costs for the single maintenance tasks, this plausibility check can also be executed for maintenance budgets.

3. Results

We collected maintenance frequencies for different technical boundary conditions from various infrastructure managers, mainly from Europe facing comparable mixed traffic networks. In this results section, we depict the main effects of changing parameter values on the maintenance frequencies and service lives. Concluding, we also calculated the average maintenance needs for an artificial network to underline these differences.

3.1 Ballast-related maintenance and renewal

As mentioned in Chapter 2, ballast maintenance frequency and service life of track cannot be analyzed separately, at least for tracks on concrete sleepers.

We start with the influence of track loading depicted in gross-tons. Note again that tonnage is only a rough estimate for the actual loading of track, but in this case, we compare mixed traffic tracks only so that the influence of different traffic segments and vehicles smooth out or average over the network. This is true for traffic volumes up to some 25 million gross-tons. Higher tonnages in mixed traffic can only be operated with an increasing amount of freight trains. This effect cannot be depicted by tonnage only [23]. If we look at the main maintenance action for the ballast, leveling-lining-tamping, we can see an almost linear increase in tamping needs, while service life drops also more or less linearly (**Figure 4**).

The curvature defines both tamping demand and service life to a certain degree, especially if curves get narrow. Below some 600 m, tamping demand increases slightly and service life starts to drop. In very narrow curves, about one-third of the service life is lost even though tamping is more than doubled (**Figure 5**). The high cant in combination with increasing lateral forces leads to a fast loss of track geometry.

The influence of substructure quality is dominant as **Figure 6** shows: poor dewatering of the track leads to a one-third higher tamping demand, and poor soil (and proper drainage) to a doubled demand. With this increased ballast maintenance at least, the service life can be assured. If poor soil condition and poor drainage condition meet, service life can drop to half even though tamping is frequent (every

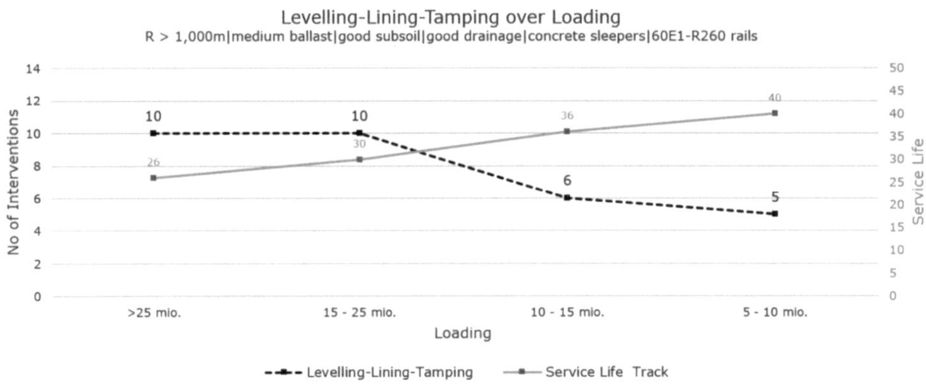

Figure 4.
Leveling-lining-tamping over loading.

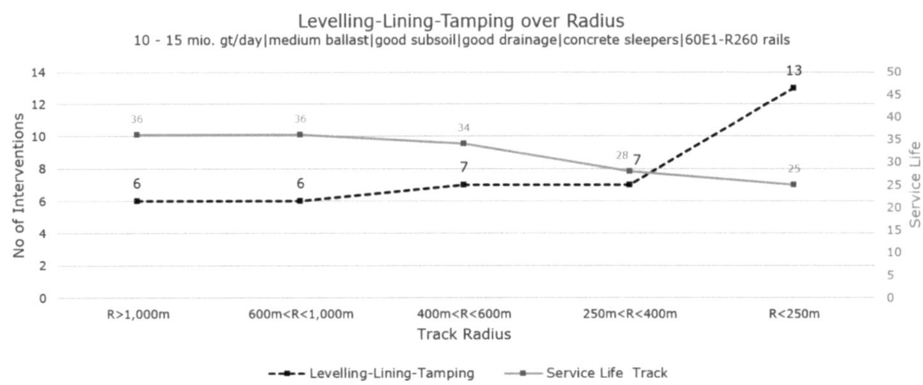

Figure 5.
Leveling-lining-tamping over track radius.

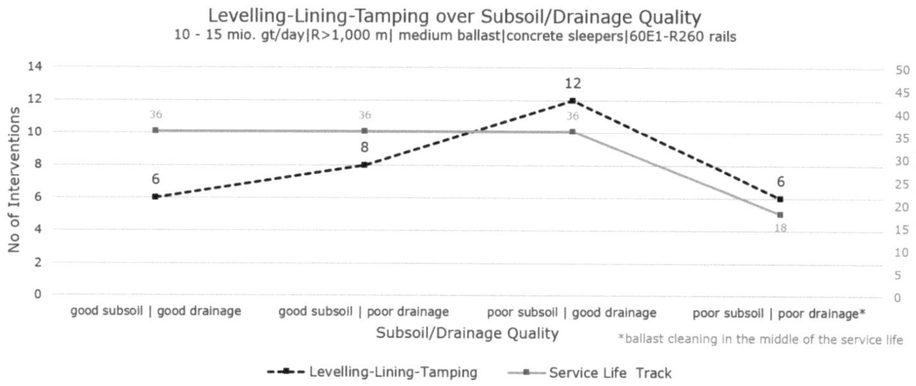

Figure 6.
Leveling-lining-tamping over subsoil/drainage quality.

third year) and ballast is to be cleaned one time. These technical consequences lead to the high economic efficiency of subsoil rehabilitation in such cases as the additional costs in track renewal are compensated by the saving in maintenance over the (prolonged) life cycle.

In ballasted track, of course, ballast quality is crucial. This is true for the composition of the ballast in terms of ballast stone size distribution (sieve curve) as well as the shaping of the single stones. The main trigger is the material quality though. This quality is assessed using the LA values usually (see Chapter 2). In case of magmatic material (e.g., basalt) with LA values below 14, the ballast is not the limiting component in track with loadings below 15 million gross-tons (mixed traffic) as also tamping demand is very low (every 10 years). Metamorphic material (LA values around 17, e.g., granite) leads to an acceptable tamping frequency (every 6 years for 10–15 mio. Gross-tons, **Figure** 7) and a reasonable service life. Poor ballast material (e.g., limestone) leads to doubled tamping needs on the one hand, while the halved service life compared to good ballast quality can only be achieved in cleaning the ballast once in the lifetime of track.

Moreover, the sleeper type in use triggers the tamping demand. Here, concrete and wooden sleeper perform similarly concerning track geometry stability (the service life of the wooden sleeper track is limited due to the limited average lifespan of wooden

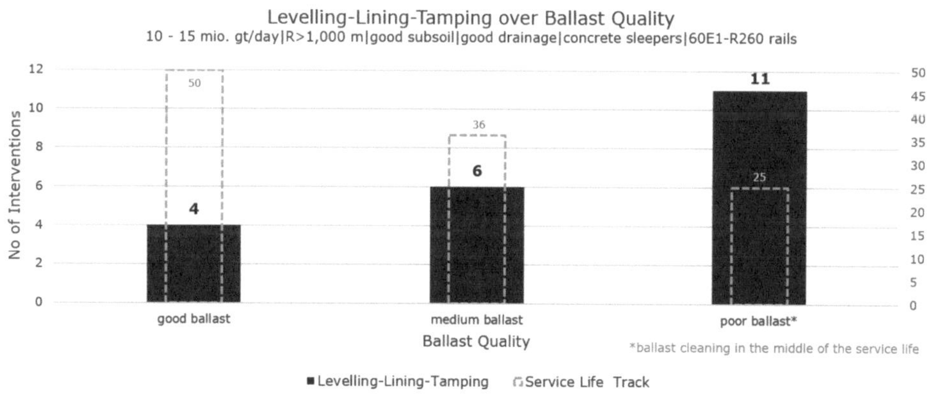

Figure 7.
Leveling-lining-tamping over ballast quality.

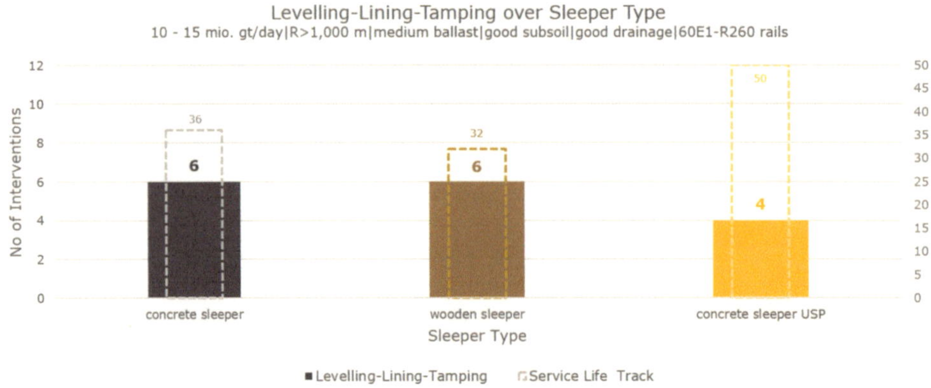

Figure 8.
Leveling-lining-tamping over sleeper type.

sleepers, **Figure 8**). A remarkable reduction of tamping needs is recorded using concrete sleepers with under sleeper pads (USP): for the traffic load used in the example in **Figure 8**, the tamping frequency drops from every 6 years to every 12 years. As the sleeper-ballast interface provides much more contact area, stresses are lower and the ballast is protected especially at the uppermost layer, leading to a significant increase in service life.

3.2 Rail-related maintenance

This subchapter depicts the maintenance necessary for rails. Accumulated tonnage may lead to a necessary exchange of the rails due to an overridden fatigue limit. This limit is about 280 mio. Gross-tons for 49 kg rails, around 500 mio. Gross-tons for 54 kg rails, and beyond 1000 mio. Tons for profiles with more than 60 kg per meter. For the rail profile 60E1 or UIC60 which is widely used in European mixed traffic railway networks, a rail exchange due to this limit is not necessary generally looking at the achievable service lives (**Figures 4–8**) of 50 years in maximum. Even if the loading reaches some 100,000 gross-tons, which is somehow the maximum possible traffic volume in mixed traffic, rail replacement as a result of exceeding the fatigue limit will not be necessary.

What keeps is maintenance linked to rail surface failures and to the side wear in the outer rails of curves where the contact between wheel flange and rail head leads to heavy wear. While the latter phenomenon can only be handled by changing the rails, surface failures can be treated by rail grinding or milling as long as executed early enough. In sharp curves, corrugation waves and re-profiling are the dominant aspects, while in wider curves rolling contact fatigue (RCF) damage in form of head checks is the trigger for the rail surface maintenance. **Figure 9** shows the number of grinding interventions for different track radii.

Rail grinding cannot be handed without looking in parallel to the rail exchange. **Figure 10** gives the amount of outer rail exchange for different track radii. In brackets, we added the exchange of the inner rail which is a consequence of rail grinding due to corrugation waves on the one hand and on the other one of achieving matching rail profiles for outer and inner rail supporting smooth track guidance.

Both rail grinding and rail exchange can be reduced significantly by using higher rail steel grades. **Figures 11** and **12** show the rail maintenance frequency over the trach

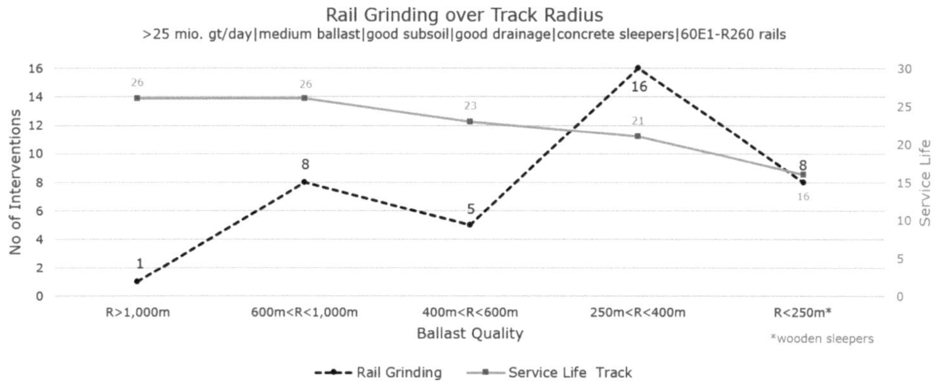

Figure 9.
Rail grinding over track radius.

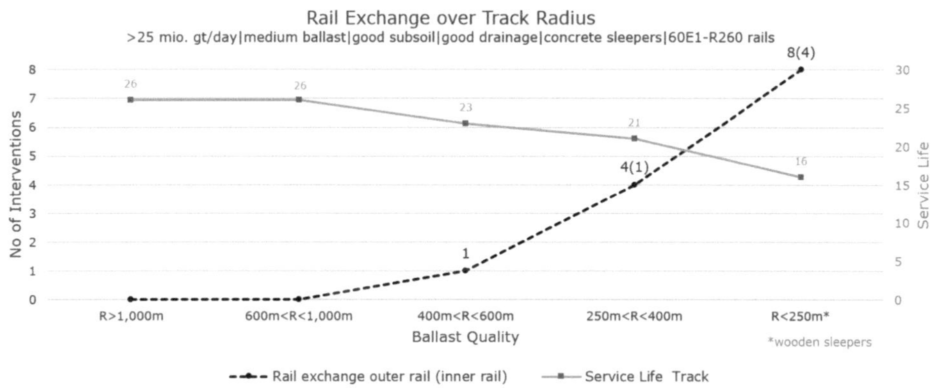

Figure 10.
Rail exchange over track radius.

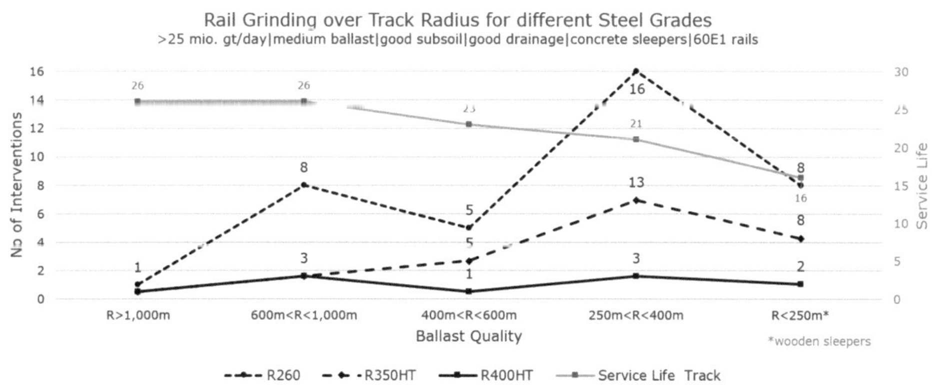

Figure 11.
Rail grinding over track radius for different steel grades.

radius for different rail steel grades. As rails with higher steel grades come along with slightly higher investment costs only, the savings in maintenance pay back at least for high transport volumes generally.

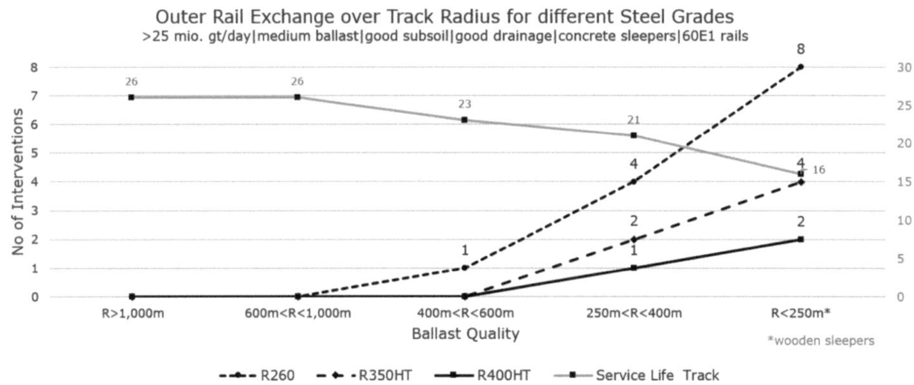

Figure 12.
Rail exchange over track radius for different steel grades.

3.3 Network-wide maintenance and renewal demand

Having described the average maintenance frequencies and the service lives as a consequence of varying boundary conditions, we can also calculate network-wide demands on this basis. This is important to guarantee sufficient budgets in order to achieve sustainable network quality and the economic optimum. The positive influence of higher-quality of track components and improved subsoil can also be depicted on this aggregated level. Of course, in a network, it takes time until all track sections are equipped with the most sustainable components, but at least such evaluations can highlight the goal and support the implementation.

The following evaluations are performed for an artificial mixed traffic network of 10,000 kilometers. The network consists of 76 percent of straight tracks (R > 1000 m), 12.5 percent of 600 m < R < 1000 m sections, 6.5 percent of tracks with radii between 400 and 600 meters, 4.5 percent of radii in the 250 m < R < 400 m class, and 0.5 percent sharp curves below 250 meters. The track loading is fixed to 10 to 15 million gross-tons per year, summing up to 125×10^6 gross-ton kilometers per year. In the basic scenario, the tracks consist of concrete sleepers with 60E1 rails with standard steel grade R260 on medium ballast and good subsoil and good drainage conditions.

This configuration leads, according to the referring Standard Elements, to a maintenance demand of 1747 km of tamping, 504 km of rail grinding, 5.6 km of rail exchange, and a renewal demand of 283 km which is equivalent to an average service life of 35.3 years.

If we look at the ballast-related maintenance and the renewal demand for different ballast and sleeper types, and varying subsoil and drainage conditions, we can see the high importance of high-quality components (**Table 2**): perfect ballast helps reducing the tamping demand by 40 percent and renewal demand by 30 percent, while poor ballast in contrary leads to m a more the double tamping demand, additional ballast cleaning needs and nevertheless to a 50 percent increased renewal demand. Moving from standard concrete sleepers to padded sleepers helps to achieve similar benefits than good ballast quality, almost 30 percent reduction of yearly renewals, and only half of the tamping demand.

The condition of subsoil and drainage is crucial. Facing poor dewatering conditions can easily increase the tamping demand by one-third, poor subsoil even by 100

Track structure	Tamping [km]	Ballast cleaning [km]	Renewal [km]
concrete sleeper | good subsoil | good drainage			
medium ballast	1.747	0	283
poor ballast	4.494	169	414
good ballast	1.049	0	201
medium ballast | good subsoil | good drainage			
concrete sleeper	1.747	0	283
concrete USP sleeper	860	0	207
concrete sleeper | medium ballast			
good subsoil | good drainage	1.747	0	283
good subsoil | poor drainage	2.331	0	283
poor subsoil | good drainage	3.479	100	283
poor subsoil | poor drainage	3.479	198	566

Table 2.
Maintenance and renewal demand for different track structures.

percent, and additionally costly ballast cleaning. Having both poor subsoil and poor drainage means doubled tamping demand (and some 2 percent of the network yearly ballast cleaned) and a doubled renewal demand. In this case and considering the transport volume, one-third of the network needs to be tamped every year and 5.7 percent of the track require track renewal.

For rail maintenance, the rail steel grade is a game changer: For the given loading of 10 to 15 mio. Gross-tons and a superstructure with concrete sleepers on medium ballast with good subsoil and drainage condition, rail grinding demand can be reduced by up to 40 percent using heat treated rails. The side wear-driven rail exchange (including the necessary exchange of the inner rail every second to third exchange) drops to 20 percent in the case of rails R350HT, to almost 0 in the case of a steel grade of R400HT (**Table 3**).

In extended mixed traffic networks, we find a mixture of these parameter values and tracks with different ages. Some—low—percentages of the networks might face any poor subsoil conditions. Ballast might be better or worse in different parts of the network or in different lines. We find different rail profiles, rail steel grades, and sleeper types. If assessing realistic samples of existing networks, it turns out that following a high-quality strategy has the potential of a 50 percent reduction of yearly maintenance expenses and up to one-third of renewal costs per year. Of course, the first step is to invest in more robust components and improved subsoil. This comes

Rails	Rail Grinding [km]	Rail Exchange [km]
60E1-R260	504	5.6
60E1-R350HT	407	1.2
60E1-R400HT	299	<0.1

Table 3.
Rail maintenance for different rail steel-grades.

not for free but pays back in the long term. Economic evaluation is not the topic of this paper but is well-published [24, 25].

As shown, the Standard Element Approach allows for predicting future demand of track work, maintenance as well as track renewal. The following examples show the net wide effects of implementing innovation. The calculation is executed on a fictive network. Different innovative track components prolonging the service life are modeled. Furthermore, an increase in the transport volume of annual 2.5 percent is assumed. Starting at **Figure 13**, we see the forecast of track renewal assuming that transport volume is constant and tracks are re-built with the same components and on unchanged subsoil condition. The five renewal waves (in different colors in **Figure 13**) smooth out over the long term to a somewhat constant renewal demand.

If following the strategy to re-invest tracks always with the optimal component mix and to rehab subsoil in case it is necessary, the renewal demand decreases after the first renewal wave. The prolongation of service lives (e.g., due to padded concrete sleepers plus 25 percent [26, 27] stretches the necessary renewals in the future (**Figure 14**).

Considering an increase in transport volume, in this example by 2.5 percent per year, track sections move from one loading class to the next higher one. This goes

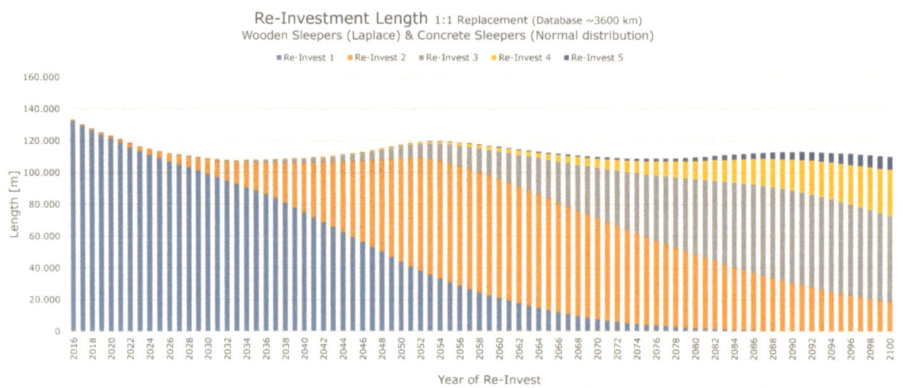

Figure 13.
Re-investment demand for several service lives with a 1:1 replacement of components.

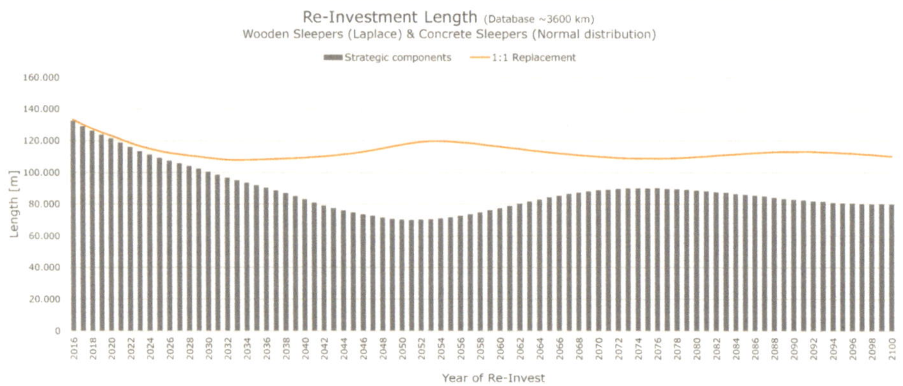

Figure 14.
Re-investment demand for several service lives with replacement of strategic components.

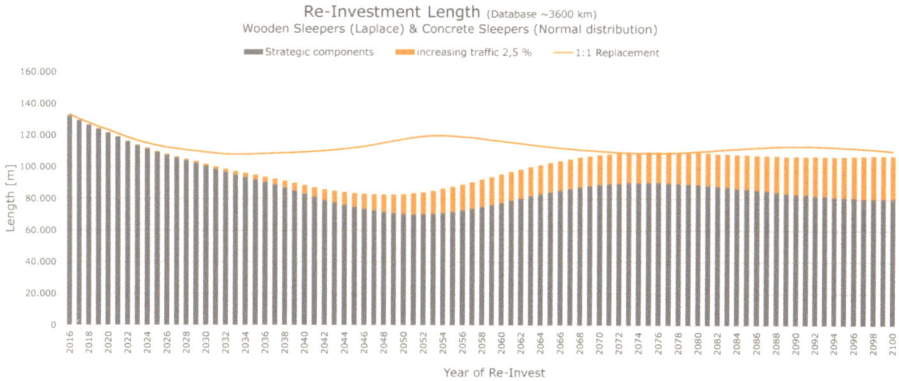

Figure 15.
Re-investment demand for several service lives with replacement of strategic components for rising transport volume.

along with decreasing service life and early renewal of course. Again, this process impacts the renewal demand in the long term only (**Figure 15**), but we see that it shifts the necessary track re-investments to the level displayed in **Figure 13**. We learn that further improving track components in order to increase the total service life of the track is not only to reduce maintenance in the short and mid term but to keep networks with growing transport volume in a balanced state.

This example underlines that Standard Elements form sound knowledge which can be used for various analyses and evaluations.

4. Conclusions

The methodology of Standard Elements is suitable for all infrastructure assets. Deriving deterministic maintenance cycles is only possible for assets with a wear-driven and thus loading-driven behavior (track, switches, bridges, catenary), in case of stochastic failure-driven assets (signaling) the working cycles depict probabilistic maintenance frequencies.

The Standard Element Approach is implemented at several railway infrastructure companies. Austrian Federal Railways, Swiss Railways [28], Scandinavian infrastructure companies, and railway companies in the Balkans, and in the Caucasus area. It always is the first step for implementing sustainable track asset management. In the past, the methodology led to new regulations for track components, specified for the different sets of parameters and thus a specific use (e.g., rail steel grades).

While the first implementations were primarily based on experience, progressive digitization, and more data in longer time series, allow for further specification and detailing of the Standard Elements. Working cycles can nowadays increasingly be based on deterioration functions derived from trend analyses of time-sequenced data. It should be noted that the basic thrust has not been lost by further detailing and that fine-tuning based on prognoses of track behavior is already taking place.

Essential specifications and extensions of the methodology are currently on the way. Thanks to new data, it is now possible to consider also the track work quality. In addition, more specific transport data is available allowing for considering wheel-rail interaction and also optimization within the parameter transport volume [23]. In addition, the model is currently being expanded to include environmental aspects.

CO_2 equivalents are calculated for track components, track work, and the end-of-life scenario and considered in the economic evaluation [29–31].

All these developments are possible due to additional and more specific data enabling more detailed working cycles. An adaption of the principles of the methodology of Standard Elements itself is not necessary.

Acknowledgements

The research was funded by various infrastructure managers, first and foremost ÖBB-Infrastruktur AG.

Author details

Stefan Marschnig* and Peter Veit
Graz University of Technology, Graz, Austria

*Address all correspondence to: stefan.marschnig@tugraz.at

References

[1] Neuhold J.et.al. Measurement data-driven life-cycle management of railway track. Transportation Research Record. 2020;**2674**(11):685-696

[2] Marschnig S, Veit P. Life cycle management in der Realität. ZEV Rail Glasers Annalen. 2012;**9**:335-341

[3] Auer F. Best Practice in Track Maintenance Vol. 1 - Infrastructure Management. Trackomedia, ISBN/EAN 978-3-96245-155-4, 2018

[4] Report from the EIM-EFRTC-CER Working Group on Market Strategies for Track Maintenance & Renewal. Available from: https://www.cer.be/publications/brochures-studies-and-reports/report-eim-efrtc-cer-working-group-market-strategies [Accessed: December 9, 2022]

[5] Lidén T. Railway infrastructure maintenance - a survey of planning problems and conducted research. Transportation Research Procedia. 2015; **10**:574-583

[6] Ouyang Y. Track maintenance production team scheduling in railroad networks. Transportation Research Part B. Methodological. 2012, **46**(10):1474-1488

[7] Lander S, Petterson J. Tamping Planning in Railway Maintenance. Thesis, Department of Civil and Environmental Engineering. Göteborg, Sweden: Chalmers University of Technology; 2012. p. 63

[8] Pouryousef H, Teixeira P, Sussman J. Track maintenance scheduling and its interactions with operations: Dedicated and mixed high-speed rail (HSR) scenarios. ASME. 2010;**2010**: 317-326

[9] Neuhold J et al. Preparing track geometry data for automated maintenance planning. Journal of Transportation Engineering Part A System. 2020;**146**(5):04020032

[10] Praneeth C et al. Train-based differential eddy current sensor system for rail fastener detection. Measurement Science Technology. 2020;**30**:125105

[11] Andrade AR, Teixeira PF. Biobjective optimization model for maintenance and renewal decisions related to rail track geometry. Transportation Research Record. 2011; **2261**(1):163-170. DOI: 10.3141/2261-19

[12] Burstow M. Whole Life Rail Model Application and Development for RSSB—Continued Development of an RCF Damage Parameter, Derby (United Kingdom). London: RSSB; 2004

[13] Burstow M. V/T SIC Annual Seminar: What We Now Know. London (United Kingdom); 2009

[14] Schoech W, Heyder R, Grohmann H-D. Contact geometry and surface fatigue—Guidelines for appropriate rail maintenance. In: 7th International Conference on Contact Mechanics and Wear of Rail/Wheel Systems(CM2006). Brisbane, Australia

[15] Loidolt M, Marschnig S. Evaluating short-wave effects in railway track using the rail surface signal. Applied Sciences. 2022;**12**:2529. DOI: 10.3390/app12052529

[16] Magel E et al. The blending of theory and practice in modern rail grinding. Fatigue & Fracture of Engineering Materials & Structures. 2003;**26**:921-929

[17] Audley M, Andrews JD. The effects of tamping on railway track geometry

degradation. Proceedings of the Institution of Mechanical Engineers, Part F: Journal of Rail and Rapid Transit. 2013;**2013**:376-391

[18] Taufan A et al. Improving the performance of railway tracks through ballast interventions. Proceedings of the Institution of Mechanical Engineers, Part F: Journal of Rail and Rapid Transit. 2018;**232**(2):337-355

[19] Przybyłowicz M et al. Comparison of the effects and efficiency of vertical and side tamping methods for ballasted railway tracks. Construction and Building Materials. 2022;**314**(Part B). DOI: 10.1016/j.conbuildmat.2021.125708

[20] Sauni M et al. Advancing railway asset management using track geometry deterioration modeling visualization. Journal of Transportation Engineering, PartA: Systems. 2022;**148**(2):04021106

[21] Marschnig S, Neuper G, Hansmann F, Fellinger M, Neuhold J. Long term effects of reduced track tamping works. Applied Science. 2022; **12**:368. DOI: 10.3390/app12010368

[22] Neuhold J. Tamping within Sustainable Track Asset Management. Monographic Series TU Graz / Railway Research, Band 6. Graz: Verlag der Technischen Universität Graz; 2020

[23] Marschnig S, Ehrhart U. Traffic load and its impact on track maintenance. In: Hessami AG, editor. New Research on Railway Engineering and Transport [Internet]. London: IntechOpen; 2022

[24] Marschnig S, Veit P, Girsch G, Eichberger P. Fahrwegstrategien für Gleis und Weiche auf Komponentenbasis: Optimierung auf Basis von Lebenszykluskosten. ZEV Rail Glasers Annalen. 2022;**130**(11–12): 500-508

[25] Marschnig S. Innovative track access charges. In: Proceedings of 6th Transport Research Arena. Warsaw, Poland; 2016

[26] Marschnig S, Berghold A. Besohlte Schwellen im netzweiten Einsatz. Eisenbahntechnische Rundschau. 2011;**5**:10-12

[27] Marschnig S, Ehrhart U, Offenbacher S. Long-term behaviour of padded concrete sleepers on reduced ballast bed thickness. Infrastructures. 2022;**7**(10):132

[28] Renggli R, Holzfeind J, Marschnig S. Netzweite Investitions- und Instandhaltungsstrategien im Bereich Fahrbahnoberbau der SBB. Eisenbahntechnische Rundschau. 2012;**6**:10-17

[29] Landgraf M, Horvath A. Embodied greenhouse gas assessment of railway infrastructure: The case of Austria. Environmental Research: Infrastructure and Sustainability. 2021;**1**:025008

[30] Zeiner, M. et.al.: Assessment and recommendations for a fossil free future for track work machinery. Sustainability. 13, 11444, 2021

[31] de Vries R, Sroba P, Magel E. Preventive Grinding Moves into the 21st Century on Canadian Pacific Railroad, Proceedings AREMA 2003 Annual Conference. Chicago, IL, USA; 2001

Chapter 11

Analysis on the Technological Needs and Requirements of SMEs in Rail

Garazi Carranza Ruiz de Loizaga, Oihane de la Rua, Olatz Amorrortu, Bérengère Nihoul and Nikolaos Matskanis

Abstract

Change is the permanent situation we live in, fostered by three main drivers identified in the new European Industrial Strategy plus by the COVID-19 pandemics, which on the one hand is pushing SMEs towards digitalisation and on the other hand is pushing us all towards the need to recover and build a resilient – and sovereign – industry for Europe. These necessary adaptations should be exploited as an opportunity to improve their environmental footprint and economic resilience. One path to achieve this is the uptake of advanced technologies. This chapter identifies the needs and barriers for rail SMEs to adopt technologies and advance their level of digitalisation in order to improve their services, production processes and products. The analysis has received funding from the European Union's Competitiveness of Small and Medium-Sized Enterprises (COSME) programme under the Grant Agreement number: 101037897 — STARS — COS-STRAT-2020-3-05.

Keywords: railway, multimodality, SMEs, recovery, resilience, innovation, Europe's rail 2030, digital maturity, demonstration, advanced technologies, Blockchain network, clusters as mining nodes

1. Introduction

Change is the permanent situation we live in, driven by the double transition we are in the midst of, centred on climate neutrality and digital leadership. According to the European Industrial Strategy, the industrial transformation is articulated around three drivers: global competition, climate neutrality and digital future [1].

The COVID-19 crisis has highlighted the importance of digitalisation and the need to accelerate the progress of technological innovations developed to date [2]. This situation, on the one hand, has pushed SMEs towards digitalisation and, on the other hand, pushes all of us towards the need to recover and build a resilient - and sovereign - industry for Europe. This is why it is important to build on the progress that has been made in the sector towards its digital transformation [3].

The rail sector is one of the most energy-efficient modes of transport as it has significant advantages compared to other modes of transport. Being responsible for

9% of passenger transport and 7% of freight transport [4], it accounts for less than 0.4% of the GHG emissions of the entire transport sector. Rail is set to be a key driver for the decarbonisation of the transport industry [5].

Additionally, after the COVID-19 crisis, a global recovery of the rail supply market is expected with a growth of 3% per year until 2027, reaching an annual volume of approximately 211 billion euros in 2027 [3]. For all these reasons, we can define the railway sector as a technological, efficient and sustainable sector.

The chapter provides an analysis identifying the needs and obstacles that SMEs face in implementing advanced technologies in their organisations, which enable them to increase their digitalisation level in order to improve their services, production processes and products. The analysis has received funding from the European Union's Competitiveness of Small and Medium-Sized Enterprises (COSME) programme under the Grant Agreement number: 101037897 — STARS — COS-STRAT-2020-3-05.

STARS project aims to boost SMEs in terms of production performances and innovation capabilities as well as sustainability and green transition thanks to the uptake of advanced technologies that can make the change possible. Advanced technologies have proven to give competitive advantages and definitely will build the "Future Mobility". Several technologies that will drive the future of the Railway and Multimodality industrial meta-ecosystem are: Advanced Manufacturing, Advanced Materials, artificial intelligence, big data, blockchain, cloud computing, augmented/virtual reality and IoT, among others.

The following sections of this chapter describe on the one hand the state of the art and on the other hand the survey design and the methodology. This data was extracted on 27 June 2022 and has been processed to perform the classification and analysis of the SMEs depending on the needs and obstacles shown.

STARS priority lies in the recovery, transition and resilience of passenger and freight mobility to overcome, adapt and boost the several challenges that Europe is facing; to name a few: (1) transition in public mobility of passengers to respond to the megatrends of urbanisation (need to move growing numbers of people in urban areas), (2) the ageing population (need to make stations, vehicles and the whole travel experience accessible to people with reduced mobility), (3) changes in users and transport behaviour (need to adapt to the needs and aspirations of end users, seamless intermodality, etc.), (4) rail freight as the backbone of European logistics, as set out in the EU Green Deal and the new EU Strategy on Sustainable and Smart Mobility and (5) promoting and completing the green transition in our sector.

2. Background

In the last two decades, the European rail sector has undergone major changes. Today, rail transport plays a key role for Europe's economy and society, but it has the potential to contribute much more. Like most sectors, the rail sector is experiencing the impact of digital transformation, but it is not doing so as fast as it should. In 2021, the European Commission adopted the Communication "The 2030 Digital Compass: the European way for the Digital Decade", which mentions the importance of digital solutions that promote a more connected and automated mobility that reduces accidents, improves the quality of life and increases the efficiency of transport, as well as reduces its environmental impact.

That is why the sector has to adapt to this new revolution to remain competitive. The development of digital technologies in the rail sector will be key to achieving the ambitions of the European Green Pact and the Sustainable Development Goals. In addition, the digitisation of rail will increase its capacity, reliability and comfort, maintaining the EU's leadership in rail transport equipment and services. New technologies such as artificial intelligence, big data, cloud computing, connectivity and autonomous driving will have a major impact on the sector. These technologies are creating a new environment in which rail operators will have to be more agile, act faster and continuously change to be successful.

This chapter's results are aligned with the main principles of the EU Green Deal and the sustainability and mobility policies, also addressing the challenges faced by the railway industry:

1. Changing customer requirements: political, demographic, technological and market trends are changing customer needs. These changes, along with disruptive events such as the COVID-19 pandemic, require rail to be more competitive.

2. Rail as the backbone of a more sustainable European mobility: The need to improve performance and capacity to deliver a more sustainable transportation mix, adapting to increase demand.

3. High cost: currently, rail is often more expensive compared to other modes of transport. To be more competitive and support greater use in the future, it must offer services at a reduced cost.

4. Climate change: rail is one of the most sustainable modes of transport. Greater use of rail is necessary to meet climate targets.

5. Legacy systems and obsolescence: rail system assets are acquired on the assumption of very long life cycles and are based on national approaches.

6. Interaction with other transport modes: rail networks and associated services link well with other transport modes. Still, improvements are needed to better serve customer needs and to make rail a central element for future mobility and to increase its overall attractiveness.

7. Increased competition: the European rail industry is a world leader. However, it faces many challenges on a global level.

This means that the railway industry in the Europe, and specifically the SMEs, must also face these challenges. The survey results served to analyse the capacity of railway SMEs in reaching these challenges and trends.

3. Methodology

3.1 Survey

The implementation of the survey was based on an online European Survey platform. The implemented survey aims to collect data about the needs of enterprises of the railways and multimodal sector. It consists of 3 parts: the introduction, the SME profiling and the SME technological assessment.

The introduction section briefly describes the project, its aims and the purpose of the survey. Additionally, users of the questionnaire are presented with the GDPR statement of the STARS survey. The GDPR statement explains in detail the reasons for collecting the data of the SMEs, how this data is going to be used and by whom and how it will be stored. Users of the survey are also requested to accept and comply with the privacy statement.

As of 27 June 2022, 128 enterprises had answered the survey. After filtering out those outside the scope, a sample of 106 SME companies working or wishing to work in the Railway and Multimodality sector and located in the COSME countries was obtained.

3.2 Questionnaire structure and deployment process

The main section of the STARS survey questionnaire is focused on the evaluation of the status, the needs and the obstacles faced by the SMEs and is structured in the following sections:

- Digitalisation Level – SME's Today's Picture. This section contained questions aiming to capture the current level of digitalisation of the SME, the currency of advanced technologies they were using or exposed to and their approach and barriers to digitalisation.

- Strategy and Vision. This section attempted to capture the digitalisation needs and priorities, as well as the digital transition and green transition strategies of the SMEs.

- Future implementation of technologies. In this section, SMEs were requested to specify the expected future degree of use for the advanced technologies in their enterprise.

- Innovate for the Market. In this section, SMEs provided information regarding their obstacles and drivers for innovation in the market.

The process of deploying the survey, result collection and analysis included the following steps: (1) a requirement analysis and specifications for the survey itself, (2) an analysis of the available survey platforms and tools in order to select the most suitable one, (3) an iterative design and authoring of the survey in collaboration with the partners of the STARS consortium, (4) dissemination of the survey through the partner network, (5) initial analysis and classification of the answers data, (6) feedback on the analysis of the answers by the focus groups at regional level and (7) final version of the analysis and grouping of SMEs.

The methodology used for designing and running the survey as well as for the analysis of the answers received can be summarised as shown in **Figure 1**.

The survey was made available online on the 8 February 2022.

The design of the survey was an iterative process between STARS consortium members as well as external experts of the Advisory Group and the supporting Digital Innovation Hubs.

The Advisory Group consists of experts from the STARS consortium, as well as external experts from European initiatives that have committed through the signature of Letters of Support and other relevant European organisations that have been invited to become part of it (IoT4Industry Partnership, ECSO, EU Blockchain

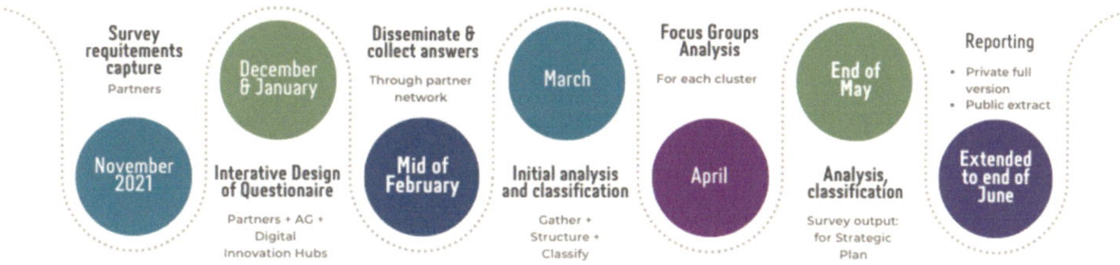

Figure 1.
Survey design methodology and timeline.

Observatory & Forum, EuroXR, ADMA and others). The role of all the entities involved in the development of the survey was to provide advice on how to structure the initial survey to SMEs and how to analyse its results.

In order to better understand the current industrial meta-ecosystem, similar questions to EU statistics were used, such as the EU Flash Eurobarometer 486 "SMEs, start-ups, scale-ups and entrepreneurship".

A first analysis of the early results took place in March by querying the database of the survey and producing statistics on the answers data.

For the purposes of the analysis and classification of the answers of the survey, we have followed the following workflow consisting of three phases:

1. Realisation of a descriptive analysis of raw data. This provided a complete picture of the survey and also provided some trends.

2. Cleaning up and transformation of data to extract knowledge from raw data

3. Data analysis computation

Cleaning consisted of: (1) deleting blank rows, that is, respondents who did not accept the analysing of their data, or all respondent who were not in the scope of the project, (2) dividing multiple choice questions into binary statements, (3) transforming scale text responses into numerical scales, (4) changing fraction to decimal and (5) replacing null answers by mean values.

After the import and cleaning of raw data, some statistics and more complex calculations were carried out: mean values by countries and classification and clustering.

The data analysis step of the workflow following the steps of data cleaning and data transformation includes the execution of an analysis algorithm that computes the distances of SMEs based on their answers and groups them accordingly. The results of both analyses are explained in detail in Section 4 of this document.

4. Results

The collected sample of SME data presents an almost balanced distribution with a third of micro enterprises (35), a third of small enterprises (49) and a third of medium enterprises (35).

Respondents were segmented as shown in **Figure 2** through the different area of the railway domain.

The top three places were engineering, supplies and manufacturing in rolling stock and the design and manufacturing of control command signalling. However, all

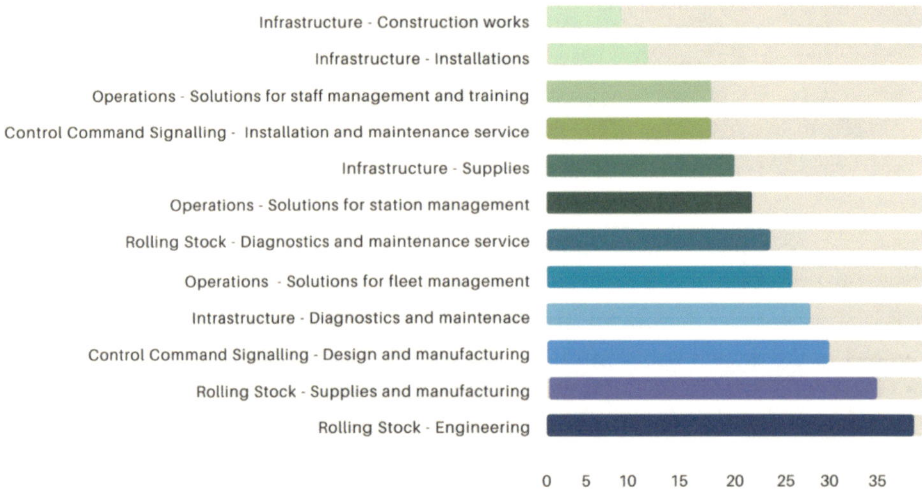

Figure 2.
Areas of the railway domain that the SMEs are operating or planning to operate.

segments were practically well represented in the sample, with a balanced distribution. Construction works and Infrastructure facilities were the least represented among the respondents.

4.1 Trends on today's picture about use of technologies

In terms of respondents' approach to digital technologies, almost 48.5% of companies said that they need to introduce advanced digital technologies, and they had already started to adopt technologies. However, it was found that around 50% of them had not yet reached this point. Of these, 30% were considering adoption, although 10% of them do not have the necessary skills or financial resources to do so. The remaining 20% think that they do not need to adopt any advanced technology.

As can be seen in **Figure 3**, there are many barriers to digitalisation. Financial resources are the main obstacle to adopting digital technologies, followed by technical risks such as security or standardisation. The lack of skills is also considered a barrier

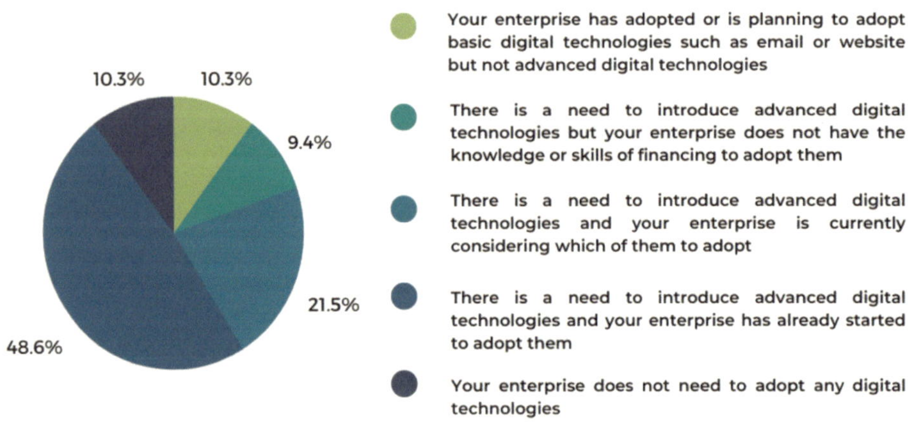

Figure 3.
Enterprises' approach to digital technologies.

to digitalisation; however, it can be assumed that there is a linkage with the lack of resources (See **Figure 4**).

In order to analyse the actual state of technology adoption, the survey asked about the level of adoption for a number of new digital technologies. As can be seen in **Figure 5** and **Figure 6**, the majority of respondents had heard about the majority of the technologies identified but did not use them. The less-known digital technology was photonics and the best-known connectivity, cloud computing, internet of things and cybersecurity.

All the identified advanced technologies can be implemented in different business areas and processes. According to **Figure 7**, analysis of operations-related data and

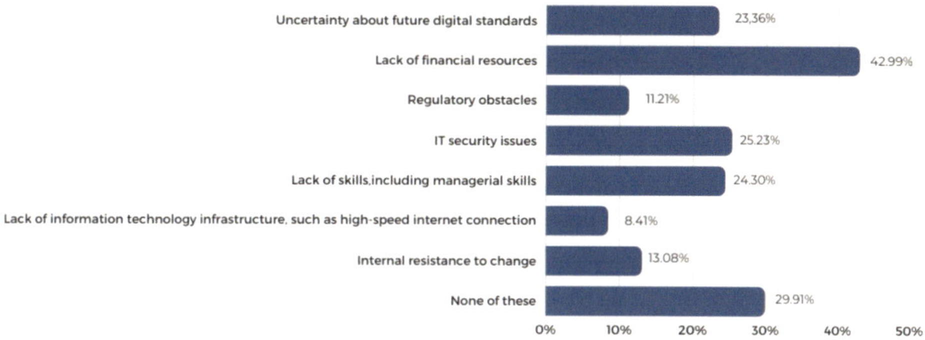

Figure 4.
Barriers to digitalisation in the enterprises.

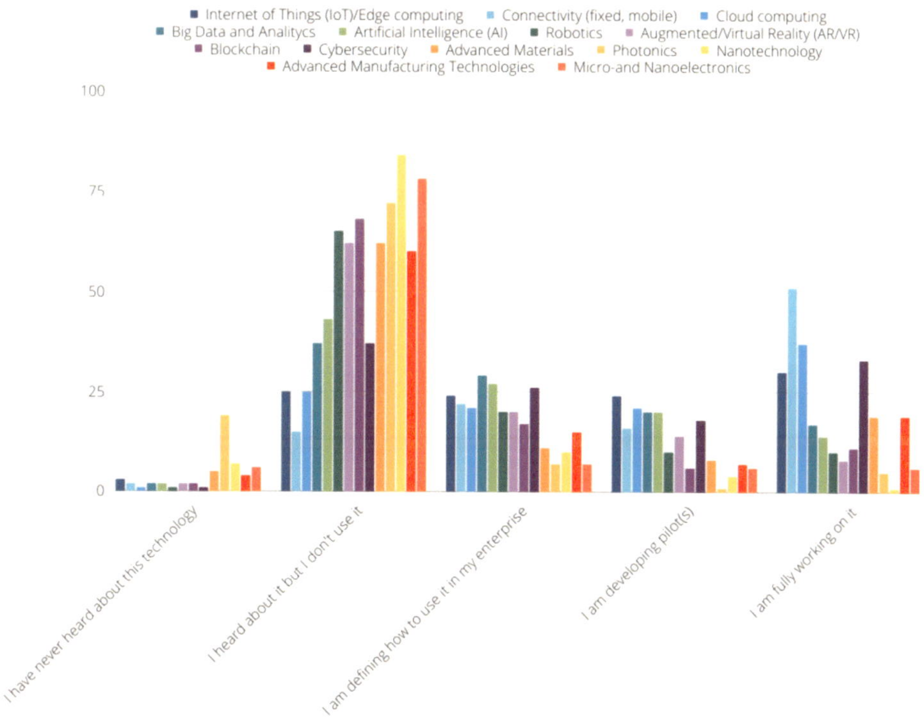

Figure 5.
Advanced technologies used by the enterprises.

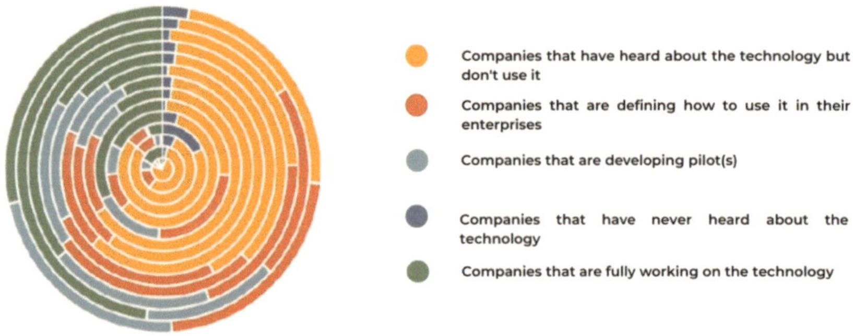

Figure 6.
Levels of knowledge of each of the technologies in the companies.

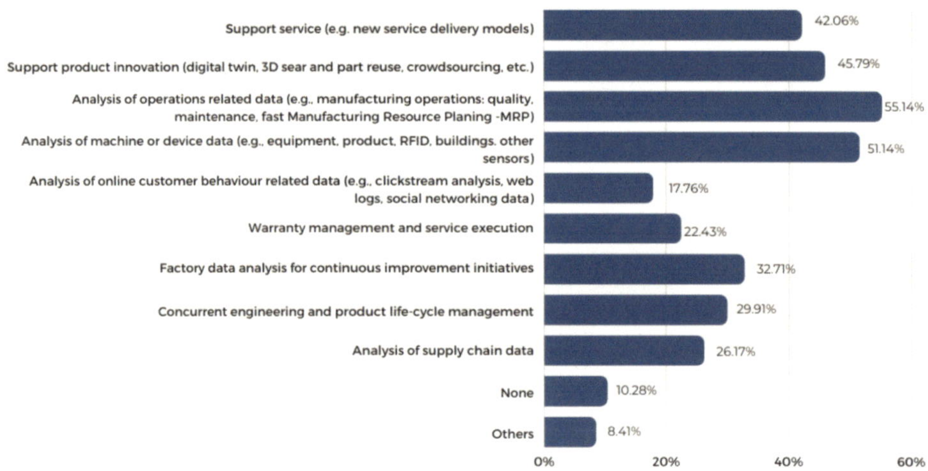

Figure 7.
Areas in which the companies implement technologies.

analysis of machine or device data are the two areas where technologies are the most needed and implemented according to the survey responses, followed by support services/products innovation. The Railway and Multimodal sectors seem to have less interest in data analysis of online customer behaviour or in technologies that could ensure warranty management and service execution.

However, as can be seen in **Figure 8**, the SMEs prefer customised solutions, and while in 14% of the cases, solutions are used in their entirety, in almost 20% of cases, future needs are taken into account in a specific solution, and for 15% of respondents, solutions are fully specific and integrated in the complete production process.

In terms of maintenance strategy (see **Figure 9**), the majority of the respondents (80%) apply predictive maintenance, while less than 10% rely solely on reactive maintenance (10% answered that they do not apply any maintenance strategy at all). 30% of companies follow fixed time schedules to ensure maintenance plan, while some (19%) manage it according to the use of current technology.

Regarding the use of standards and digital tools to ensure the security and quality of production processes in the companies, 68% of them stated that they were applying engineering standards for products, processes and services and that they relied on

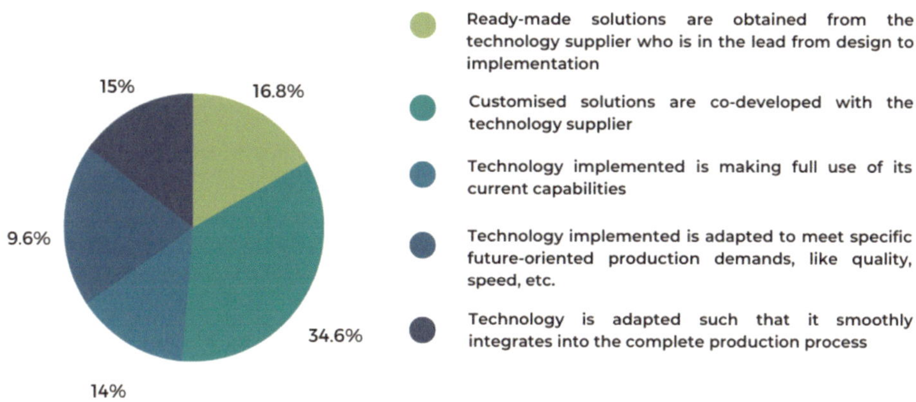

Figure 8.
Advanced technologies sources implementation in the companies.

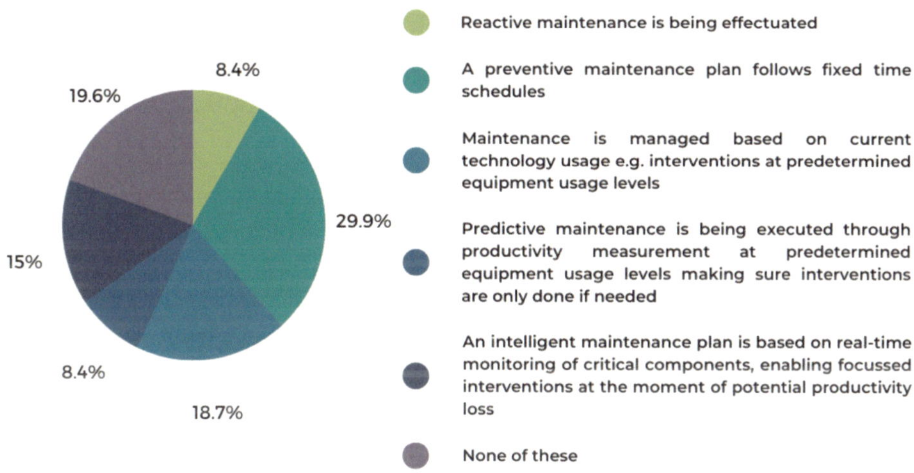

Figure 9.
Maintenance strategy of advanced technologies and tools.

digital means in order to ensure at least one of the following requirements: reliability, security or predictability. However, 20% of the answers were negative to the question.

Only 28% of enterprises collected data from sensors or machinery to monitor production and infrastructure. As for cybersecurity, it is fairly well considered (92%), in half of the cases, only by IT departments, but in the other half, in all areas of the company. In this last case, in 70%, cybersecurity standards and tools are installed and maintained internally by the company's cybersecurity team, while in the other 30%, cybersecurity is outsourced.

4.2 Strategy and vision

After collecting information on the current status of the implementation of digital technologies, it can be seen whether companies have a clear vision of their needs and how they could become more competitive by adopting technologies.

It was found that most companies have identified their technology needs; one-third of them have identified possible use cases that require the adoption of one or more advanced technologies but had not yet implemented them, and another third had not yet identified use cases.

In terms of priorities for the use of advanced technologies by companies, the most prominent would be the optimisation of operations, the development of innovative features in existing products/services and the development of entirely new products/services. There is no clear ranking across the board; each of these three needs seems to be considered of equal priority; however, the development of innovative features has a small lead in preference, followed by optimisation and thirdly by the development of new products or services.

A small majority of respondents do not have a clear digitalisation strategy; almost all of those who do have a clear digitalisation strategy have a time horizon of at least 2 years, and a considerable part of them even have more than 3 years in their time horizon (see **Figure 10**).

In the case of investment policies, 50% focus on the cases of the replacement of obsolete machines or software. However, the other half of the SMEs are less advanced in their investment: 13% have plans for a state of the art; 14% have a road map that includes the evaluation of new technologies through feasibility studies, and 17% have a road map that includes a research and development approach for advance relevant technologies to higher maturity levels.

Looking at each country's average about innovation strategies (see figure table above), the results are balanced, and all countries seem to have identified their needs.

In terms of green transition, almost 65% of respondents said that there is not a clear transition strategy in their company. Respondents rated the different areas to which they aim to contribute to make the railway and multimodality sector greener as below (the score in parentheses represents the average obtained by each item proposed).

- Energy consumption (4.11/5)
- Material (3.87/5)
- Waste (3.85/5)

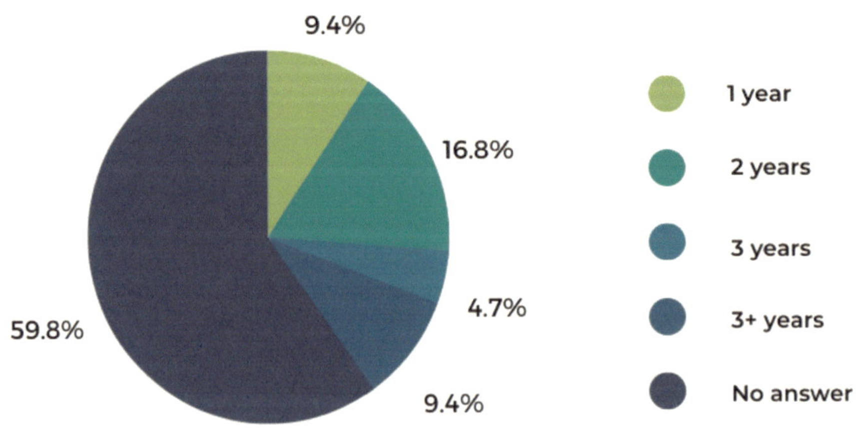

Figure 10.
Time horizon for delivering the strategy.

Location of the SMEs	Has your enterprise already identified its development needs and how to address them?	Using advanced technologies to optimise my enterprise's operations (e.g., manufacturing process, supply chain management, relations with customers, staff management, etc.)	Using advanced technologies to develop innovative features in my current products/ services	Using advanced technologies to develop brand-new products-services	Has your enterprise adopted a clear Digitalisation Transition Strategy?	Do you have an investment policy that matches your Digital Transition Strategy?	SUM by country
Austria	2.25	1.25	2.50	2.25	0.00	1.50	11.85
Belgium	2.33	2.33	2.33	2.22	0.44	3.11	14.61
France	2.29	2.29	2.00	2.43	0.71	2.86	14.60
Germany	2.10	2.50	2.40	1.80	0.30	2.30	13.67
Italy	2.08	2.33	2.38	2.29	0.50	2.29	13.92
Serbia	2.00	2.40	2.40	2.60	0.40	2.60	14.66
Slovenia	2.00	3.00	3.00	1.00	0.00	2.00	**13.10**
Spain	2.17	2.28	2.28	2.33	0.39	2.67	14.11
Türkiye	2.13	2.20	2.47	2.33	0.53	3.13	**15.04**
United Kingdom	2.23	2.15	2.46	2.31	0.08	2.77	14.16

- Hazardous (3.48/5)
- Noise and vibration (3.39/5)
- GHC emissions (3.36/5)
- Land use, biodiversity (3.13/5)

In order to analyse the companies' contribution to a greener rail and multimodal sector, a list of related actions was proposed. This showed that 51% of the companies already implement a strategy to reduce energy consumption and that recycling and the use of more sustainable and advanced materials are also strategies implemented by almost 40% of the respondents, as well as waste reduction. Other avenues proposed were new ways of working, such as hybrid and home working, the production and use of renewable energy or the use of IoT and data analytics for a more robust engineering process (see **Figure 11**).

Despite all these achievements, a big majority of respondents were not compliant with standards or certification, even if 75% of them intended to apply in the future, and 13% had a clear deadline for achieving a targeted eco-efficiency level. Only 12% already had a low footprint certification or a green transition standard like ISO (9001, 14,000, 14,001, 45,001), SA80003, EMAS4 and ECOVADIS5.

Finally, half of the SMEs do not advertise about this point, and from the other half, only 12% follow a specific communication strategy. For the rest, communication is

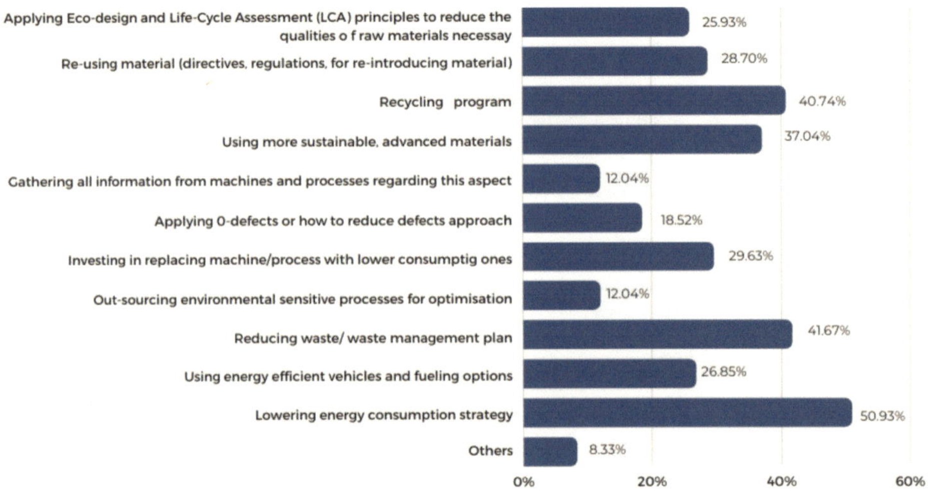

Figure 11.
Contributions of the companies.

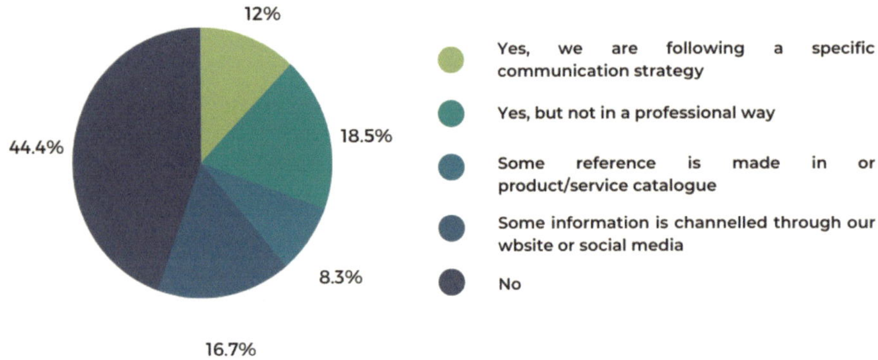

Figure 12.
Enterprises advertising green products/services.

summed up to non-professional advertising or reference on product/services or on websites or social media (see **Figure 12**).

4.3 View on future implementation of technologies

After understanding the actual state of implementation of technologies and the strategic vision for the coming years, SMEs were asked about the advanced technologies they expect to use in the future and the degrees of each of them (see **Figure 13**).

As we can see in **Figure 13**, on the one hand, there are advanced technologies which are clearly already implemented or will be in the short term (less than one year). It is the case for internet of things (IoT)/edge computing, connectivity (fixed, mobile), cloud computing and cybersecurity. On the other hand, there are some technologies that seem to be less appealing for SMEs to be implemented, according to those who have answered the survey: blockchain, advanced materials, photonics, nanotechnology, advanced manufacturing technologies, micro- and nanoelectronics and, to a lesser extent, augmented/virtual reality (AR/VR). Big data and analytics and

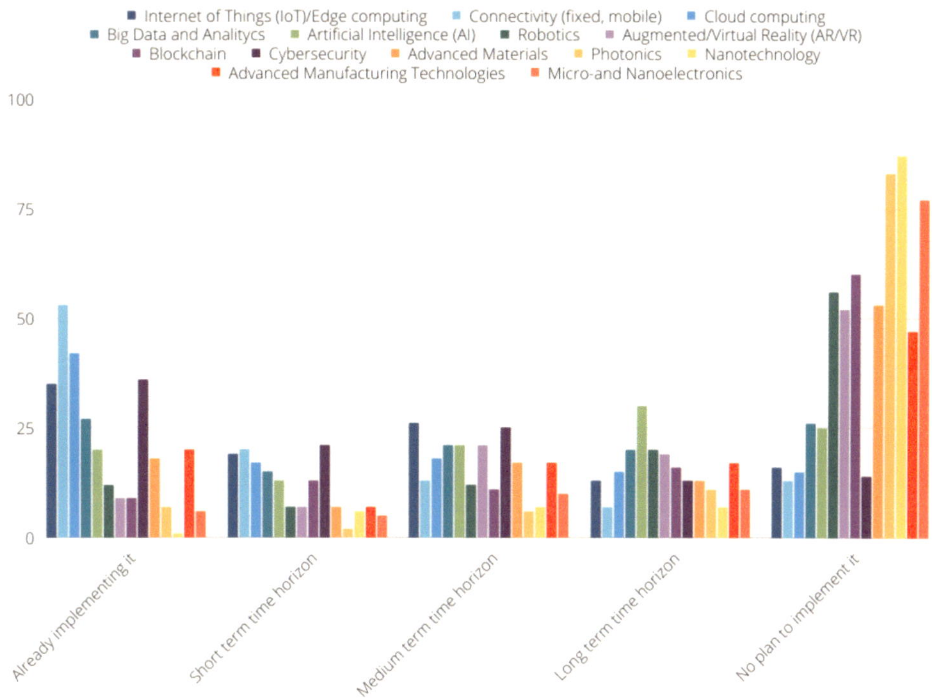

Figure 13.
Expected future degree of use of the advanced technologies in the companies.

artificial intelligence (AI) will take at least 2 years to be deployed, but they are of interest for the companies.

The main risks or obstacles when fostering innovation in the railway and multimodality market are: regulatory aspects, legacy technologies, internal lack of economic resources for investments, no innovation demand in my market segment clients with low digital capacities.

5. Conclusions and future work

The rail sector is one of the most energy-efficient modes of transport, but being responsible for only 9% of passenger transport and 7% of freight transport, it needs to increase its competitiveness and further reduce its carbon footprint. One way of achieving these goals is by the uptake of advanced technologies for boosting digitalisation and green transition of SMEs' industrial processes. This chapter presents the work of surveying the technological needs of the SMEs in the domain and their classification and grouping according to those needs. The results of this survey will be the basis of our future work, for advocating the appropriate technologies to the SME groups, the business areas to focus and identifying the use cases and technological solutions for each SME.

Acknowledgements

We thank respondents who provided insight through the online survey and gave us the opportunity to analyse the results offering conclusions.

Funding

This work has received funding from the European Union's Competitiveness of Small and Medium-Sized Enterprises (COSME) programme in the context of the STARS project under the Grant Agreement number: 101037897 — STARS — COS-STRAT-2020-3-05.

Availability of data and material

The datasets generated during and/or analysed during the current study are available from the corresponding author on reasonable request.

Author details

Garazi Carranza Ruiz de Loizaga[1]*, Oihane de la Rua[1], Olatz Amorrortu[1], Bérengère Nihoul[2] and Nikolaos Matskanis[2]

1 MAFEX- Spanish Rail Industry Association, Spain

2 CETIC – Centre of Excellence in ICT, Charleroi, Belgium

*Address all correspondence to: garazi@mafex.es

References

[1] E. Commission. A New Industrial Strategy for Europe. Brussels. 2020

[2] G. C. R. d. Loizaga. Innovation Methodologies to Activate Inclusive Growth in the Organisation. de Advances in Decision Making. 2021

[3] Berger R. World Rail Market Study. Retrieved from Roland Berger on behalf of UNIFE, the European Rail Industry Association. 2022

[4] Tattini J. Tracking report IEI - International Energy Agency. 2020

[5] ERRAC. Rail 2050 Vision. Rail the backbone of Europe's mobility. 2017

Section 6

Planning and Management

Chapter 12

Traffic Load and Its Impact on Track Maintenance

Stefan Marschnig and Ursula Ehrhart

Abstract

Transport volume is often addressed as gross-tonnage when it comes to track degradation, maintenance needs and maintenance costs. Tonnage and thus weight are insufficient to address track loading: The vehicle properties, mainly axle load, unsprung masses and bogie stiffness, as well as train speed have a major impact on track maintenance demand. This chapter gives an overview on vehicle-specific track deterioration models and describes the maintenance needs of different track as well as the influence on track maintenance cost of different vehicles and trains. The approach proposed is both simple enough to be used and sufficiently detailed to address the main technical aspects. The differences of track maintenance necessary for mixed traffic lines, high-speed lines and heavy haul freight lines can be derived from the vehicles used and the existing train speeds. In passenger transport, the difference between push-pull loco hauled trains (with wagons) and multiple units are a major aspect.

Keywords: track maintenance, traffic load, gross-tonnage, track deterioration model, vehicle properties, track access charges

1. Introduction

Transport volume is one major trigger of track maintenance [1]. For all structures, the loading drives damage, wear and overall system deterioration. In the case of railway track, the loading can be described in different ways. The number of trains gives information on the track utilization. This indicator is important in terms of capacity and linked issues like timetabling, but does not directly influence the maintenance as trains range from heavy freight trains to fast long-distance passenger trains to regional trains. The number of trains does have influence on the maintenance costs as most maintenance tasks need closed-down tracks. Line utilization specifies that the duration of track closures which in turn defines the possible length maintenance tasks can be carried out and thus impacts the cost per meter [2]. Additionally, so-called cost of operational hindrances emerges as trains are delayed, re-routed or simply do not run [3, 4]. Alternatively, the accumulated weight of trains can be applied which describes the intensity of track usage. The calculated gross-tonnage is widely used as an indicator for track loading and used for both classifying tracks [5] and defining maintenance frequencies [6–8]. Consequently, track maintenance cost is allocated to gross-ton-kilometers [9, 10]. Also charging is based on this unit [11, 12]. On the other hand, tracks are designed, constructed and classified for a maximum permissible static axle load and a maximum allowed speed [13–15]. Forces are generally used as loading for the design of structures, and the static axle load at least approximates the vertical loading. Summarizing, the gross-tonnage is a feasible indication for the vertical loading and thus works as an approximation for ballast maintenance and possible rail fatigue. However, the gross-tonnage is definitely not sufficient for all damage mechanisms that are additionally triggered by lateral forces, slip and/or applied traction forces and thus for determining maintenance requirements for rails and turnouts. Moreover, gross-tonnage does not cover axle loads and speed.

There are some approaches covering those—and even more—aspects in calculating so-called equivalent gross-tons. In [16, 17] speed and high axle loads act as increasing factors, so does the aggressiveness of powered axles. [18, 19] additionally address the unsprung mass. Those models improve the description of vertical impacts, but still do not address lateral aspects. The approach of Burstow adds these in providing a damage index for the rails [20].

Finally, there are existing track deterioration models combining all the mentioned aspects [18, 21]. These models have been the inspiration for the proposed model in this paper which is close to the SBB-model. We specified the loading for crossings in turnouts and also added an additional damage term for the entire track structure being restored by a track renewal only. In the result section, we give several examples using this model for both the prediction of track maintenance and the effects of different train service and vehicles. These examples leave out the damage mechanisms for turnouts and the entire track and turnout structures so that one can follow the examples well.

2. Methodology

2.1 Track deterioration model

A track deterioration model (TDM) enables damage assessment by using one mathematical formula. The TDM presented is therefore an analytic approach. The model is based on three input categories: track characteristic, rolling stock and track maintenance. **Figure 1** gives an overview of the input categories that have an impact on the TDM.

The first category **track characteristics** provides components of the track's superstructure such as rail type (profile and steel grade), sleeper type and ballast bed condition. Furthermore, the category track characteristic includes information about the condition of substructure and functionality of drainage. Radius and maximum permissible speed (alignment caused) can be found as well in this category. In Section 2.3.1, the track characteristic is discussed in more details.

In the second category **Maintenance** track works are documented by their type, frequency and costs. This category is strongly linked to the track characteristic. Types of maintenance work considered in the model can be found in the subchapters of Section 2.2.

Vehicle parameters that are necessary for the TDM and therefore those which can cause damage to the track can be allocated to **Rolling Stock**. Parameters like maximum vehicle speed, number of powered/unpowered axles, static and dynamic vertical wheel load, etc., are collected in the third category. This is discussed in Section 2.3.2.

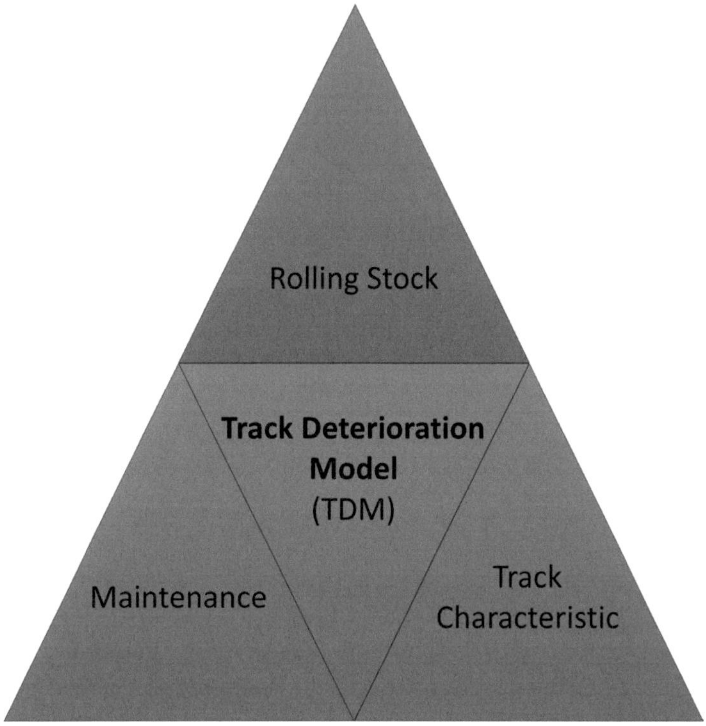

Figure 1.
Input categories for the track deterioration model.

Linking all parameters of those three input categories together enables setting up a track deterioration model. The core of the TDM is a developed mathematical wear formula that is composed of damage terms (D_n for damage) and cost calibration factors (c_n for costs). The wear formula consists of seven damage terms which range from D_1 to D_7. Each of the seven damage terms describes a partial damage that occurs due to a vehicle's run through the track and the arising vehicle track interaction. As those damage terms do differ in their units and do cause different maintenance costs, the damage terms are multiplied by cost calibration factors. Eq. (1) depicts the calculation method and general composition of the wear formula, in which the products of damage terms and cost calibration factors are added up to give total costs per vehicle kilometer.

$$C_{Veh_{S,R}} = \sum_{n=1}^{7} c_n \times D_n \tag{1}$$

$C_{Veh,S,R}$—costs per vehicle kilometer depending on speed and radius (costs/km); c_n—cost calibration factors (n = 1, 2, 3, 4.1, 4.2, 5, 6, 7) (costs/(unit km*)); D_n—damage term (n = 1, 2, 3, 4.1, 4.2, 5, 6, 7)(unit*).

**Unit of the damage term:* kN^3 *(D_1, D_6 and D_7),* $kN^{1.2}$ *(D_2),* kW/mm^2 *(D_3), kN (D_5) and $D_{4.1}$ and $D_{4.2}$ are unitless*

Damage terms D_1, D_2, D_5, D_6 and D_7 represent physical forces that are expressed in kN. However, D_1, D_6 and D_7 are weighted by the exponent 3 (kN^3), while damage term D_2 is weighted by the exponent 1.2 ($kN^{1.2}$). D_3 describes the rail deterioration caused by the physical power per wheel in kW/mm^2. The contact patch frictional energy described in D_4 is expressed in Nm/m. By the fact that the damage terms differ in their physical units, the cost calibration factors correlate to the track damage terms in costs/unit-kilometer.

Precising the general Eq. (1) by means of its damage terms D_n leads to the track deterioration formula set up by Graz University of Technology based on the existing SBB-formula [21]. The track deterioration model and its detailed approach for each damage term are depicted in Eq. (2).

$$\begin{aligned} C_{Vehs_{S,R}} &= c_1 \times {P_{2,S}}^3 + c_2 \times {P_{2,S}}^{1.2} + c_3 \times TPV + c_{4.1} \times D_{4.1} + c_{4.1} \times D_{4.2} + c_5 \\ &\times \sqrt{\left(0.5 \times {P_{2,S}}^2 + 0.5 \times {Y_R}^2\right)} + c_6 \times {P_{1,S}}^3 + c_7 \\ &\times \sqrt{\left(f_{7_{1,R}} \times {P_{2,S}}^2 + f_{7_{2,R}} \times {Y_R}^2\right)}^3 \end{aligned} \tag{2}$$

$C_{Veh,S,R}$—costs per vehicle kilometer depending on speed and radius (costs/km); c_n—cost calibration factors (n = 1, 2, 3, 4.1, 4.2, 5, 6, 7) (costs/(unit km*); $P_{2,S}$—dynamic vertical wheel force (long-waved) depending on speed (kN); $P_{1,S}$—dynamic vertical wheel force (short-waved) depending on speed (kN); Y_R—lateral force of the guiding wheel on the outer rail within radius R 190 m (kN); TPV—traction power value (kW/mm²); $D_{4.1}$—damage index for rolling contact fatigue (RCF); $D_{4.2}$—damage index for plastic deformation/rail abrasion; $f7_{1,R}$—weighting factor for the vertical dynamic wheel force depending on radius R; $f7_{2,R}$—weighting factor for the lateral wheel force depending on radius R.

**Unit of the damage term: kN³ (D_1, D_6 and D_7), $kN^{1.2}$ (D_2), kW/mm² (D_3), kN (D_5) and $D_{4.1}$ and $D_{4.2}$ are unitless*

The track deterioration model consists of seven damage terms, whose calculation is based on one of the four physical parameters vertical or lateral force, power or friction work. These four parameters lead to wear mechanism in the track and therefore to maintenance work. A more detailed explanation of each damage term can be found in the following Section 2.2, while Section 2.3.2 describes the four physical parameters.

2.2 Damage terms

2.2.1 D1: Track geometry deterioration

Damage term D_1 describes track geometry deterioration and ballast destruction on the basis of dynamic vertical wheel contact force P_2. [22] and [23] are the foundation for the approach of the P_2 force (see Section 2.3.2 for more details). This force is not only vehicle-dependent but also a function of speed and represents the long-waved force influence that is caused by track joints/isolated defects. As shown in Eq. (3), the dynamic vertical wheel contact force P_2 is weighted by the exponent 3. The approach of the over linear influence (exponent 3) of the representative axle load bases on [24]. In 1987, the dynamic effects due to increasing axle load from 20 to 22.5 metric tons were investigated on the railway test circuit in Velim (Czechia).

$$D_1 = {P_{2,S}}^3 \tag{3}$$

D_1—damage term 1 (kN³); $P_{2,S}$—dynamic vertical wheel force (long-waved) depending on speed (kN).

In the calculation of damage term D_1, each vehicle's wheelset is classified as damage-relevant. Ballast bed cleaning, line and spot tamping are types of maintenance work that are related to this damage term.

2.2.2 D2 and D3: Rail surface damage (straight tracks)

The damage terms D_2 and D_3 are discussed together in this chapter because both describe rail surface failures in straight tracks, however, due to different impacts. Damage term D_2 describes rail surface failures due to vehicle's dynamic vertical wheel force. The wheel force P_2 used in D_2 corresponds to the force applied in damage term D_1. As depicted in Eq. (4), the P_2 force is weighted by the exponent 1.2. The approach of the power 1.2 does also form the basis of [24, 25], in which the influence of increasing axle load from 20 to 22 metric tons was investigated. In damage term D_2, every wheelset is relevant for determining the surface failure due to vertical force impact.

$$D_2 = {P_{2,S}}^{1.2} \tag{4}$$

D_2—damage term 2 ($kN^{1.2}$); $P_{2,S}$—dynamic vertical wheel force (long-waved) depending on speed (kN).

As not only dynamic vertical forces affect the rail surface, the model does also consider longitudinal forces on the rail. Longitudinal forces are induced by the traction power of the rolling stock [26]. Damage term D_3 describes the influence of traction power on the rail surface by means of the traction power value (TPV). This value is based on the vehicle's power density. The power density is related to the cumulated contact area between rail and wheel of powered axles. Hence, a multiplication with the number of axles (like it is done, e.g., in D_1 and D_2) is not necessary. Furthermore, only powered axles are considered in damage term D_3. Eq. (5) describes the content of damage term D_3 with its unit. A more detailed description of the traction power value can be found in Section 2.3.2.

$$D_3 = TPV \tag{5}$$

D_3—damage term 3 (kW/mm^2); TPV—traction power value (kW/mm^2).

Both damage terms D_2 and D_3 describe rail surface fatigue in straight tracks, where head checks, squats and corrugation occur. The vertical (D_2) and lateral (D_3) force impact cause maintenance work in form of rail surface treatment, such as grinding and milling.

2.2.3 D4.1 and D4.2: rail surface damage and wear (curved tracks)

As damage term D_2 and D_3 describe the rail surface damage in straight tracks, damage terms $D_{4.1}$ and $D_{4.2}$ do so for curved tracks. In curved tracks, a distinction between three damage characteristics can be drawn: rolling contact fatigue (RCF), rail abrasion/plastic deformation and a mixture of both effects. The three damage characteristics are evaluated due to the contact patch frictional energy Tγ (T-Gamma) that is calculated by a multiple-body simulation. The evaluating function that describes the relationship between the contact patch frictional energy (Tγ) and the fatigue damage is based on findings by Burstow [18]. This function with its different wear areas (A to D) is depicted in **Figure 2**. In the function's area A ($T\gamma < 15$ Nm/m), no rail

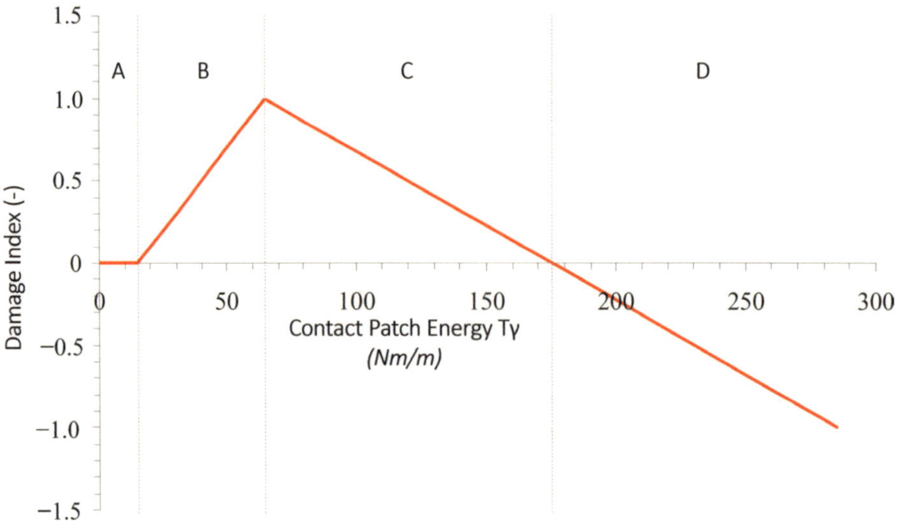

Figure 2.
Burstow's evaluating damage function [18] that distinguishes in no wear (A), RCF (B), RCF and abrasion/deformation (C) and abrasion/deformation (D).

surface wear occurs, whereas in area B RCF takes place ($15 \leq T\gamma < 65$ Nm/m). RCF as well as abrasion/deformation appears in function area C ($65 \leq T\gamma < 175$ Nm/m). Isolated rail abrasion/deformation is described in area D ($T\gamma > 175$ Nm/m).

As the TDM distinguishes between rail surface treatments and rail exchange, Burstow's evaluating damage function is split up into descriptive functions for damage term $D_{4.1}$ as depicted in Eq. (6) and damage term $D_{4.2}$ as shown in Eq. (7). These two equations reflect different scenarios (a to e). Scenarios (a)-(c) describe rail surface wear that is connected to rail surface treatments in curved tracks as it is meant by damage term $D_{4.1}$. Scenarios (d) and (e) are linked to damage term $D_{4.2}$ that represents rail exchange in curved tracks.

$$\begin{aligned} &(a)\ D_{4.1} = 0 && \text{for } T_\gamma < 15\text{Nm/m and } T_\gamma \geq 175\text{Nm/m} \\ &(b)\ D_{4.1} = n \times (0.02 \times T_\gamma - 0.3) && \text{for } 15 \leq T_\gamma < 65\text{Nm/m} \\ &(c)\ D_{4.1} = n \times \left(\frac{-T_\gamma + 175}{110}\right) && \text{for } 65 \leq T_\gamma < 175\text{Nm/m} \end{aligned} \tag{6}$$

$$\begin{aligned} &(d)\ D_{4.2} = 0 && \text{for } T_\gamma < 65\text{Nm/m} \\ &(e)\ D_{4.2} = n \times \left(\frac{T_\gamma - 65}{110}\right) && \text{for } T_\gamma \geq 65\text{Nm/m} \end{aligned} \tag{7}$$

$D_{4.1}$—damage term 4.1; $D_{4.2}$—damage term 4.1; n—number of leading wheelsets of a bogie; Tγ—contact patch energy (Nm/m).

The total damage potential of a vehicle is obtained by multiplying each term $D_{4.1}$ and $D_{4.2}$ by the number of all leading wheelsets in a bogie, since only these are considered as relevant for damage. The contact patch friction that appears in curved tracks not only leads to rail surface treatments but also to rail exchanges of the outer rail. Damage terms $D_{4.1}$ and $D_{4.2}$ therefore enable an evaluation of curve-friendly vehicles. Nerlich and Holzfeind [27] shows that actively controlled wheelsets that enable a better radial position in curved track sections lower the contact patch friction by over 60%. A more detailed description of the contact patch energy $T\gamma$ is discussed in Section 2.3.2.

2.2.4 D5: Wear of turnout components

Damage term D_5 describes wear of turnout components with the exception of the crossing nose. The wear of the crossing nose is specified in an extra damage term (D_6). As can be seen in Eq. (8), term D_5 is described by the vertical (P_2) and lateral (Y_R) forces. P_2 is thereby analogous to the use in D_1. Whereas Y_R is estimated by a multiple-body simulation and represents the lateral force of the guiding wheel on the outer rail that occurs while driving through a s-curve. As a reference for the simulation, a turnout deviation at a radius of 190 m and speed of 40 kmph was chosen as representative for turnouts in the Austrian railway network. This turnout geometry not only represents the vehicle's characteristic significantly, and this geometry also occurs frequently in the network of the Austrian Federal Railways (appearance of still 15% on the mainlines). With regard to [24] and [25], a linear damage approach is selected for D_5, in which the vertical and lateral forces are weighted in each case by 50%.

$$D_5 = \sqrt{\left(0.5 \times {P_{2,S}}^2 + 0.5 \times {Y_R}^2\right)} \tag{8}$$

D_5—damage term 5 (kN); $P_{2,S}$—dynamic vertical wheel force (long-waved) depending on speed (kN); Y_R—lateral force of the guiding wheel on the outer rail within radius R 190 m (kN).

Calculating damage term D_5, each wheelset is concerned for the vertical force. For the lateral force the number of leading axles in a bogie is considered. Damage in D_5 correlates to maintenance work in turnouts, such as exchange of turnout parts (switch and guard rail), grinding, welding, singular sleeper exchange and deburring. As mentioned in the beginning of this chapter, D_5 does not deal with crossing noses. This is done in the following chapter.

2.2.5 D6: Wear of crossing nose

Damage term D_6 describes wear of the turnout's crossing nose. The short-waved dynamic vertical wheel force P_1 is used to estimate the wear, as depicted in Eq. (9).

$$D_6 = {P_{1,S}}^3 \tag{9}$$

D_6—damage term 6 (kN^3); $P_{1,S}$—dynamic vertical wheel force (short-waved) depending on speed (kN).

The approach of P_1 is based on [22] and is justified due to the fact that a wheel running through a crossing nose results in an immediate impact stress. This short-waved force impact leads to wear in the crossing nose. The damage influence of the P_1 force on the crossing nose is evaluated super linearly with the exponent 3 and is therefore comparable with the damage term D_1. The power of 3 is also based on [24, 25]. As every wheelset has its impact on the crossing nose during a drive through the turnout, each wheelset is classified as damage relevant in the calculation of damage term D_6. The described wear in D_6 results in maintenance work, such as exchange of the crossing nose and surface-layer welding/build-up welding. Section 2.3.2 gives a more detailed description of the vertical wheel force P_1.

2.2.6 D7: track renewal (reinvestment)

Damage term D_7 was invented to describe the damage to ballast, sleeper and rail components and implies the renewal of tracks and turnouts. Eq. (10) shows the composition of damage term D_7 and its parameters.

$$D_7 = \sqrt{\left(f_{7_{1,R}} \times P_{2,S}^{\ 2} + f_{7_{2,R}} \times Y_R^{\ 2}\right)}^{\,3} \qquad (10)$$

D_7—damage term 7 (kN^3); $P_{2,S}$—dynamic vertical wheel force (long-waved) depending on speed (kN); Y_R—lateral force of the guiding wheel on the outer rail within radius R (kN); $f_{71,R}$—weighting factor for the vertical dynamic wheel force; $f_{72,R}$—weighting factor for the lateral wheel force.

The speed-dependent vertical force P_2 and the speed and radius-dependent lateral force Y_R are used in D_7 similarly to damage term D_5. Again, the ORE report for question D161 [24] is the basis for the over-linear approach (3rd power) to the component damage in this term. The weighting factors ($f_{7.1}$ and $f_{7.2}$) represent the damage allocation of the vertical and lateral forces depending on the radius. In straight track sections, the lateral force between wheel and rail is negligible, which is why only the vertical force P_2 contributes to the component damage in straight sections. Furthermore, the smaller the radius, the higher the share of lateral force Y_R. **Table 1** includes the radius-dependent weighting factors for the vertical and lateral force. The classification of radii is explained in Section 2.3.1 in more detail.

For calculating the component renewal of tracks and turnouts, each vehicle's wheelset is considered in the calculation for the vertical stress. The leading wheelsets of bogies are used for the lateral stress in curved tracks.

2.3 Input data

Applying the TDM needs input data. On the one hand these input data refer to the characteristics of rolling stock. The vehicle's information about their characteristics is the foundation for calculations and simulations regarding stresses on the track. On the other hand, the input data refer to the track alignment radii and speeds. Classifying the track in radii and speed classes allows distinguishing between wear in straight and

Radii class	**Track alignment** *in m*	$f_{7.1,R}$ *unitless*	$f_{7.2,R}$ *unitless*
R_5	$R > 1000$	1.0	0.0
R_4	$600 < R \le 1000$	0.8	0.2
R_3	$400 < R \le 600$	0.7	0.3
R_2	$250 < R \le 400$	0.6	0.4
R_1	$R \le 250$	0.5	0.5

Table 1.
Weighting factors for vertical and lateral forces in damage term D7.

curved tracks. One and the same vehicle running once through a straight and once through a curved track section causes therefore wear to a different extent. Furthermore, maintenance work that appears due to vehicle–track interaction needs to be defined and its strategic costs determined. In this TDM, the three categories, namely rolling stock, track characteristics and maintenance are strongly connected to each other (see **Figure 1**). The following chapters expand on these impact categories.

2.3.1 Track characteristics: Radii and speed classes

Based on the differences between alignment of track sections, a classification of radii and speeds was done. For the TDM 5 radii classes (R) were defined. These 5 radii classes correspond to those used in the Standard Elements [1]. Further, defining different radii classes comprises not only the law of superelevation of the outer rail in curved tracks, but also the wear of this outer rail.

The fifth radius class (R_5) represents track sections that show a radius of more than 1000 m. These sections are classified as straight track sections. Radius class R5 is again divided into eight speed classes (S_1 to S_8). For each radius and speed class one reference speed is allocated with which the vehicle parameters are calculated. The reference speeds of radius class R_1 to R_4 each represent the rounded median value of the line speeds of the Austrian core network. The reference speed of 60 kmph of radius class R_1 thus indicates that on the Austrian main lines in curves with radii less than 250 m a median speed of 60 kmph is driven. The calculation of the reference radii is done analogously. **Table 2** lists the radius and speed classes with their reference radius and speed.

In the Austrian network vehicles used in freight traffic are always treated with a maximum speed of 100 kmph, in the model 90 kmph is therefore assumed as a

Classification of radius *in m*	**Reference radius** *in m*	**Classification of speed** *in kmph*	**Reference speed** *in kmph*
R_1 $R < 250$	215	S_{R1} -	60
R_2 $250 < R \leq 400$	315	S_{R2} -	70
R_3 $400 < R \leq 600$	500	S_{R3} -	90
R_4 $600 < R \leq 1000$	800	S_{R4} -	110*
R_5 $R > 1000$	—	S_1 $S \leq 80$	75
		S_2 $80 < S \leq 100$	90
		S_3 $100 < S \leq 120$	110*
		S_4 $120 < S \leq 140$	130*
		S_5 $140 < S \leq 160$	150*
		S_6 $160 < S \leq 200$	190*
		S_7 $200 < S \leq 230$	220*
		V_8 $230 < S \leq 250$	240*

**For fright rolling stock 90 kmph is assumed as a reference speed.*

Table 2.
Speed and radii classification.

reference speed for freight traffic. This speed limitation applies to both freight wagons and also freight locomotives. For speed classes S_3 to S_8 of radius class R_5 (straight tracks) and radius class R_4, the reference speed is applied in the model, as the reference speed exceeds 90 kmph in these classes. The speed classes S_2 to S_8 are thus identical and further lead to equal calculated parameters for freight rolling stock. For universal locomotives, it therefore makes sense to differentiate according to type of traffic (passenger or freight traffic).

Additionally, it should be mentioned here that the reference speed is a reference value that is based on the maximum permissible speed of track sections due to their track alignment (S_{Track}). If the maximum permissible speed of the vehicle/train ($S_{max,Veh}$) exceeds the speed due to track alignment ($S_{max,Veh} > S_{Track}$), in the calculation of the vehicle parameters (Section 2.3.2), the reference speed (S_{Track}) of **Table 2** is applied. This also applies analogously vice versa, as shown in Eq. (11).

$$S = \min[S_{max,Veh}, S_{Track}] \tag{11}$$

S—relevant speed (kmph); $S_{max,Veh}$—maximum permissible speed due to rolling stock (kmph); S_{Track}—maximum permissible speed due to track alignment (kmph).

Using the minimum of both speeds $S_{max,Veh}$ and S_{Track} reflects the operational reality. Regarding the speed the determined damage that is speed dependent (D_1, D_2, D_5, D_6 and D_7) is thus also close to reality.

2.3.2 Rolling stock parameters

The TDM and its damage terms described in Section 2.2 are based on the following four parameters:

- Dynamic vertical force: P_1 and P_2
- Lateral force: Y_R
- Contact patch energy: $T\gamma$
- Traction power value: TPV.

The parameters P_1 and P_2 stand for the dynamic vertical force, while Y describes the lateral force that occurs in a track curve on the outer wheel of a bogie in the leading wheelset. The traction power value (TPV) indicates the physical power per wheel due to traction. The fourth physical input parameter for the deterioration formula is $T\gamma$ that stands for the friction work due to longitudinal and lateral slip. The following sections include a discussion of the four vehicle parameters in more details.

2.3.2.1 Dynamic vertical force (P_1 and P_2)

Due to [23] and [22], the P_1 and P_2 force is applied in the TDM. Eqs. (12)-(16) depict the composition and calculation scheme of the vertical impact forces P_1 and P_2.

$$\text{Iteration}: P_{1,S} = P_0 + S \times 2\alpha \times \sqrt{\frac{K_H \times m_e}{1 + {m_e}/{m_u}}} \tag{12}$$

$$m_e \approx 0.4 \times \left(m_r + \frac{m_s}{l}\right) \tag{13}$$

$$K_H = \frac{P_{1,est} - P_0}{G \times \left(P_{1,est}^{2/3} - P_0^{2/3}\right)} \tag{14}$$

$$G = \frac{3.86}{R_{Wheel}{}^{0.115}} \times 10^{-8} \tag{15}$$

$$P_{2,S} = P_0 + S \times 2\alpha \times \sqrt{\frac{m_u}{m_u + m_t}} \times \left(1 - \frac{c_t \times \pi}{4 \times \sqrt{K_t \times (m_u + m_t)}}\right) \times \sqrt{K_t \times m_u} \tag{16}$$

$P_{1,S}$,$P_{2,S}$—vehicle dynamic wheel forces (N); P_0—vehicle static wheel force (N); S —relevant speed (limited by vehicle or track alignment) (m/s); 2α—total joint angle (rad); m_u_unsprung mass per vehicle wheel (kg); m_t—effective vertical track mass per vehicle wheel (kg); c_t—effective track damping per vehicle wheel (Ns/m); K_t—effective vertical track stiffness per vehicle wheel (N/m); m_e—effective track mass per vehicle wheel (kg); m_r—rail mass per unit length (kg/m); m_s—mass of half a sleeper (kg); l—sleeper spacing (m); $P_{1,est}$—estimated vertical dynamic wheel force (N); G—Hertzian flexibility constant (for worn tyre profiles) ($m/N^{2/3}$); R_{Wheel}—wheel radius (m); K_H—linearized Hertzian contact stiffness per vehicle wheel (N/m).

The total vertical force that arises due to the interaction between rail and wheel comprises not only the static gravitational loading (P_0) but also the dynamic forces activated by speed (S), unsprung mass (m_u) and the rail's alignment (2α). Both calculation schemes of the vertical wheel forces P_1 (Eq. (12)) and P2 (Eq. (16)) are based on this approach. While P_2 does also depend on track parameters (track stiffness K_t, track damping c_t and track mass m_t per wheel), P_1 is additionally conditioned by the effective track mass (m_e) and the linearized Hertzian contact stiffness (K_H) per wheel. The Hertzian contact stiffness K_H is subject to an iterative calculation between Eqs. (12) and (14) and therefore depending on the P_1 force. P_1 can be summarized to be the high-frequency portion of the impact force that is primarily responsible for surface wear and local stress peaks of the rail material. P_2 is the low-frequency and long-waved force component that mainly stresses the sleepers and ballast bed [22, 28].

In the TDM the included constants m_t, K_t, c_t and 2α are applied according to [23]. Further, rail mass (m_r) belongs to UIC60 rail, the mass of half a sleeper (m_s) to concrete sleepers and the sleeper spacing (l) to the standard spacing in Austria. The included constants in the TDM have the following values:

- 2α = 0.02 rad
- m_t = 245 kg
- c_t = 55,400 Ns/m
- K_t = 62,000,000 N/m
- m_r = 60 kg

- m_s = 150 kg
- l = 0.6 m

As the approach for P_1 and P_2 depends on the unsprung mass (m_u), the model differs between the wheels of powered and unpowered axles. Wheels of powered and unpowered axles therefore show varying values in both, P_1 and P_2. Section 2.4 gives an example calculation of the P_1 and P_2 force for a universal locomotive at a speed of 90 kmph.

2.3.2.2 Lateral force (Y_R)

The lateral force Y_R describes the horizontal force component that arises in a track curve on the outer wheel of a bogie in the leading wheelset. On the basis of a multiple-body simulation (MBS) Y_R can be determined. Representing the case of a typical turnout (for Austria EW190-1:9: single turnout at an inclination 1:9 and radius of the junction of 190 m) and calculating Y_R values for damage term D_5 at 40 kmph, a s-curve with radii 190 m was used in the MBS. In this s-curve no cant or transition curves, but an intermediate straight line of 6 m were used to ensure a representative simulation of a vehicle's run through a turnout. Due to the s-curve, both the left and right wheel run on the outside of the curve. Two Y_R values (left and right) are therefore produced within one simulation. The higher value of both is considered for the TDM [21].

As not only damage term D_5 (turnout wear) depends on the lateral force but also D_7 (track renewal), an MBS is done also for tracks representing radius class R_1 to R_4. In this MBS a cant was implemented, so that the lateral acceleration is constantly 0.85 m/s^2 in each radius class. As a reference for the classification (R_1 to R_4) of the track curves, radii in the amount of 215, 315, 500 and 800 m were chosen. The Y_R was simulated at the reference speed of each radii class and for both, powered and unpowered wheelsets. **Table 3** gives an overview of the input parameters that were chosen for the MBS and calculating Y_R.

2.3.2.3 Contact patch energy ($T\gamma$)

Not only the lateral force is determined by a multiple-body simulation (MBS), but also the contact patch energy $T\gamma$. $T\gamma$ represents the input parameter for Burstow's [20] evaluating function for rail surface wear in curved tracks. For determining $T\gamma$ on the

Radii class	Reference Radius *in m*	Reference Speed *in kmph*	Superelevation *in mm*	Lateral acceleration *in m/s²*
R_1	215	60	67.6	0.85
R_2	315	70	53.6	0.85
R_3	500	90	61.2	0.85
R_4	800	110*	48.5	0.85

**For fright rolling stock 90 kmph is assumed as a reference speed.*

Table 3.
Input parameters for MBS of Y_R.

basis of MBS, again a s-curve has been used. The applied curves in the simulation belong to the radii classes R_1 to R_4 and their reference radius (215 m for R_1, 315 m for R_2, 500 m for R_3 and 800 m for R_4). Further, the simulation includes superelevation and transition curves. **Table 3** in Section 2.3.2.3 depicts the input parameters for the MBS of Tγ. As the simulation is not only done for powered but also unpowered wheelsets up to 8 Tγ values can be assigned to one specific vehicle due to 4 radii classes. The contact patch energy Tγ is given in Nm/m for the TDM. A detailed specification for the MBS is described in the SBB guidance for vehicle pricing [21].

2.3.2.4 Traction power value (TPV)

As damage term D_3 describes rail surface wear due to the impact of vehicles power, the traction power value (TPV) is calculated. TPV describes the power of a vehicle (P_{Wheel} in kW) related to the contact area between rail and wheel (A_{eff} in mm^2) as shown in Eq. (17). The TPV therefore only exists for powered wheelsets.

$$\text{TPV} = \frac{P_{Wheel}}{A_{eff}} \tag{17}$$

TPV—traction power value (kW/mm^2); P_{Wheel}—power per wheel (kW); A_{eff}—effective contact area between rail and vehicle wheel (mm^2).

The vehicle's power is a value that should be available from the data sheet of the vehicle, whereas the contact area between rail and wheel can be calculated. Basis for this is the methodology due to the Hertzian contact area for arbitrarily curved surfaces as it is described in [29]. However, the Hertzian contact area is downsized by a factor of 2/3 to include system-related uncertainties. Eq. (18) depicts the general approach of the downsized Hertzian contact area due to the major (a) and minor (b) radius. In Eq. (19) the major and minor radius of the ellipse are specified as well as the auxiliary angle (ϑ). Due to the formula of Hertz, the major and minor radius are functions dependent on static wheel force (P_0), Poisson's ratio (ν), modulus of elasticity (E) and the wheel and rail radius (R_{Wheel}, R_{Rail}). The coefficients η and ξ are functions of ϑ. Summarizing the general Eq. (18) of Hertz with the parameters in Eq. (19) leads to a simplified Hertzian formula, as depicted in Eq. (20).

$$A_{eff} = \frac{2}{3}\pi\, a\, b \tag{18}$$

$$a = \sqrt[3]{\frac{3\,\xi^3\,(1-\nu^2)\,P_0}{E\left(\frac{1}{R_{Wheel}}+\frac{1}{R_{Rail}}\right)}} \times 10^6 \text{ and } b = \sqrt[3]{\frac{3\,\eta^3\,(1-\nu^2)\,P_0}{E\left(\frac{1}{R_{Wheel}}+\frac{1}{R_{Rail}}\right)}} \times 10^6$$

$$\vartheta = \arccos\left(\frac{R_{Wheel}-R_{Rail}}{R_{Wheel}+R_{Rail}}\right) \tag{19}$$

$$A_{eff} = \frac{2}{3}\pi\,\xi\,\eta\left(\frac{3\,(1-\nu^2)}{E}\right)^{2/3\times}\left(\frac{P_0}{\left(\frac{1}{R_{Wheel}}+\frac{1}{R_{Rail}}\right)}\right)^{2/3\times 10^6} \tag{20}$$

A_{eff}—effective contact area between rail and vehicle wheel (mm^2); a—major radius of ellipse (mm); b—minor radius of ellipse (mm); P_0—vehicle static wheel force (kN); R_{Wheel}—wheel radius (m); R_{Rail}—rail radius (m); ϑ—auxiliary angle (rad); ξ, η—coefficients; ν—Poisson's ratio; E—modulus of elasticity (kN/m^2).

For the analytic calculation of the Hertzian contact area, the vehicle is considered to be standing on a straight track. The rail head radius (R_{Rail}) is therefore expected to be 0.3 m. Further, the Poisson's ratio (ν) is uniformly set to 0.3 and the modulus of elasticity (E) to 2.1×10^8 kN/m^2. These parameters are constant values in the TDM and do not get changed. The coefficients ξ and η can be summarized (approximation) in a function for each, as shown in Eq. (21) due to [21] and [29].

$$\xi(\vartheta) = 1.5281739 \times \vartheta^{-0.8571601}$$
$$\eta(\vartheta) = 0.4724037 \times \vartheta + 0.2366389 \quad (21)$$

ϑ—auxiliary angle (rad); ξ, η—coefficients.

Implementing the functions of the coefficients ξ and η into Eq. (20) gives the final description of the effective contact area due to the Hertzian methodology, as shown in Eq. (22). This formula is valid for the Austrian TDM due to constant values for R_{Wheel}, E and ν.

$$A_{eff} = \frac{8.3593707 \times \vartheta + 4.1874191}{\vartheta^{0.8571601}} \times \left(\frac{P_0}{\frac{1}{R_{Wheel}} + \frac{1}{0.3}} \right)^{2/3}$$
$$\text{with } \vartheta = \arccos\left(\frac{R_{Wheel} - 0.3}{R_{Wheel} + 0.3} \right) \quad (22)$$

A_{eff}—effective contact area between rail and vehicle wheel (mm^2); P_0—vehicle static wheel force (kN); R_{Wheel}—wheel radius (m); ϑ—auxiliary angle (rad).

The determined TPV in kW/mm^2 corresponds directly to the damage term D_3. A multiplication by the number of driven axles is not necessary, since the vehicle's power per wheel is related to the Hertzian contact area between wheel and rail.

2.4 Exemplary calculations

In the following subchapters, exemplary calculations are done for a sample vehicle. A universal locomotive is chosen as a sample vehicle that goes through a track segment of radius class R_3. The maximum permissible speed of this vehicle is 230 kmph at an axle load of 22 t. In **Table 4**, all input parameters are given that are necessary to calculate the rolling stock parameters and the damage increments for the sample vehicle.

2.4.1 Rolling stock parameters

In this chapter, the rolling stock parameters of the sample vehicle are calculated due to the described formula in Section 2.3.2. The vertical wheel forces P_1 and P_2 and the TPV are determined for powered axles as the sample vehicle has four powered axles and no unpowered axles. Furthermore, $T\gamma$ and Y_R are not treated in this chapter as they are estimated by MBS. Determining the dynamic P_2 force for radius class R_3 at 90 kmph as shown in Eq. (23) and for 40 kmph (damage term D_5) in Eq. (24):

Universal locomotive: axle load = 22 t | $S_{max,Veh}$ = 230 kmph
R_3: 400 < R ≤ 600 m | $S_{reference}$ = 90 kmph | $R_{reference}$ = 500 m

	Parameter	Value (per wheel)	Unit
P_0	Vehicle static wheel force	22 × 9.81 × 1000/2 = 107,910	N
S	Vehicle speed (limited by vehicle or track alignment)	90 × 3.6 = 25	m/s
S_{40}	Vehicle speed for D_5	40 × 3.6 = 11.11	m/s
2α	Total joint angle	0.02	rad
m_u	Unsprung mass per vehicle wheel	1247.5	kg
m_t	Effective vertical track mass per vehicle wheel	245	kg
c_t	Effective track mass per vehicle wheel	55,400	Ns/m
K_t	Effective vertical track stiffness per vehicle wheel	62,000,000	N/m
R_{Wheel}	Wheel radius	1.15	m
m_r	Rail mass per unit length (rail UIC60/60E1)	60	kg/m
m_S	Mass of half a sleeper	300/2 = 150	kg/m
l	Sleeper spacing	0.6	m
$P_{1,est}$	Estimated dynamic vertical wheel force	Iteration (367,283.51)	N
n_p	Number of powered wheelsets	4	—
n_{up}	Number of unpowered wheelsets	0	—
n_p‘	Number of leading powered wheelsets in a bogie	2	—
n_{up}‘	Number of leading unpowered wheelsets in a bogie	0	—
P_{Wheel}	Power per wheel	6400/4/2 = 800	kW
$S_{max,Veh}$	Maximum permissible speed of vehicle	230	kmph

Table 4.
Input parameter of the sample vehicle.

$$P_{2,90} = \left(107{,}910 + 25 \times 0.02 \times \sqrt{\frac{1{,}247.5}{1{,}247.5 + 245}} \times \left(1 - \frac{55{,}400 \times \pi}{4 \times \sqrt{62{,}000{,}000 \times (1{,}247.5 + 245)}}\right) \times \sqrt{62{,}000{,}000 \times 1{,}247.5}\right) \times 10^{-3} = \underline{216.8\ \text{kN}} \quad (23)$$

$$P_{2,40} = \left(107{,}910 + 11 \times 0.02 \times \sqrt{\frac{1{,}247.5}{1{,}247.5 + 245}} \times \left(1 - \frac{55{,}400 \times \pi}{4 \times \sqrt{62{,}000{,}000 \times (1{,}247.5 + 245)}}\right) \times \sqrt{62{,}000{,}000 \times 1{,}247.5}\right) \times 10^{-3} = \underline{156.3\ \text{kN}} \quad (24)$$

Determining the dynamic P_1 force for radius class R_3 as shown in Eq. (25). to Eq. (28):

$$m_e \approx 0.4 \times \left(60 + \frac{150}{0.6}\right) = 124 \text{ kg/m} \tag{25}$$

$$G = \frac{3.86}{1.15^{0.115}} \times 10^{-8} = 3.798 \times 10^{-8} \text{ m/N}^{2/3} \tag{26}$$

$$K_H = \frac{P_{1,est}{}^{*} - 107{,}910}{3.798 \times 10^{-8} \times \left(P_{1,est}{}^{2/3} - 107{,}910^{2/3}\right)} = 2.386 \times 10^{9} \text{N/m} \tag{27}$$

$$P_{1,90} = \left(107{,}910 + 25 \times 0.02 \times \sqrt{\frac{2.386 \times 10^{9} \times 124}{1 + {}^{124}\!/_{1{,}247.5}}}\right) \times 10^{-3} = 367.3 \text{ kN} \tag{28}$$

**At the end of the iteration process $P_{1,est}$ is expected to be ~367,283.5 N.*
Determining the TPV as shown in Eq. (29):

$$A_{eff} = \frac{8.3593707 \times \vartheta + 4.1874191}{\vartheta^{0.8571601}} \times \left(\frac{107{,}910 \times 10^{-3}}{\frac{1}{1.15} + \frac{1}{0.3}}\right)^{2/3} = 110.4 \text{ mm}^2$$

$$\text{with } \vartheta = \arccos\left(\frac{1.15 - 0.3}{1.15 + 0.3}\right) = 0.94442737 \tag{29}$$

$$TPV = \frac{P_{Wheel}}{A_{eff}} = \frac{800}{110.4} = 7.243 \text{ kW/mm}^2$$

Table 5 summarizes the parameter values for the powered wheels of a universal locomotive in a track segment of radius class R_3. This locomotive does not have unpowered wheelsets. In the case of vehicles with both powered and unpowered wheelsets there exist another table with vehicle parameters for unpowered wheels. In the following table, the parameters Y_R and $T\gamma$ that belong to MBS are added too for further calculations. P_2 and Y_R values at 40 kmph are needed later on for damage term D_5.

For every vehicle type (such as locomotives, freight wagons, passenger wagons or multiple units) that appears in a network these five physical parameters shown in **Table 5** need to be estimated. As not every parameter occurs in every radius and/or speed class in the following their appearance is summarized:

- P_1: R_1 to R_4 and values for R_5 S_1 to S_8 (12 input values)
- P_2: R_1 to R_4 and values for R_5 S_1 to S_8, one value at 40 kmph (13 input values)
- Y_R: R_1 to R_4 and one value at 40 kmph (5 input values)

Universal locomotive: axle load = 22 t | $S_{max,Veh}$ = 230 kmph
R_3: 400 < R ≤ 600 m | $S_{reference}$ = 90 kmph | $R_{reference}$ = 500 m

Parameters are related to one wheel

$P_{1,90}$ *in kN*	$P_{2,40}$ *in kN*	$P_{2,90}$ *in kN*	TPV *in kW/mm²*	$Y_{R,90}$ *in kN*	$Y_{R,190}$ *in kN*	$T\gamma$ *in Nm/m*
367.3	156.3	216.8	7.243	40.0	69.0	291.0

Table 5.
Vehicle parameters for powered wheels of a sample vehicle (universal locomotive).

- TPV: one input value
- Tγ: R_1 to R_4 (4 input values).

If those input parameters are available for all vehicles in the network, the seven damage increments of the TDM can be estimated for each vehicle. This is illustrated for the sample vehicle in the following Section 2.4.2.

2.4.2 Damage increment (D_n)

In this chapter exemplary calculations for the seven damage increments of the TDM are given. Required data and information to do so are.

- the estimated vehicle parameters (**Table 5**),
- number of powered and unpowered wheelsets and the number of leading wheelsets of bogies (**Table 4**) and
- the maximum permissible vehicle speed (**Table 4**).

The basic formula of the vehicle-specific damage increments D_n can be seen in Eq. (30). Since the vehicle parameters are given per wheel, they must be multiplied by the number of wheelsets (powered and unpowered).

$$\begin{aligned} Dn_{Veh,Si|Rj} &= \text{formula for damage termn}\left[VehPn_{Si|Rj_p(S_{min})}\right] \times n_{Veh_p} \\ &+ \text{formula for damage termn}\left[VehPn_{Si|Rj_{up}(S_{min})}\right] \times n_{Veh_{up}} \\ &\text{with} : S = \min[S_{max,Veh}, S_{Track}] \end{aligned} \tag{30}$$

$D_{Veh,Si|Rj}$ —damage increment Dn of the sample vehicle at speed (i) and radius class (j) (with i=1-8 und j=1-4); n_{Veh} —number of powered (p) or unpowered (up) wheelsets of the sample vehicle; $VehP_{Si|Rj(S_{min})}$—vehicle parameter of powered/unpowered wheelsets at R/S-class, depending on relevant speed; $S_{max,Veh}$ —maximum permissible vehicle speed in kmph; S_{Track}—reference speed at a certain speed and radius class in kmph; S—relevant speed for determining the vehicle parameter in kmph.

Calculating the vehicle-specific damage increments, it is important to distinguish between powered and unpowered wheelsets. On the other hand, attention must be paid to the relevant speed due to the vehicle and the radius/speed-class. The minimum of both is considered in the calculation.

Exemplary calculation for damage increment D_1 is given in Eq. (31).

$$\begin{aligned} &\text{Formula for } D_1 : {P_{2,S}}^3 \\ &S = \min[230 \text{ kmph}, 90 \text{ kmph}] = 90 \text{ kmph} \\ &D_{1,R3} = 216.8^3 \times 4 + 0 = 40{,}793{,}916 \text{ kN}^3 \end{aligned} \tag{31}$$

$D_{1,R3}$—damage increment D1 of the sample vehicle in radius class R3 (kN^3/vehicle); n_{Veh_p}—4; $n_{Veh_{up}}$ —0; $P_{2,R3_p}$ —216.8 (kN/wheel); $P_{2,R3_{up}}$ —0 (kN/wheel).

Exemplary calculation for damage increment D_2 is given in Eq. (32).

$$\text{Formula for } D_2 : {P_{2,S}}^{1,2}$$

$$S = \min[230 \text{ kmph}, 90 \text{ kmph}] = 90 \text{ kmph}$$

$$D_{2,R3} = 216.8^{1.2} \times 4 + 0 = 2{,}543.7 \text{ kN}^{1.2} \tag{32}$$

$D_{2,R3}$—damage increment D2 of the sample vehicle in radius class R3 ($kN^{1.2}$/vehicle); n_{Veh_p}—4; $n_{Veh_{up}}$ —0; $P_{2,R3_p}$ —216.8 (kN/wheel); $P_{2,R3_{up}}$ —0 (kN/wheel).

Exemplary calculation for damage increment D_3 is given in Eq. (33).

$$\text{Formula for } D_3 : TPV = 7.243 \text{ kW/mm}^2 \tag{33}$$

D_3—damage increment D_3 of the sample vehicle (kW/mm^2); TPV_{Veh}—7.243 (kW/mm^2)

Exemplary calculation for damage increment $D_{4.1}$ and $D_{4.2}$ is given in Eq. (34).

$$\text{Formula for } D_{4.1}(\text{area a}) : \quad D_{4.1} = 0 \qquad \text{for } T_\gamma \geq 175 \text{Nm/m}$$

$$D_{4.1,R3} = 0.00$$

$$\text{Formula for } D_{4.2}(\text{area e}) : D_{4.2} = \left(\frac{T_\gamma - 65}{110}\right) \qquad \text{for } T_\gamma \geq 65 \text{Nm/m}$$

$$D_{4.2,R3} = 2 * \left(\frac{291 - 65}{110}\right) + 0 = 4.11 \tag{34}$$

$D4.1_{R3}$—damage increment D4.1 of the sample vehicle in radius class R_3 (-/vehicle); $D4.2_{R3}$—damage increment D4.2 of the sample vehicle in radius class R_3 (-/vehicle); n'_p —2; n'_{up} —0; $T_{\gamma,p}$ —291 (Nm/m); $T_{\gamma,up}$ —0 (Nm/m).

Exemplary calculation for damage increment D_5 is given in Eq. (35).

$$\text{Formula for } D_5 : \sqrt{\left(0.5 \times {P_{2,S=40}}^2 + 0.5 \times {Y_{R=190}}^2\right)}$$

$$D_{5,R3} = \sqrt{\left(0.5 \times 156.3^2 + 0.5 \times 69.0^2\right)} * 2 + 0 = 241.6 \text{ kN} \tag{35}$$

$D_{5,R3}$—damage increment D5 of the sample vehicle in radius class R_3 (kN/vehicle); n'_p —2; n'_{up} —0; $P_{2,40,p}$ —156.3 (at 40 kmph) (kN/wheel); $P_{2,40,up}$ —0 (at 40 kmph) (kN/wheel); $Y_{R=190,p}$ —69.0 (at 40 kmph and radius 190 m) (kN/wheel); $Y_{R=190,up}$—0 (at 40 kmph and radius 190 m) (kN/wheel).

Exemplary calculation for damage increment D_6 is given in Eq. (36).

$$\text{Formula for } D_6 : {P_{1,S}}^3$$

$$S = \min[230 \text{ kmph}, 90 \text{ kmph}] = 90 \text{ kmph}$$

$$D_{6,R3} = 367.3^3 \times 4 + 0 = 198{,}208{,}728 \text{ kN}^3 \tag{36}$$

$D_{6,R3}$—damage increment D6 of the sample vehicle in radius class R_3 (kN^3/vehicle); n_{Veh_p}—4; $n_{Veh_{up}}$—0; $P_{1,R3_p}$—367.3 (kN/wheel); $P_{1,R3_{up}}$—0 (kN/wheel).

Exemplary calculation for damage increment D_7 is given in Eq. (37).

$$\text{Formula for } D_7: \sqrt{\left(f_{7_{1,R}} \times P_{2,S}{}^2 + f_{7_{2,R}} \times Y_R{}^2\right)}^{-3}$$

$$S = \min[230 \text{ kmph}, 90 \text{ kmph}] = 90 \text{ kmph}$$

$$D_{7,R3} = 2 * \sqrt{\left(0.7 \times 216.8^2 + 0.3 \times 40.0^2\right)}^{-3} + 0 = 12{,}207{,}944 \text{ kN}^3 \tag{37}$$

$D_{7,R3}$—damage increment D7 of the sample vehicle in radius class R_3 (kN^3/vehicle); n'_p—2; n'_{up}—0; $P_{2,R3_p}$—216.8 (kN/wheel); $P_{2,R3_{up}}$—0 (kN/wheel); $Y_{R3,p}$—40.0 (kN/wheel); $Y_{R3,up}$—0 (kN/wheel); $f_{7_{1,R}}$—0.7; $f_{7_{2,R}}$—0.3.

2.4.3 Damage per vehicle-kilometer ($D_{n,km}$)

In a further step the damage increments of every vehicle are related to length by means of unit-kilometer. The absolute value of the damage increment stays the same; however, the unit changes (see Eq. (38)).

$$D_{n,km} = D_n \times km \tag{38}$$

$D_{n,km}$—damage increment per vehicle of one unit kilometer (unit*km/vehicle); D_n—damage increment (n = 1, 2, 3, 4.1, 4.2, 5, 6, 7) (unit*/vehicle); km—length (km).

**Unit of the damage term: kN^3 (D_1, D_6 and D_7), $kN^{1.2}$ (D_2), kW/mm^2 (D_3), kN (D_5) and $D_{4.1}$ and $D_{4.2}$ are unitless*

Due to Eq. (38), a length reference is established in the TDM and vehicle-kilometer are therefore included. This approach is similar to the conventional vehicle gross-tonnes that turn into vehicle gross-ton-kilometer when length is considered.

3. Results

In the first part of the result section, we assess the influence of traffic on track maintenance. We start with a given situation in an artificially configurated network with a mixed traffic load of some 30,000 gross-tons per day (~10 mio. Gross-tons per year). We analyze the influence of changes in the traffic mix preliminary with focus on the different traffic segments, long-distance passenger traffic, regional passenger traffic, and freight transport. We also include by way of example the influence of different superstructure and substructure conditions (see more details in [1]).

In the second part, the influence of different vehicles is highlighted (see Section 3.2). This influence is much smaller at the network level of course, but might be significant in some parts, line segments or whenever entire vehicle fleets are replaced. We compare different vehicles at different speed levels and also powered and running axles. Estimating future track maintenance needs the assessment of future vehicles on the network. New vehicles might be track-friendly or track-harming and train operators focus on many other aspects than track maintenance consequences. Again, in these evaluations, we see the strong impact of the technical configuration of track, as different tracks indicate a higher or lower importance of the one or the other vehicle characteristics.

3.1 The impact of loading on track maintenance

In order to highlight the main difference between treating track loading as gross-tonnage or as specific damage for different track components, this paper delivers results based on an artificial network. Note, this network is also used in [1] dealing with the impact of different track-related boundary conditions on overall track maintenance and renewal. We add the line speed as a parameter and use the modulation done in [1] concerning superstructure components.

The network covers 10,000 km with a given radii distribution and line speed distribution for the straight sections given in **Table 6**. To keep the focus on the loading, we model the maintenance for a given, constant track consisting of heavy superstructure (concrete sleepers, wooden sleepers in the sharp curves R ≤ 250 m 60E1-R260 rails) on medium ballast quality, good subsoil and good drainage condition. The tracks are loaded with 30,000 gross-tons per day everywhere in the network. Both assumptions are never true for any existing network, but were chosen for illustration purposes.

According to the track maintenance assessment provided by [1], this network characteristics deliver a certain track maintenance and renewal demand depicted in **Table 7**.

For applying the alternative description of track loading proposed in Section 2, we need to have a closer look at the transport load. The traffic mix shows long-distance passenger trains (LDP), regional passenger trains (RP) and freight trains (F) consisting of different vehicles (**Table 8**).

The train configuration of the different market segments is depicted in **Table 9**. The long-distance train is a loco-wagon train with a maximum speed of 160 kmph (and thus reaching the maximum line speed). For the regional passenger traffic, the traffic mix consists of two different types of electric multiple units (EMU), one (EMU1) with a total weight of 160 tons consists of four powered bogies and two trailer bogies (axle scheme Bo'Bo'2'2'Bo'Bo'), the other one (EMU2) is a lighter and shorter trainset with two powered bogies and three trailer bogies (axle scheme

Radii Class	Track Alignment *in m*	Radii Distribution *in %*	Line Speed* *in kmph*	Line Speed Distribution *in %*
R_5	R > 1000	76.0	140 < S ≤ 160	12
			120 < S ≤ 140	31
			100 < S ≤ 120	30
			80 < S ≤ 100	19
			S ≤ 80	8
R_4	600 < R ≤ 1000	12.5	140	—
R_3	400 < R ≤ 600	6.5	110	—
R_2	250 < R ≤ 400	4.5	90	—
R_1	R ≤ 250	0.5	80	—

**Line Speed is not reached by all trains/vehicles.*

Table 6.
Reference network.

Track work	**Amount** *in km*
Leveling-Lining-Tamping	1747
Ballast Cleaning	0
Rail Grinding	504
Rail Exchange	5.5
Renewal	283

Table 7.
Maintenance and renewal demand for the reference network.

	No. of Trains *unitless*	**[relative]** *in %*	**Loading** *in gross- tons*	**[relative]** *in %*
Long-distance Passenger	13	[20]	6284	[21]
Regional Passenger	35	[53]	5072	[17]
Freight	18	[27]	18,814	[62]
Sum	66	[100]	30,170	[100]

Table 8.
Traffic mix.

	No. of vehicles *in Vehicle/ Train*	**Vehicle weight** *in tons*	**Train weight** *in tons*
Long-distance Passenger			
4ax-Universal-Loco	1	88	483.36
4ax-2Bogie-Passenger Car	7	56.48	
Regional Passenger			
EMU1	19	160	160.00
EMU2	16	127	127.00
Sum/Average	**35**		**144.91**
Freight			
4ax-Universal-Loco	1	88	1045.23
4ax-2Bogie-Y25 Freight Car "empty" (Axle-load 6.5 t)	7	25.56	
4ax-2Bogie-Y25 Freight Car "medium" (Axle-load 14 t)	8	56.65	
4ax-2Bogie-Y25 Freight Car "full" (Axle-load 20 t)	4	81.28	

Table 9.
Train configuration.

Bo'2'2'2'Bo') and a total weight of 127 tons. The freight trains are modeled as some 1000 ton-trains with one four-axle loco and a mix of empty, medium-loaded and full-loaded freight wagons (detailed loads see **Table 9**), all with two Y25-bogies.

3.1.1 Ballast maintenance

The transport volume for the entire network can be addressed as gross-ton-kilometers or as damage-kilometer following Section 2. For the latter, for the ballast maintenance we need to calculate the P_2-forces, specifically for the different line speed sections. Thereby it is to be considered that the train speed may be limited by the allowed maximum vehicle speed and not by the given line speed. For freight trains, the maximum speed assessed for this calculation is 100 kmph. Thus, for all sections with higher line speeds the damage increments D_1 (P_2-force) are calculated with 100 kmph only. What we can see directly from **Table 10** is that the alternative description of "loading" moves relative damage shares toward long-distance passenger traffic. This effect is mainly driven by the higher speed as axle loads for these two traffic segments are about the same on average. The share of regional passenger traffic remains almost the same as speeds are higher than in freight transportation, but axle loads lower.

We know that this loading leads to 1747 kilometers of necessary tamping in the network or a tamping interval of 5.7 years (6 years in the straight sections). If we allocate this tamping demand to the loading, we can calculate an incremental tamping demand for the unit gross-ton-km and the alternative unit kN^3km.

	Gross-ton-kilometers per year		**D_1: kN^3-km per year**	
Long-distance Passenger	2,29E+10	21%	9,83E+15	28%
Regional Passenger	1,85E+10	17%	5,57E+15	16%
Freight	6,87E+10	62%	1,94E+16	56%
Sum	1,10E+11	100%	3,48E+16	100%

Table 10.
Track loading.

	R_5	**R_4**	**R_3**	**R_2**	**R_1**
LDP gross-ton-km	1,75E+10	2,86E+09	1,48E+09	1,03E+09	1,13E+08
RP gross-ton-km	1,41E+10	2,30E+09	1,19E+09	8,28E+08	9,15E+07
F gross-ton-km	5,23E+10	8,55E+09	4,43E+09	3,07E+09	3,39E+08
Sum gross-ton-km	8,38E+10	1,37E+10	7,10E+09	4,93E+09	5,44E+08
Tamping demand [km]	1269	208	133	112	26
incremental tamping demand ton-km	**1,51E-08**	**1,51E-08**	**1,87E-08**	**2,27E-08**	**4,72E-08**
LDP kN^3-km	7,99E+15	1,18E+15	4,35E+14	2,06E+14	1,85E+13
RP kN^3-km	4,52E+15	6,74E+14	2,48E+14	1,17E+14	1,05E+13
F kN^3-km	1,49E+16	2,49E+15	1,29E+15	6,10E+14	5,46E+13
Sum kN^3-km	2,74E+16	4,34E+15	1,97E+15	9,34E+14	8,36E+13
Tamping demand [km]	1269	208	133	112	26
incremental tamping demand kN^3-km	**4,62E-14**	**4,78E-14**	**6,73E-14**	**1,20E-13**	**3,07E-13**

Table 11.
Incremental tamping demand.

		R_5	R_4	R_3	R_2	R_1
Sum gross-ton-km		8,38E+10	1,37E+10	7,10E+09	4,93E+09	5,44E+08
Tamping demand	*in km*	1269	208	133	112	26
LDP kN^3-km		3,87E+16	5,70E+15	2,11E+15	1,00E+15	8,95E+13
Tamping demand LDP only	*in km*	1791	273	142	120	28
	in %	**141**	**131**	**107**	**107**	**107**
RP kN^3-km		2,69E+16	4,01E+15	1,48E+15	6,98E+14	6,23E+13
Tamping demand RP only	*in km*	1244	192	99	84	19
	in %	**98**	**92**	**75**	**75**	**75**
F kN^3-km		2,41E+16	4,01E+15	2,08E+15	9,83E+14	8,80E+13
Tamping demand F only	*in km*	1112	192	140	118	27
	in %	**88**	**92**	**105**	**105**	**105**

Table 12.
Tamping demand for different track loading scenarios.

As both tamping demand and the damage according to D_1 vary over the track radius, we split this calculation into radii classes. All values are presented in **Table 11**. The incremental tamping demand in **Table 11** can also be seen as the calibration of the damage model D_1.

If we now change the transport volume, the differences between those two approaches become visible, as shown in **Table 12**. In the gross-ton-approach, it does not matter which trains generate the ton-km. Therefore, the tamping demand calculated with the incremental tamping demand based on ton-km according to **Table 11** stays constant for constant gross-ton-kilometers. In the case of the specified loading derived from the damage function D_1, the results change significantly: In the straight sections, the higher speeds of long-distance trains lead to 40% higher taping demands, whereas the lower speeds decrease the demand for the "freight trains only" scenario (note: on average, freight transport delivers lower axle loads than long-distance passenger traffic).

These results are well in line with common experience and still count, even if we consider the extremes on both ends of railway operation: slow freight traffic needs less ballast maintenance, even in case of heavy haul operation, while high-speed train operation leads to very frequent interventions in order to keep track geometry on the necessary level [30].

In mixed traffic though, these effects occur at a much lower level. To show the limited, but nevertheless existing effects, we double the transport volume in our example. For the gross-ton-approach, this simply gives double amount of ballast related maintenance. Increasing the transport volume to some 60,000 gross-tons per day by adding trains of one market segment only delivers different maintenance needs using the specific damage function D_1 (**Table 13**).

Again, we learn that maintenance demands for increasing transport volumes can be estimated sufficiently well by using gross-tonnage as long as this increasing volume consists mainly of regional passenger trains (the difference to specific damage function is less than 3%) or freight trains (the difference is some 5%). Adding faster long-distance trains add much more tamping needs than estimated by the simplified gross-ton-approach with a resulting 16% higher total tamping demand.

		R_5	R_4	R_3	R_2	R_1
Doubled gross-ton-km		1,68E+11	2,74E+10	1,42E+10	9,85E+09	1,09E+09
Tamping demand	*in km*	2538	415	266	224	51
	in %	**200**	**200**	**200**	**200**	**200**
	SUM in km			*3493*		
Doubling with LDP only kN^3-km		6,56E+16	9,95E+15	4,05E+15	1,92E+15	1,72E+14
Tamping demand LDP only	*in km*	3032	476	272	230	53
	SUM in km			*4062*		
Doubling with RP only kN^3-km		5,46E+16	8,38E+15	3,46E+15	1,64E+15	1,46E+14
Tamping demand RP only	*in km*	2524	401	233	196	45
	SUM in km			*3399*		
Doubling with F only kN^3-km		2,41E+16	4,01E+15	2,08E+15	9,83E+14	8,80E+13
Tamping demand F only	*in km*	2381	399	273	230	53
	SUM in km			*3335*		

Table 13.
Tamping demand with increasing track loading.

In order to deepen these findings, we performed another variation: instead of simply increasing the amount of tonnage of long-distance passenger trains, we introduced an additional long-distance passenger service with a speed up to 200 kmph in the sections with maximum line speed of 160 kmph (12% of the straight line-section according to **Table 6**) covering half of the trains (scenario LDP+ only, see **Table 14**). Of course, this assumes that technically the maximum line speed can be increased to this level. In this case, the tamping demand in the radii class R_5 rises by another 250 kilometers.

Usually this effect is not significantly recognized even though higher speeds in passenger long-distance services are introduced quite intensively in European mixed traffic networks. The reason for this is that introducing faster passenger services is often accompanied by establishing new lines or the total rehabilitation of existing lines. This comes along with tracks on perfect substructure and robust components

Doubling with LDP only kN^3-km		**6,56E+16**	**9,95E+15**	**4,05E+15**	**1,92E+15**	**1,72E+14**
Tamping demand LDP only	*in km*	3032	476	272	230	53
	SUM in km			*4062*		
Doubling with LDP+ only kN^3-km		7,10E+16	9,95E+15	4,05E+15	1,92E+15	1,72E+14
Tamping demand LDP+ only	*in km*	3282	476	272	230	53
	SUM in km			*4312*		
Doubling with LDP+ only kN^3-km		7,10E+16	9,95E+15	4,05E+15	1,92E+15	1,72E+14
Tamping demand LDP+ only improved Track Quality	*in km*	2894	476	272	230	53
	SUM in km			*3925*		

Table 14.
Tamping demand with increasing track loading—Scenario LDP+ only.

		Rail Surface Damage[1]	Rail Wear[2]
Doubled gross-ton-km		1,37E+11	1,37E+11
Grinding[1]/Rail Exchange[2] demand	*in km*	584	11.0
	in %	**200**	**200**
Doubling with LDP only D_4-km		1,07E+09	8,76E+08
Grinding[1]/Rail Exchange[2] demand LDP only	*in km*	653	14,6
	in %	**224**	**265**
Doubling with RP only D_4-km		1,16E+09	7,81E+08
Grinding[1]/Rail Exchange[2] demand RP only	*in km*	707	13,0
	in %	**242**	**236**
Doubling with F only D_4-km		9,13E+08	5,48E+08
Grinding[1]/Rail Exchange[2] demand F only	*in km*	558	9,1
	in %	**191**	**165**

[1]*According to damage term $D_{4.1}$.* [2]*According to damage term $D_{4.2}$.*

Table 15.
Rail maintenance demand with increasing track loading.

such as padded concrete sleeper. Such tracks generally perform better, and ballast-related maintenance is reduced by 50%. Considering this for our example, the scenario "LDP+ only, improved Track Quality" even delivers a lower tamping demand (**Table 14**).

Summarizing, we can state that the influence of the specific loading is considerably high even though it is hard to extract in top-down figures for extended mixed traffic networks.

3.1.2 Rail maintenance

Similar to the ballast maintenance, we can analyze the rail maintenance in more detail using the damage function according to Section 2, the damage mechanism D_4 for both rail surface damage in wider curves and rail side wear in tighter curves. Note: Rail grinding in tangent track is not assessed, as this is driven by the damage functions D_2 and D_3.

Differently to ballast maintenance, regional passenger trains contribute significantly to both rail surface damage and rail side wear (**Table 15**). Again, doubling the transport volume (and thus the rail maintenance following the gross-ton-approach) by increasing one market segment only, we see distinct differences.

This result is not linked to train speeds, but mainly to the vehicles in use. Rail maintenance is mainly driven by the type of bogie and the longitudinal stiffness of the car body. Therefore, we need to look deeper into different vehicles.

3.2 The influence of vehicle properties on track maintenance

As shown in Section 2, the alternative description of track loading as specific damage goes along with vehicle properties such as axle load, unsprung mass, traction power or longitudinal stiffness and also for some damage mechanisms with vehicle speed. Since the total damage is calculated as a sum of those impacts, it is in turn also possible to allocate track maintenance to single vehicles. This chapter gives examples for the influence of different vehicle properties in combination with operational scenarios. In order to sum up different track maintenance works, for this task, we use the maintenance costs. Thus, the following results are track maintenance cost per vehicle-kilometer, sometimes re-calculated to track maintenance cost per gross-ton-kilometer. As the absolute cost level varies from infrastructure manager to infrastructure manager and are moreover not of importance for this investigation, the costs are normalized. The 100% base level is a fully laden 4-axle freight car with Y25-bogies (axle load 20.5 tons) running at 90 kmph on a straight track. The track is according to the network configuration in Section 3.1 a ballasted concrete sleeper track on good subsoil with 60E1-R260 rails. Note: In order to keep results comprehensible, we do not add any costs of turnouts so that the damage mechanisms D_5 and D_6 are not added. Also, D_7 is not addressed as we model track maintenance only, without incorporating renewals.

To better understand the results, it is essential to know the track maintenance expenses allocated to the damage mechanisms. According to the network configuration and the superstructure parameters, track maintenance costs split into the percentages displayed in **Figure 3**. We see that the overwhelming part of the costs (85%) is triggered by the damage mechanism D_1, the dynamic vertical forces. These costs

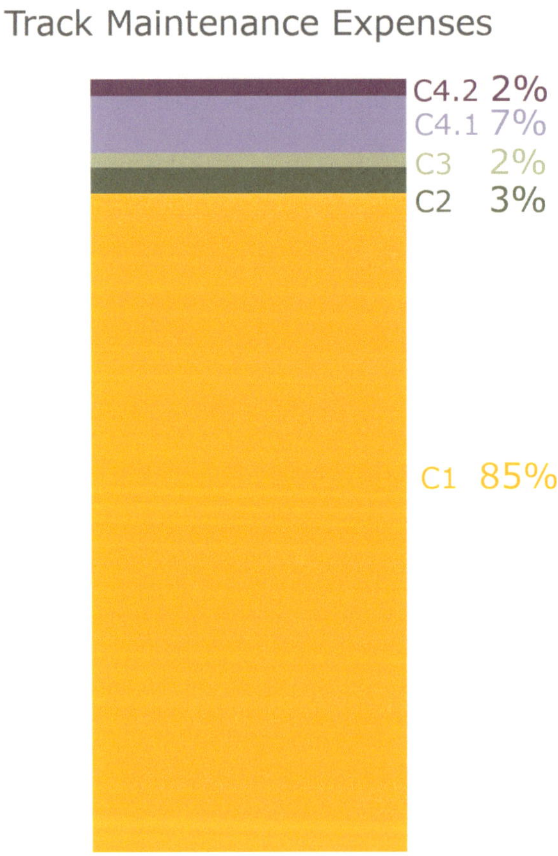

Figure 3.
Track maintenance expenses.

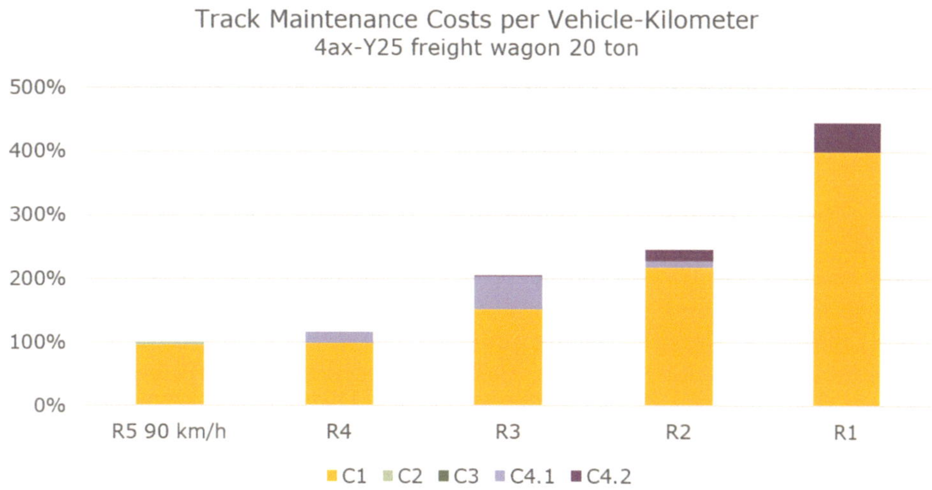

Figure 4.
Track maintenance costs per vehicle-kilometer—4ax-Y25 freight wagon_20 t.

contain mainly ballast-related maintenance and small reactive maintenance. In this network, some 12% of track maintenance costs are due to rail surface damage (D_2, D_3 and $D_{4.1}$) and thus rail grinding or milling costs. Only 2% of the costs are the consequence of rail side wear in curves.

Looking at the reference vehicle (**Figure 4**), the heavy freight car, we learn that ballast-related maintenance costs increase with decreasing track radius. Moreover, rail surface maintenance costs in straight tracks are very low and increase with decreasing track radius to be finally replaced by rail exchange costs (rail wear) in the smallest radius (250 m).

Locomotives and generally powered axles come along with higher unsprung masses and in addition with traction forces. The latter deliver a significant contribution to the rail surface damage in straight sections (damage D_3). **Figure 5** shows both

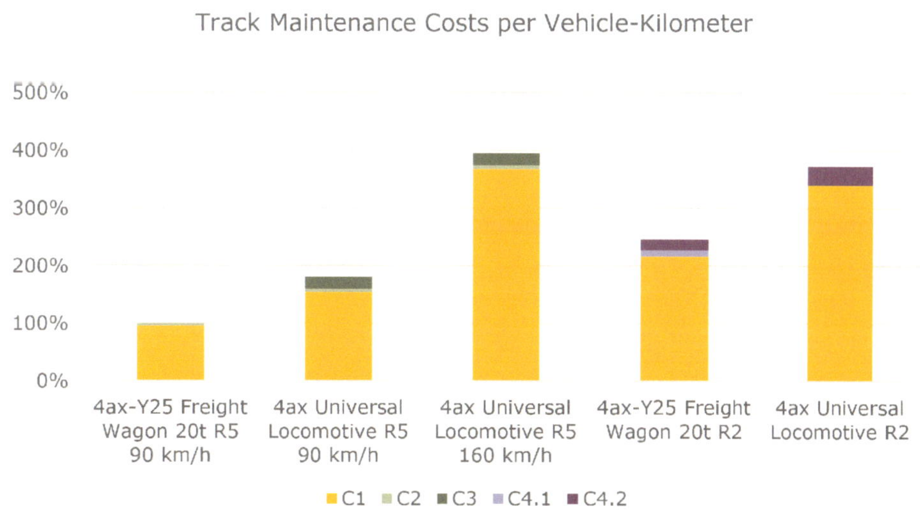

Figure 5.
Track maintenance costs per vehicle-kilometer—Powered and unpowered axles.

effects directly compared to the unpowered axles of the freight wagon. Both vehicles have the same tonnage but contribute very differently to the track maintenance needs. This is in accordance with [16]. The third column shows the influence of speed: Due to the high additional dynamic forces, the allocated track maintenance costs double when operating the same loco at 160 kmph instead of 90 kmph.

The high differences between the allocated maintenance costs in straight and curved track originate mainly in the maintenance needs themselves. Track maintenance in tight curves reaches much higher levels than in straight sections (see [1]). In a network, curves form only a small part. Looking at the network used in this paper as reference (**Table 6**), only 24% of the track show radii below 1000 m. The wear formula allows for calculating associated track maintenance costs as a consequence of track radius and speed level. For this example, we use nine different operational situations (four radii classes and five speed levels according to **Table 6**). In our simplified example, all vehicles run along all lines exactly in the distribution of these operational situations. Thus, we can calculate the average track maintenance cost per average vehicle-kilometer (vehicles according to **Table 9**) and sum up how the single vehicles to trains deliver results as shown in **Figure 6**.

These trains have different weights and cannot be compared directly. This counts especially for the light electrical multiple units and the heavy freight trains. We thus re-calculate the train costs to gross-ton-kilometers, simply by dividing by the train weight (see **Table 9**). **Figure 7** shows the damage impact of one gross-ton-kilometer on the track: Again, the long-distance passenger train is the 100% level. The regional trainsets deliver 20 to 40% less track damage. This is due to the lighter axles, less unsprung masses and the lower speeds compared to the fast intercity-train. The freight train has similar axle loads on average compared to the long-distance passenger train and the same loco, but runs at 100 kmph maximum so that the impact per gross-ton-km is 31% lower. **Figure 7** also shows the gross-ton-kilometer approach with the dashed light gray bars: In this approach, all these trains are meant to cause the same track damage.

These first examples compare very different trains and/or vehicles. It is not surprising that the track damage caused by these vehicles differ widely. But these differences occur for similar vehicles in the same way. We looked at locos with similar total weights (which is always close to the maximum allowed weight of 90 tons), but very

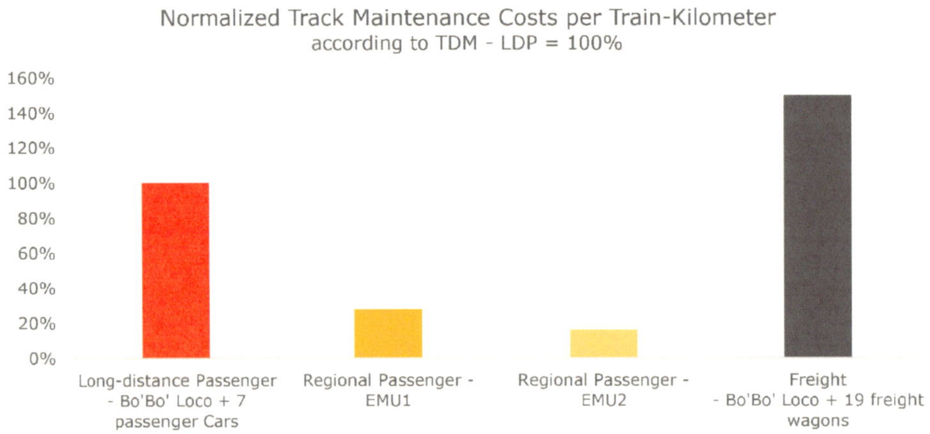

Figure 6.
Track maintenance costs per train-kilometer for different trains.

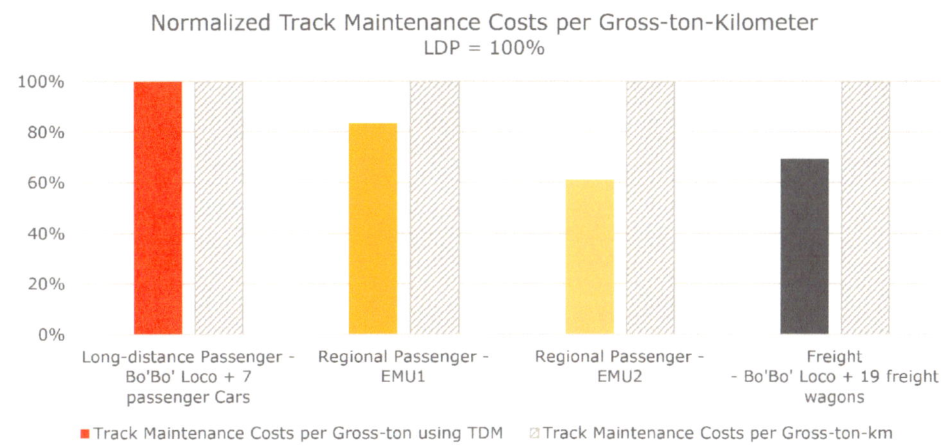

Figure 7.
Track maintenance costs per gross-ton-kilometer for different trains.

differing unsprung masses, and bogie stiffnesses. In **Figure 8**, we see differences up to 65% for operating the locos on passenger trains with speeds up to 160 kmph in the artificial network. In freight operation, the differences are much lower, 20% maximum.

From a track maintenance point of view, the use of track-friendly vehicles is of course preferable. However, the selection of vehicles or even train concepts focuses on many other aspects. We highlight the use of different train concepts as our next example: As an alternative choice for the long-distance passenger train consisting of the loco and seven wagons, an electrical multiple unit enabling higher speeds is analyzed. This trainset comes along with distributed powered axles, comparably high axles loads as the trainset is a double-decked unit and a total weight of some 400 tons. In **Figure 9**, we see three different options for performing a comparison. The red line demonstrates the allocated track maintenance costs of the loco hauled passenger train for the different radii and speed ranges (100% is the cost at speed level S_5, 150 kmph).

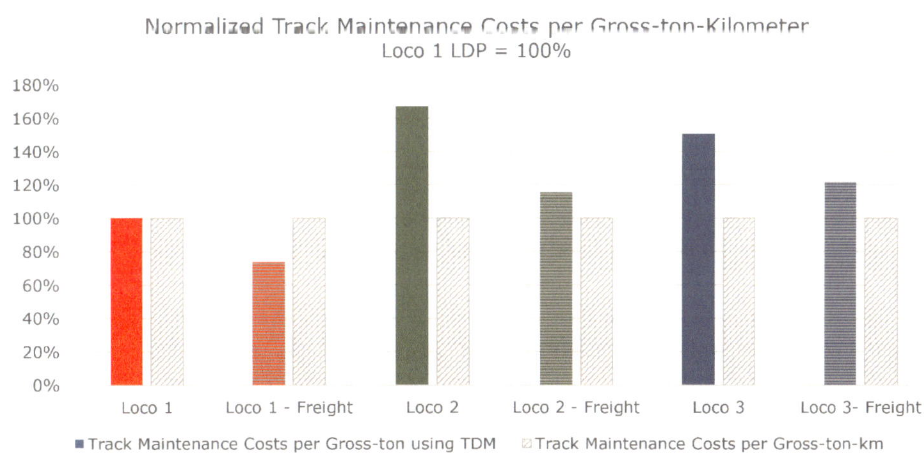

Figure 8.
Track maintenance costs per gross-ton-kilometer for different locos.

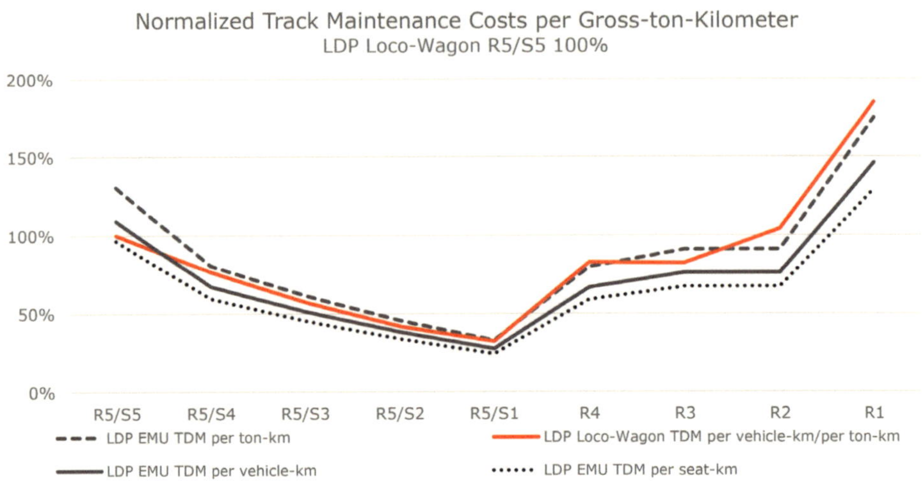

Figure 9.
Track maintenance costs per gross-ton-kilometer for different long-distance passenger trains.

This red line is the same for all three comparisons. Initially, we look at the track maintenance cost per vehicle-kilometer (black solid line). The EMU causes some 20% lower damage in curves, only slightly lower damage in straight sections with moderate speeds, and only at the highest speed (velocity range 5) higher track maintenance cost (plus 9%). As the trainset is lighter than the push/pull train, we again re-calculate the cost to gross-ton-level (dashed black line). We learn that the impact of these two trains is very similar on gross-ton-level; in the highest speed range the inserted damage of the EMU is 30% higher. The third possibility for comparing different passenger train configurations is to refer the track maintenance cost caused to seat-kilometers. In this option, the double-decked trainset option performs best: As the dotted line in **Figure 9** indicates, the difference is remarkable and reaches values up to 35% less track maintenance costs caused.

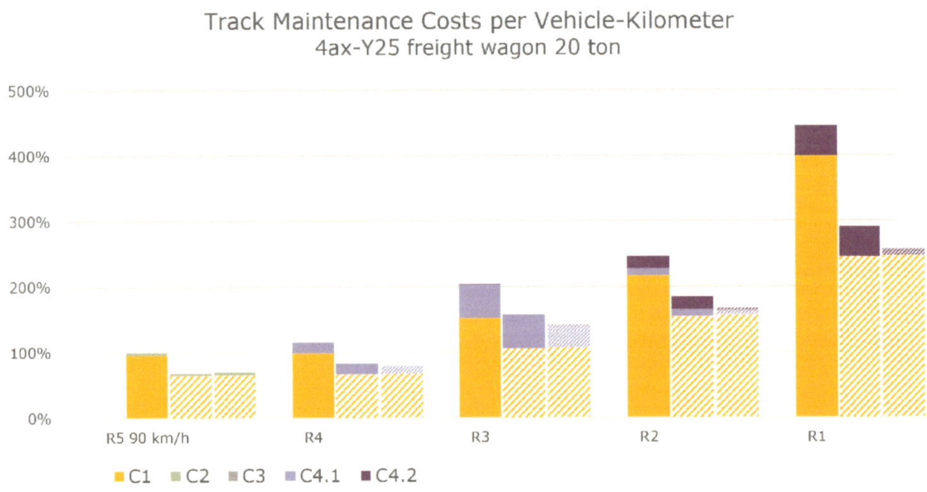

Figure 10.
Track maintenance costs per vehicle-kilometer—4ax-Y25 freight wagon_20 t with varying track components.

Concluding, we want to underline that not only loading triggers track maintenance. Improving track structure and installing innovative, robust components helps to drop the maintenance demands—much more than track-friendly vehicle concepts. **Figure 10** shows, for example, the impact of using concrete sleepers with under sleeper pads [31, 32] on the ballast-related maintenance costs (dashed yellowish bars) and on top, for example, a rail steel grade of R350HT in curves (dashed purple bars—rail grinding and rail exchange). The track maintenance cost decreases by more than 40% using these components in curves and still by one third in straight sections where higher rail steel grades do not reduce maintenance demands.

4. Conclusions

Track loading is a main trigger for track maintenance needs. Gross-tonnage is widely used, but cannot cover the different damage mechanisms properly. Damage processes on rails and in ballast are the most relevant causes for track maintenance. Dynamic vertical forces created by unevenness of track can be seen as loading for the ballast-related deterioration process, while rail surface damage and rail wear needs a more holistic description of the loading. Existing track deterioration models still need improvement, especially for new rail surface failures and for turnouts, but deliver a much more detailed insight on both track maintenance demand for changing traffic compositions and vehicle properties triggering track maintenance. Following this approach of a track deterioration model based on vehicle properties will help to forecast necessary track maintenance and the associated budgets much more precisely than the simplified gross-ton approach. The possibility to calculate track maintenance costs on vehicle basis enables of course also to allocate these costs much better in track access charge schemes [21, 33, 34]. This would support the construction and use of track-friendly vehicles and thus to reduce the costs of railway operation.

Acknowledgements

The research was funded by ÖBB-Infrastruktur AG and SBB.

Author details

Stefan Marschnig* and Ursula Ehrhart
Graz University of Technology, Graz, Austria

*Address all correspondence to: stefan.marschnig@tugraz.at

References

[1] Marschnig S, Veit P. Assessing average maintenance frequencies and service lives of railway tracks—the Standard Element Approach. In: Hessami AG, editor. New Research on Railway Engineering and Transport. London: IntechOpen; 2023

[2] Bente H, Kindler T, Wittmeier K. (BSL Management Consultants): Network Rail: Rail Infrastructure Cost Benchmarking. Hamburg/London (UK); 2008

[3] Loidolt M. Costs of operational hindrances: Reduction of railway system costs by means of a holistic approach. Internationales Verkehrswesen. 2020;**72** (3):48-50

[4] Marschnig S, Veit P. Optimierte Einheitskosten - Sperrpausendauer und Baustellenlänge. ZEVrail. 2008;**134**:10

[5] Praneeth C et al. Train-based differential eddy current sensor system for rail fastener detection. Measurement Science Technology. 2011;**30**:125105

[6] Andrade AR, Teixeira PF. Biobjective optimization model for maintenance and renewal decisions related to rail track geometry. Transportation Research Record. 2011;**2261**(1):163-170. DOI: 10.3141/2261-19

[7] Taufan A et al. Improving the performance of railway tracks through ballast interventions. Proceedings of IMechE Part F: Journal of Rail and Rapid Transition. 2018;**232**(2):337-355

[8] Schoech W, Heyder R, Grohmann H-D. Contact geometry and surface fatigue – Guidelines for appropriate rail maintenance. In: 7th International Conference on Contact Mechanics and Wear of Rail/Wheel Systems (CM2006); Brisbane, Australia

[9] Munduch, G. et al.: Estimating Marginal Costs for the Austrian Railway System; Vienna (Austria), 2002

[10] Andersson M. Marginal Cost of Railway Infrastructure Wear and Tear for Freight and Passenger Trains in Sweden; Borlänge (Sweden), 2011

[11] European Commission: Commission implementing Regulation (EU) 2015/909 of 12 June 2015. On the modalities for the calculation of the cost that is directly incurred as a result of operating the train service; Brussels (Belgium), 2015

[12] Marschnig S. iTAC – Innovative Track Access Charges. Monographic Series TU Graz / Railway Research, Band 1. Graz: Verlag der Technischen Universität Graz; 2021

[13] Federal Railroad Administration: Track and Rail and Infrastructure, Integrity Compliance Manual, Volume II Track Safety Standards, Chapter 1 Track Safety Standards Classes 1 through 5, 2014

[14] Federal Railroad Administration: Track and Rail and Infrastructure, Integrity Compliance Manual, Volume II Track Safety Standards, Chapter 1 Track Safety Standards Classes 6 through 9, 2014

[15] EN 15528:2021: Railway applications—Line categories for managing the interface between load limits of vehicles and infrastructure, 2021

[16] International Union of Railways (UIC): UIC Code 714 E—Classification of lines for the purpose of track maintenance, Paris (France), 2009

[17] Group Standards Railway Technical Centre. Group Standard Railway Group Standard GC/RT5023 Track Standards

Manual - Section 1: Basic Track Category Matrix; London (United Kingdom), 1995

[18] Worley B, Burstow M. The Variable Usage Charge (VUC) in CP5; London (United Kingdom), 2014

[19] Serco Commercial: VTISM Analysis to Inform the Allocation of Variable Usage Costs to Individual Vehicles; Cheshire (United Kingdom). 2013

[20] Burstow MC. A Whole Life Rail Model Application and Development for RSSB – Continued Development of an RCF Damage Parameter, AEATR-ES-2004-880 Issue 2; 2004

[21] Bundesamt für Verkehr BAV Abteilung Finanzierung: Basispreis Verschleiss im Trassenpreis, Anleitung für die Fahrzeugpreisbestimmung. 2022. Available from: https://www.bav.admin.ch/bav/de/home/verkehrsmittel/eisenbahn/fachinformationen/trassenpreis.html

[22] Jenkins HH, Stephenson JH, Claxton GA, et al. The effect of track and vehicle parameters on wheel/rail vertical dynamic forces. REJ. 1974;**1974**:1

[23] GM/TT0088, Permissible track forces for railway vehicles, Issue 1, Revision: A. 1993, 11 p

[24] ORE: D 161.1 RP4 the Dynamic Effects Due to Increasing Axle Loads from 20 to 22.5 t, Utrecht (Nederland). 1987

[25] ORE D141: Einfluss der Erhöhung der Radsatzlast von 20 auf 22 t auf den Oberbau. RP5: Untersuchung der technischen und wirtschaftlichen Auswirkungen einer Erhöhung der Radsatzlast von 20 auf 22 t; Utrecht (Niederlande). 1982

[26] Zhao X, Li Z, Dollevoet R. The vertical and the longitudinal dynamic responses of the vehicle–track system to squat-type short wavelength irregularity. Vehicle System Dynamics. 2013;**51**(12):1918-1937. DOI: 10.1080/00423114.2013.847466

[27] Nerlich I, Holzfeind J. Wie wirtschaftlich ist gleisschonendes Rollmaterial ... wie einen Verkehr von morgen in der zukünftigen Auswirkung auf Gleisinstandhaltungen beziffern? ZEVrail. 2014;**2014**:138

[28] Steenbergen M. Quantification of dynamic wheel–rail contact forces at short rail irregularities and application to measured rail welds. Journal of Sound and Vibration. 2007;**312**(4):606-629

[29] Grote K-H, Feldhusen J. Taschenbuch für den Maschinenbau. 21st ed. Berlin: Springer; 2005

[30] Marschnig S, Vidocic I. Estimating future tamping demands using the Swiss Wear Factor. In: Proceedings WCRR 2019; Tokyo, Japan. 2019

[31] Marschnig S, Berghold A. Besohlte Schwellen im netzweiten Einsatz. Eisenbahntechnische Rundschau. 2011;**5**:10-12

[32] Marschnig S, Ehrhart U, Offenbacher S. Long-term behaviour of padded concrete sleepers on reduced ballast bed thickness. Infrastructures. 2022;7:132. DOI: 10.3390/infrastructures7100132

[33] Marschnig S. Innovative track access charges. In: Proceedings of 6th Transport Research Arena, April 18–21 2016; Warsaw, Poland. 2016

[34] De Vries R, Sroba P, Magel E. Preventive grinding moves into the 21st century on Canadian Pacific Railroad. In: Proceedings AREMA 2003 Annual Conference; Chicago, IL. 2001

Section 7

Control Command and Communications

Chapter 13

Distributed Optical Fiber Sensing in Railway Engineering

Muhammad Adeel, Aadil Raza and Muhammad Muaz

Abstract

There are many technologies associated with optical fiber sensing (OFS) and depending upon the type of application, a specific OFS technology plays a crucial role in the associated application as compared to the use of conventional sensing technologies with these applications. The same is true with the railway industry and the two most suitable OFS technologies in the railway sector are distributed acoustic sensing (DAS) and fiber Bragg grating (FBG). The two mentioned technologies in association with the railway industry are explained briefly in this chapter.

Keywords: distributed optical fiber sensing, distributed acoustic sensing, Brillouin optical-fiber time domain reflectometry, Brillouin optical-fiber time domain analysis, Raman-based distributed temperature sensing

1. Introduction

The services, such as cargo and mass traveling using the railway infrastructure, have undergone tremendous progress in recent years due to their reliability and safety in comparison to other modes of transportation. However, there is a possibility of degradation in both the railway vehicles (trains/trams) and infrastructure (track lines and the associated hardware) due to an increase in the speed limit, especially in the high-speed railway sector. This degradation normally leads to an increase in accidents. In under and non-developed countries, these accidents are inevitable due to the high cost of maintenance requirements for railway vehicles and infrastructure. Therefore, both the rolling stock and vehicles require reliability and low-cost operational monitoring.

In all of the mentioned metrics, infrastructure and vehicle are the two subjects, the attributes of which can be measured. The conventional operational monitoring systems for the mentioned subjects are usually cost-effective, but this cost-effectiveness is at the expense of unreliability if compared to the current optical fiber sensing (OFS) system. A dedicated monitoring (either operational or traffic management) technique can be used to measure one of the mentioned subjects which is more expensive than using a single method to measure both subjects with reliability. With these factors in mind, conventional sensing techniques in the railway sector can be regarded as expensive. Other than addressing a single metric and unreliability, there are many other aspects of the conventional techniques that make these methods expensive when compared to OFS. For example, a very high amount of electricity is consumed with the track circuits (TC) technique, or a very large number of sensors are normally installed to cover the operational monitoring of both the vehicles and infrastructure in the case of communication-based train control (CBTC).

A reliable sensing system refers to a number of factors, including the avoidance of dead sensing points, ease of installation, resistance of electromagnetic interference (EMI), and avoidance of safety hazards. Compared with the distributed optical fiber sensing (DOFS) technologies, the conventional railway-oriented sensing techniques are unreliable because there is no such method used by these techniques for measuring the distributed sensing, and hence there is a possibility of a large number of dead points leading to the provision of unreliable services. To avoid the possibility of dead points, there are chances of installing a myriad number of wireless sensors, but the energy requirements are challenging to meet the objectives. If wired sensors are installed instead, the number of power and communication lines will be very large, and the system installation may be impractical. Moreover, a long-term testing and assessment procedure of a fiber optic-based railway infrastructure is possible due to the possibility of high spatial resolution based on distributed and quasi-distributed nature of sensing. Such a distributed sensing system removes the necessity of power and data cables because it serves a dual purpose; sensing system with no active power requirement and as a communication system, thus reducing the cost tremendously for tens of kilometers distance span. Therefore, the difficulty of installing a myriad number of sensors in conventional sensing is less competitive than the OFS-based solution for both cost and reliability. EMI is another issue in electrical-based sensing systems, especially in very high voltage pantographs, which can be avoided with OFS as the light is the only signal passed through optical fibers. To avoid the difficulty of providing dedicated power and communication lines, TC provides a reasonable solution to cut off the additional power lines. However, this technique does not meet the standard safety requirements, which leads to an extreme sense of unreliability in case of avoidance of safety hazards.

Any railway system requires technology to help in the operational monitoring, train traffic management, and an additional amount of data for postaccident investigation. Operational monitoring is actually the structural health monitoring of a railway system, and it is based on the investigation of both the vehicle and its infrastructure. In the case of railway vehicle traffic management, any sensing solution may provide the instantaneous location, instantaneous speed detection, and live tracking of the vehicle, and these parameters are enough to control and manage the overall traffic. As the number of accidents is increasing in economically poor countries, the third benefit that any railway system can take from the technology is the acquisition of enough data, which can be beneficial in case of postaccident investigation, which is helpful to know the potential reason for the incident to avoid in the future. Other than manual inspection, several methods have been adopted so far to automate the operational monitoring of railway vehicles and infrastructure. These methods include CBTC, TC, wheel counters, track recording cars, and onboard operational monitoring.

The data, acquired in the case of conventional railway-based sensing systems, is discrete with a relatively large interval among spatial data samples if compared with any OFS-based technology. Any technology associated with DOFS sensing is normally distributed in nature, whereas there are also discrete sensing technologies in OFS such as fiber Bragg gratings (FBGs) and interferometry-based sensing solutions. However, there is a possibility of alleviating the distance between adjacent point sensors in the case of discrete or quasi-distributed OFS in a handful of ways if compared to the conventional sensing solution. Therefore, acquiring data with minimal distance among spatial locations is possible with OFS, and this large number of data has numerous other benefits.

A discrete sensing with a relatively large interval among sensing points may help in the case of operational monitoring but in the case of traffic management and postaccident analysis, this is surely a poor solution. The reason that continuous sensing or discrete sensing with relatively short intervals among sensing points well suits traffic management and postaccidental analysis is that data is critical at every possible point in these applications. For example, instantaneous speed and position determination are critical in the case of traffic management due to the high speed of railway vehicles. Missing data due to dead points may create severe problems, which may lead to a train-to-train collision. Moreover, postaccidental analysis requires data separated spatially in close vicinity in order to make the investigation possibly easy. Therefore, the sensing points, located spatially in close vicinity, can be used for any of the application that falls under any of the three groups of applications, as discussed. However, keeping the reliability factor in mind one may target only the applications that fall under the category of operational monitoring if the sensing points are not closed enough. This conclusion leads to the fact that the selected OFS technologies provide an edge due to their capability of spatially close sensing points or distributed sensing nature if compared with conventional sensing technologies.

Over the last two decades, the OFS-based smart railway infrastructure has made tremendous progress in view of research and development as optical fiber is always preferred due to its lightweight, reliability, and cost-effectiveness. The additional benefits of these systems include their lightweight nature and the possibility of distributed sensing solutions. In the case of train tracks, there is no possibility of data disruption while the vehicle passes within the tunnels as it normally occurs in case of the navigation systems such as the Global Positioning System (GPS) and Global Navigation Satellite System (GLONASS), etc. Unlike the conventional sensing technologies in the railway sector, which measure only the frequency spectrum, the technologies associated with OFS not only measure the frequency spectrum but also the true phase of the time-based signals. Therefore, to prevent catastrophic failures and early failure detection, the real-time sensing nature of OFS-based systems is a better option if compared with the conventional sensing system. In short, OFS can be regarded as the only reliable and cost-effective solution for smart railways. Other than management system, operational inspection, and providing ease of postaccidental investigation, some technologies in OFS can also provide intrusion detection and trespassing monitoring.

In summary, OFS-based sensing systems are not only capable of targeting both the subjects (railway infrastructure and vehicles) with a single sensing system but also provide cost-effectiveness and reliability in addition to other benefits. Therefore, in comparison to the conventional operational monitoring techniques, OFS-based systems can be considered to be more suitable with the provision of long-term solutions for several types of maintenance and other miscellaneous monitoring for both the vehicle and infrastructure.

2. Optical fiber sensing

Light travels within the optical fiber and constrains itself in the medium due to the total internal reflection mechanism of light signals. This total internal reflection is possible if the light tries to pass from the highly dense medium to a lower one and the concept of this principle is depicted in **Figure 1**. **Figure 1** shows that a deflection away from the reference occurs if the light signal passes from a highly dense medium to a low dense medium. If the incident signal surpasses the critical angle (θ_3 in our case), the total internal reflection occurs. The same figure shows the total internal reflection occurs at the incident angle θ_5. Within an optical fiber, the light traveling follows the principle of total internal reflection. The two main components of an optical fiber are a core and a cladding, as depicted in **Figure 2**. The core is the highly dense medium, whereas the cladding is a low dense medium. Light is injected from one end of the optical fiber and due to total internal reflection, the light retains within the core until it reaches the end of the optical fiber.

The conventional vehicle and track monitoring systems are mostly based on electrical principles. The main use of optical fibers is in the telecommunication industry. However, the idea of OFS in railway systems was engendered to avoid EMI, which is considered to be a more challenging impairment than any other drawback of electrical-based systems. Later on, many other OFS-based solutions were used in comparison to the conventional sensing techniques in the railway industry. By that time, the OFS solution got popular, and many challenging issues of the railway infrastructure were resolved with the OFS solutions, especially in the last two decades.

DOFS is a specialized group of technologies under OFS, that provide sensing at any location of the fiber, where needed and the sensing can be done simultaneously at

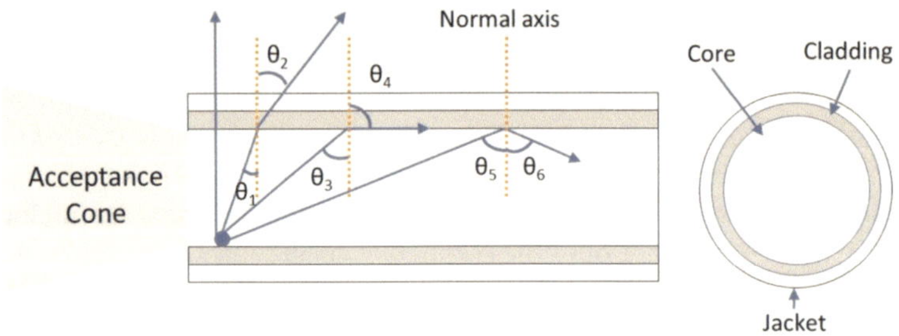

Figure 1.
A demonstration of total internal reflection.

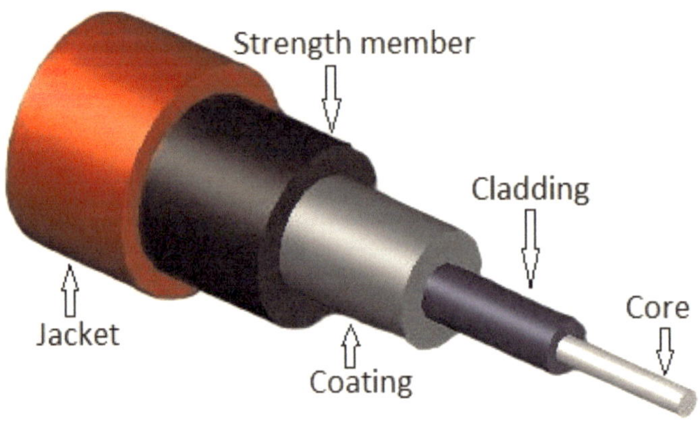

Figure 2.
Cross-section of an optical fiber cable.

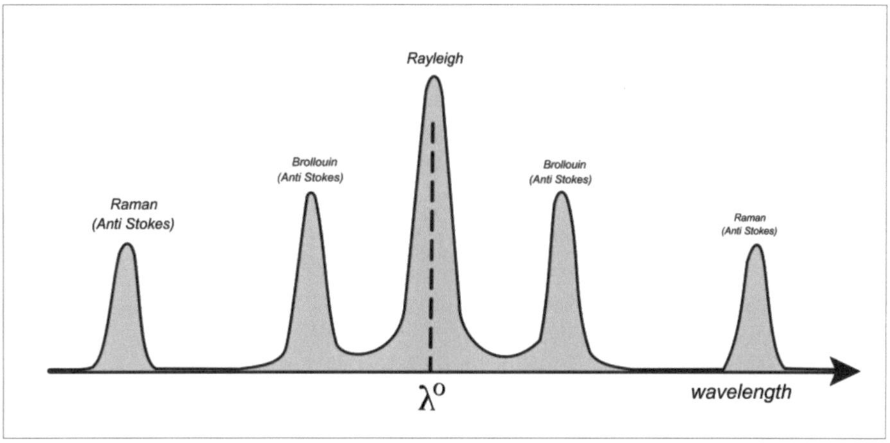

Figure 3.
Typical operating spectrum of different distributed optical fiber sensing technologies.

multiple locations. The DOFS-based technologies are grouped based on three types of light wave scattering. These are Rayleigh scattering, Brillouin scattering, and Raman scattering [1]. The frequency band of each of these scattering types is different, and an overview of these bands is depicted in **Figure 3**.

All of these scattering types are comprised of many other sensing technologies. For example, the technologies associated with the Rayleigh scattering include distributed acoustic sensing (DAS) or phase optical time-domain reflectometry (ϕ-OTDR), and polarization-OTDR. In regard to the railway industry in the last decade, the focus in the railway sector is mainly on DAS as compared to other Rayleigh-based scattering technologies.

DAS can be used in distributed and quasi-distributed nature sensing, and there is a possibility of dynamic stress sensing for both the railway vehicles and the infrastructure. Moreover, due to the possibility of supporting a very long range, train tracking, and operational monitoring have achieved a high level of maturity in the research field associated with DAS technology for its use within the railway sector. On the other hand, Brillouin scattering-based DOFS technologies include Brillouin optical time domain reflectometry (BOTDR), Brillouin optical time domain analysis (BOTDA), and Brillouin optical frequency domain analysis (BOFDA). However, due to the hybrid sensing capability of Brillouin scattering-based technologies can be rarely used in railway infrastructure monitoring as the high temperature of the railway tracks may affect the measured stresses. Beside, Raman scattering-based sensing technologies are not very popular in the railway industry due to their only sensing capability for temperature and relatively less response to a rapid change in a measured quantity. Moreover, Raman scattering-based sensing technologies only measure the temperature in a distributed manner, but this technology can be expensive for use in the railway sector. Many other OFS-based technologies, such as interferometry-based sensing techniques, can be regarded as unreliable in the railway industry due to their nonlinear nature, and hence cannot be used to meet the required objectives. One of the advanced interferometry-based sensing technologies is the optical frequency domain reflectometer (OFDR), which is not popular in the railway sector due to its short-distance applications. It is due to these reasons that the use of FBGs and DAS technologies is dominant in the railway sector, and more. than 95% of the OFS-based research articles are based on these two technologies. Therefore, this chapter provides insight into both the FBGs and DAS technologies for their extensive use within the railway industry.

2.1 Fiber Bragg gratings in railways

FBGs are the best quasi-distributed alternative in the OFS field, and their use has become quite popular within the railway industry in the last two decades. As the FBGs can support a very long range of distances, and hence this technology has attracted the researchers' attention in the railway industry. Moreover, the use of any FBG type can provide the dynamic sensing capability, which has attracted its interest in the railway industry.

With distributed sensing, the whole optical fiber can act like an array of continuous sensors at virtually no gap among the sensing points. Moreover, the same fiber can be used as a communication medium, and hence the ease of installation make these sensors a favorite choice. On the other hand, the quasi-distributed sensing points at a relatively low gap among the sensing points can be achieved with ease of installation with the help of FBG. These sensors are normally the best alternative to the conventional sensing used within the railway sector. FBGs are formed with the help of an intense optical interference pattern within a fiber core. These patterns are formed such that the grating acts like a periodic perturbation of the refractive index. The gratings can perform many functions such as filtering, diffraction, and reflection. However, the most important property of these sensors is the reflection of incident light waves according to a predefined wavelength. The FBGs can be embedded within the optical fiber at discrete positions, as shown in **Figure 4**. These sensing points are generally comprised of a periodic modulation of the refractive index, which is normally embedded in the core of a single-mode optical fiber. There are two types of these gratings, uniform and nonuniform grating. The phase fronts of the uniform grating are usually vertical to the longitudinal axis of the fiber. After the light strikes along the grating plane, it gets scattered along the core part of the fiber, as shown in **Figure 4**. The period of modulation index plays a very important role in controlling

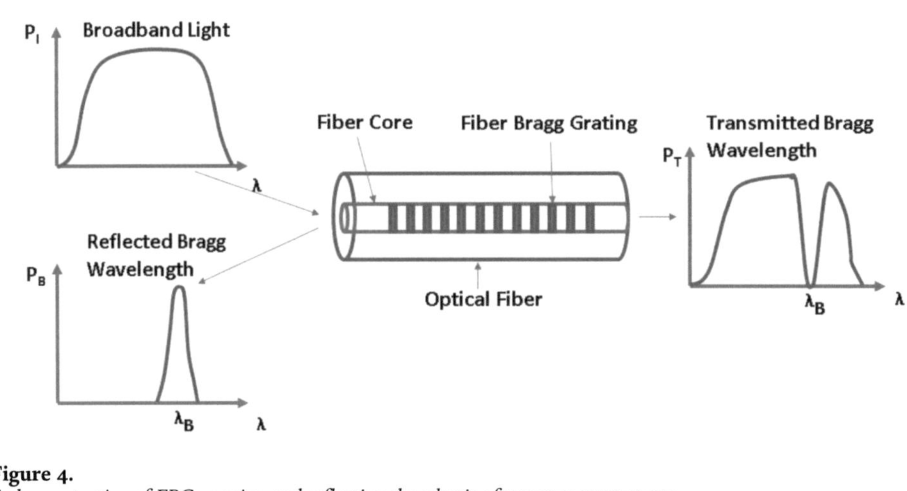

Figure 4.
A demonstration of FBG, passing and reflecting the selective frequency components.

the width of the frequencies scattered by the grating, and this index is represented by Λ, which can be defined as

$$\Lambda = \frac{\lambda_B}{2n_{eff}} \tag{1}$$

Where λ_B is the Bragg wavelength of the input light, which is back-reflected from the grating plane, and n_{eff} is the effective index of the optical fiber.

The important aspect of the grating period is that the wavelength of the reflected light is modified in accordance with the grating period due to a change in external environmental effects such as temperature or applied stress. It is due to this reason that a change in the external environmental conditions is proportional to a change in the refractive index of the core and the modulation index Λ, and this, in turn, provides a proportional change in temperature and applied stress. In other words, a change in both the external variables (stress and temperature) brings an offset to the Bragg wavelength λ_B, and the whole process is demonstrated in **Figure 4**. This figure depicts that the light waves are reflected according to the predefined wavelength gratings, and hence pass the remaining wavelengths of light. The reflected wavelength is processed such that the environmental effects such as strain or temperature are numerically acquired, and hence detection of a change in wavelength provides useful information about the measured variable.

The use of FBGs is becoming quite popular in the railway industry in regard to the monitoring and inspection of both the vehicle and track and the previous research shows a great deal of work in this area than DOFS as far as the railway sector is concerned. FBGs are popular in railway operational monitoring due to their small weight and volume, ease of handling, high spatial resolution, high precision and accuracy in the numerical results, and the capability to multiplex multiple signals simultaneously due to the wave division multiplexing capability of the fiber optics. By using a spacing between each FBG sensing point of less than 1 km, it is possible to extend the measurement range up to 100 km.

Literature reveals that FBGs in the railway sector have been used for the first time in the year 2004 for derailment detection [2] and axle counting [3]. Since then, these sensors have been used in a myriad of different applications in the railway infrastructure and the associated vehicles. The derailment detection was used as a metric for identifying different train types, whereas axle counting was used as a metric to detect the vehicle speed. After the initial steps were taken, the railway sector used FBGs in a number of applications such as the detection of defective wheels [4, 5]. These sensors were installed in the vicinity of the sleepers for health monitoring of both the rails and the vehicle's wheels status and proved the metrics such as "infrastructure measuring infrastructure metric" and "infrastructure measuring vehicle metric." The two metrics measurements were possible with the help of elegant signal processing techniques, which was otherwise impossible without the use of FBGs. Moreover, the train moving direction and axle load were measured along with many other important parameters in [6]. Beside the rail and infrastructure monitoring, the FBGs were utilized by bonding them to the railway tracks in order to measure speed and track the moving vehicles along the whole distance [6–8]. Beside, the innovative FBG interrogator has been designed to measure the train speed and axle load with a minimum possible number of FBG sensors [6]. Due to the ease of installation of FBG sensors and systems, the applications of these sensors in the railway sector did not end till the stage of limited monitoring as we can see in the case of conventional sensing systems.

These sensors can be installed in brack blocks, wheels, axles, and bogies in order to complete a composite sensing system, as installed by Mi in [9]. Such a composite sensing system can provide a complete monitoring system, which was otherwise impossible without the use of FBGs. Aside from these, different force types such as longitudinal and vertical forces on the railway tracks are possible with FBG [10, 11]. To filter out the impact of temperature from the strain measurements, efforts were made in [12], and hence there is a possibility of a provision of pure strain measurements even in the harsh environment where a large variation in temperature is possible. The benefits of FBGs in the railway industry are not limited to the mentioned applications. Vibration measurement in geogrid-reinforced ballast and unreinforced ballast along with the lateral displacement were made [13]. Moreover, FBGs can be utilized for the differential settlement of railway tracks [14]. Additionally, the operation monitoring/inspection of switchblades, fishplates, and stretcher bars is possible with the help of FBGs [15]. An important part of the railway infrastructure is the railway bridges. The effects of transverse vibration, dynamic load bending, and vertical deflection of the railway bridges were inspected in [16]. FBGs were proven to replace the electrical based sensing systems such as strain gauges, and this was verified by inspecting the vertical acceleration and contact force in pantograph catenary [17]. **Table 1** refers to different railway-specific applications associated with the use of FBGs.

S. no	References	Year	Work purpose
1	[18]	2018	Fiber-optic Bragg sensors for the rail applica-tions
2	[19]	2022	Smart railway traffic monitoring using fiberBragg grating strain gauges
3	[10]	2015	Longitudinal force measurement in continuouswelded rail with bidirectional FBG strain sensors
4	[20]	2016	The longitudinal force measurement of CWRtracks
5	[21]	2011	Strain measurements and axle counting inhigh-speed railway applications
6	[22]	2022	Discriminative monitoring of seamless railforce
7	[2]	2004	Derailment detector
8	[3]	2004	Axle counting
9	[8]	2016	Railway track operational monitoring
10	[11]	2012	Structural health monitoring of railway tracks
11	[12]	2017	Monitoring and early warning system of high-speed railway track operational monitoring
12	[6]	2016	Railway infrastructure operational monitoring
13	1	2016	High-speed railway operational monitoring
14	[23]	2016	High-speed railway operational monitoring
15	[24]	2012	Real-time railway traffic monitoring
16	[4]	2013	Wheel flat detection in high-speed railwaysystems
17	[7]	2007	Railway infrastructure operational monitoringand train tracking
18	[25]	2018	Railway operational monitoring system
19	[26]	2005	Bridge load measurement under the movingtrain load
20	[27]	2007	Railway track operational monitoring

S. no	References	Year	Work purpose
21	[28]	2012	Performance analysis of peak tracking tech-niques
22	[9]	2014	Composite railway health monitoring system
23	[13]	2015	Rail track deformations monitoring
24	[14]	2016	Railway track differential settlement monitor-ing
25	[15]	2016	Novel liquid-based FBG design for sensitivityand temperature range enhancement
26	[29]	2012	Design of a novel FBG-based sensing systemfor train vibration and weight measurements
27	[30]	2014	First work to conduct continuous structuralhealth monitoring systems on railway pantographs
28	[5]	2015	Rail and wheel wear monitoring by acquiringthe data for a very long time (6 months)
29	[17]	2013	FBG-based sensing system was proved tooutperform in comparison with conventional sensing systems in high voltage pantograph-catenary monitoring systems.

Table 1.
Applications of FBGs in references to previous work.

2.2 Distributed acoustic sensing in railways

DAS works under the principle of ϕ-OTDR, and it is a Rayleigh scattering-based DOFS technology used to sense the vibrations and perturbations at a regular or selected spatial point along the entire length of the optical fiber. Each sensing point is analogous to thousands of hydrophones connected in series. According to the working principle of DAS, the light signals from a laser source (around 1500 nm wavelength) are modulated with rectangular pulses, and these pulses are sent to the fiber under test (FUT). Each segment of the FUT reflects the Rayleigh backscattered signals, and a circulator is used to divert these back-reflected signals to the direction other than the one from which these pulses originate. At the intended port of the circulator, the photodetector is installed, which converts the modulated pulses into electrical signals. A positive aspect of the DAS system is that the simultaneous perturbed signals can be retrieved without any specialized signal processing system. The two most commonly used configurations of a DAS system are directed detected and coherent detected systems, as shown in **Figures 5** and **6** respectively. The configuration in **Figure 5** is the direct detected system that implies there is no reference light source at the photodetection stage. A coherent detected ϕ-OTDR system is one in which the laser source acts as the reference signal to provide additional phase information at the photodetection stage. Converting the direct detected system to a coherent detected system requires a modification such that the laser source is divided into two parts with the help of a coupler. One of the branches of this signal is injected into the optical modulator, whereas another part of the same signal leads to the balanced photodetector, as depicted in **Figure 6**.

In both the direct and coherent detected systems, the optical modulator is considered to be any type of acousto-optic modulator (AOM) or electro-optic modulator (EOM) with a high extinction ratio (a ratio between transmitted one and transmitted zero). The modulation drops the signal strength many folds, and therefore an Erbium-

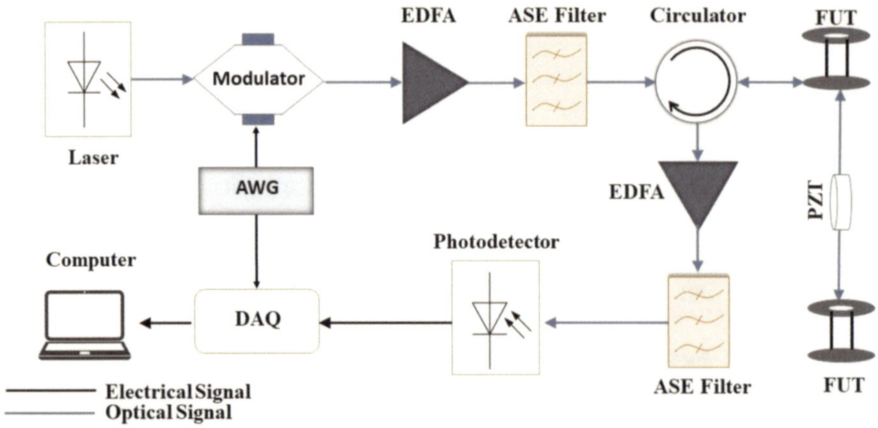

Figure 5.
A direct detected ϕ-OTDR system.

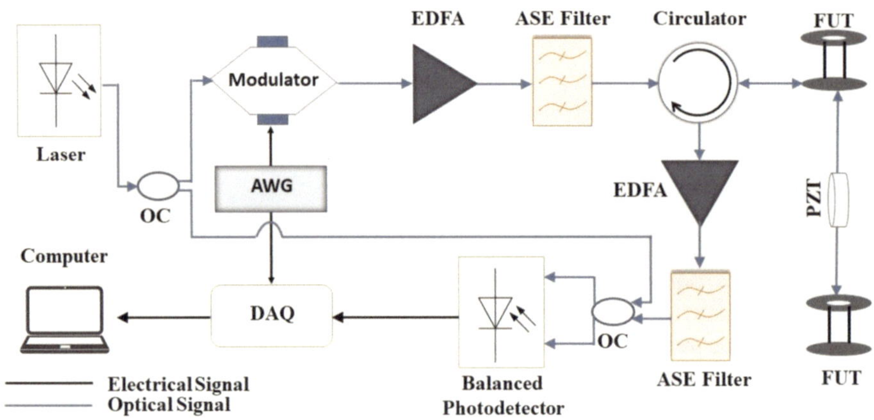

Figure 6.
A coherent detected ϕ-OTDR system.

doped fiber amplifier (EDFA) is used to amplify these signals. The EDFA and filters are optional and these devices are installed if needed in the very large L_{FUT} (length of FUT) applications. Each injected pulse with a predefined pulse width (PW) within the fiber defines the spatial resolution (SR) of the measured distance within which the perturbation is felt. These pulses are injected within the fiber with a specific pulse repetition rate (f_{PRR}), and the backscattered signals are acquired with the sampling frequency of f_{DAQ}. The maximum frequency (f_{PER}) from the frequency components generated due to applied perturbation is another factor, that is. restricted due to the associated parameters. Here is a list of compromised parameters that depend on another parameter and are mentioned below with examples (**Table 2**).

Two types of sampling rates are defined in the abovementioned discussion. One of these types relates to the spatial sampling rate, and its speed depends on the sampling frequency of the data acquisition (DAQ) card, termed f_{DAQ}. The second type relates to the temporal sampling rate, and its speed depends on PRR. From the length of a fiber, one can decide the maximum allowable frequency of the perturbation. The smaller the length of the fiber, the higher will be the frequency at which the perturbation can be

Compromise between L_{FUT} and f_{PRR} A typical length of the FUT restricts the maximum f_{PRR} at which the pulses are injected. Formula: $f_{PRR} = c/(2 \times n_{eff} \times L_{FUT})$, where c = velocity of light.	
Example-1	L_{FUT} of 1 km fiber can allow a maximum of $f_{PRR} = 100$ kHz against each sensing points.
Example-2	L_{FUT} of 500 m fiber can allow a maximum of $f_{PRR} = 200$ kHz against each sensing point.
Compromise between f_{PRR} and f_{PER} PRR restricts the maximum frequency of the applied perturbation. $f_{PER} < f_{PRR}/2$	
Example-1	f_{PRR} of 10 kHz can permit all frequencies under $f_{PRR} =$ 5 kHz of the applied perturbation
Example-2	f_{PRR} of 5 kHz can permit all frequencies under $f_{PER} =$ 2.5 kHz of the applied perturbation
Compromise between SR and f_{DAQ} **Spatial resolution (SR) restricts the sampling rate of a DAQ card (f_{DAQ})** **$SR = c/(3 \times f_{DAQ})$, where c = velocity of light and the unit for SR is in meters.**	
Example-1	SR of 1 m (10 ns pulse width of pulse) requires the DAQ card with a minimum sampling rate of $f_{DAQ} = 100$ MHz.
Example-2	SR of 0.1 m (1 ns pulse width of pulse) requires the DAQ card with a minimum sampling rate of $f_{DAQ} = 1$ GHz.
Compromise between PW and SR PW controls the SR. $SR = c \times PW/(2 \times n_{eff})$, where c = velocity of light (m/sec), n_{eff} is the effective refractive index, and PW is the pulse width (in seconds).	
Example-1	A PW of 2 ns (Function generator frequency of 500 MHz) can provide a minimum of SR = 20 cm spacing between two sensing points.
Example-2	A PW of 1 ns (Function generator frequency of 1 GHz) can provide a minimum of SR = 10 cm spacing between two sensing points.

Table 2.
Threshold of the limiting parameters.

detected. Both the SR and the DAQ sampling rates are required to define the minimum spacing between two sensing points. The lower the SR and DAQ sampling rate, the smaller the spacing between two sensing points. Moreover, the lower the SR and DAQ speed, the smaller the spacing between two sensing points. After sending a single pulse, the fiber provides a response signal and each of these samples correspond to a distance 1/PRR in the time domain, where the mentioned response signal is the result of Rayleigh backscattered signals along the whole FUT. The correspondence distance of this response in meters can be obtained using the formula: $L_{FUT} = c\ 2\ n_{eff}\ f_{PRR}$. The response against the whole FUT is called a single trace, which is a stationary random process, and this fact can be observed with the help of observing each received trace provided that the fiber is not disturbed with the assumption of the very long linewidth of the laser source. Differentiating one received from its previous counterpart is termed the differential data-trace. One can easily observe the phase change due to the applied perturbation with the help of a differential data-trace. In case of any perturbation, multiple traces are normally acquired, and the difference of each subsequent trace is taken in order to determine the point of perturbation.

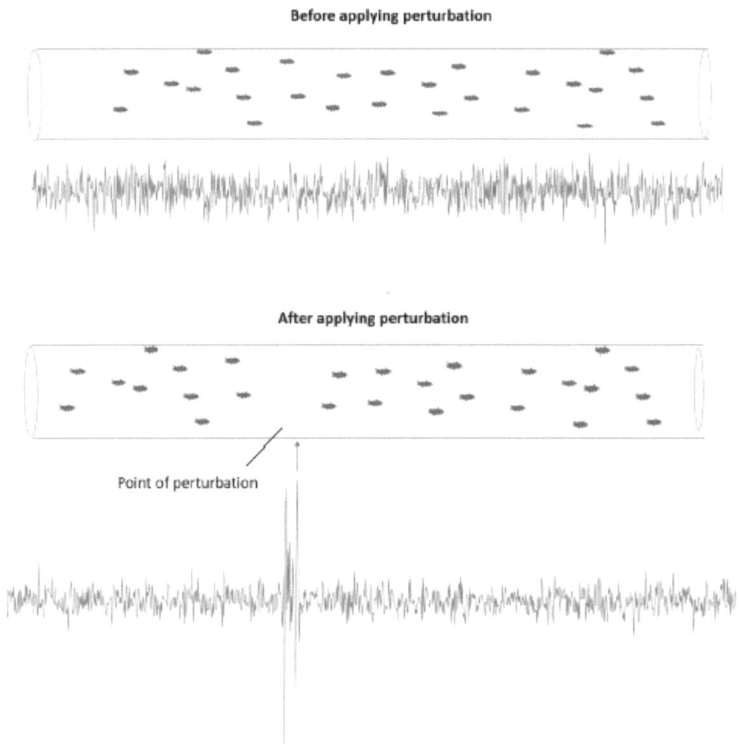

Figure 7.
Perturbation demonstration using differential signals of ϕ-OTDR system.

A schematic diagram showing the behavior of differential signals due to a stretch of the fiber by an applied perturbation is shown in **Figure 7**.

DAS is a revolutionary photonic sensing technology that exploits the use of a standard communications fiber into a linear array of discrete vibration sensors. Activities such as people walking/running, hot-tapping pipelines, pipeline leakage detection, perimeter intrusions, moving vehicles, industrial operations, failing mechanical components, firing direction detection, and many applications are responsible for generating vibrations with distinct acoustic characteristics. DAS technology monitors these vibrations and accurately detects, classifies, and reports on the vibration events. With DAS, there is no need to install the conventional sensors, and a simple G.652 type single-core fiber optic cable is enough to sense the whole distance in a distributed fashion, and hence it saves a huge amount of cost among hundreds of kilometer distance spans. The efficient algorithms requiring very few data traces [31–33] can help to calculate the instantaneous speed of the trains. As we know, terrorist acts along railway tracks, derailments, and train collision accidents were quite common in the history of railway-based accidents; therefore, DAS can play a very important role to avoid these accidents in the future. The use of DAS technology in the railway sector is not only a cost-effective solution but due to the provision of enormous data at each spatial location also there is a possibility of involving artificial intelligence to automate both the security and train tracing along the whole railway track, which is normally in hundreds of kilometers range. Fiber response or traces received against a few injected pulses are mostly insufficient to detect a certain perturbation as most differential data traces do not describe the effect of external perturbations. Most often, the machine learning algorithms applied to these differential data traces do not provide essential information for classifying different types of intrusions unless a myriad number of data traces are acquired. Normally, a single data trace is received with a certain delay after the injection of a single pulse within an optical fiber. A large percentage of the received data traces are irrelevant, and these irrelevant data traces are also received with the same delay. Hence, DAS can be successful in many applications for which the delay is not important. However, in the case of high-speed moving trains, where instantaneous intrusion detection, instantaneous speed determination of the trains, and the instantaneous location of each wheel of the respective bogies of the whole train are solicited, the lag in the DAS system is unacceptable. There are many applications of the DAS system, including fence, border, and pipeline security systems. Several signal processing methods applied directly on differential data traces include time-series-based algorithms [34–38], and frequency-based algorithms [39–41] approaches were suggested to provide a better probability of detection and classification accuracy in perturbation detection and event recognition applications respectively without imparting emphasize on utilizing a smaller number of data-traces efficiently. The time-frequency-based approaches such as discrete wavelet transform [42–44], Hilbert-Huan transforms [45], or similar algorithms [46–48] are best suited for trace-to-trace fluctuation-based noise alleviation [49, 50] than the frequency-based or time-series-based approaches. A drawback of the time-frequency-based approach is that these techniques are intensively parameter-dependent. For example, choosing a very selective mother wavelet and a vanishing moment in DWT-based algorithms for each specific event in a perturbation recognition application may not be possible, although, these algorithms can be suitable for the applications such as perturbation detection. The issues relevant to sampling data relevant to high-speed vehicles were suggested for the first time in [31–33], and hence now it is possible to consider DAS for very high-speed vehicles.

A detailed insight into DAS technology with respect to the railway infrastructure and the associated railway-based vehicles, whether trains or trams, is discussed here. DAS was first introduced by Juarez in the year 2005 while demonstrating the concept of intrusion detection, and this class of distributed sensing became popular afterward. Initial work was carried out on the ballastless track structure in the year 2013 [51] followed by speed and position detection in the year 2014 [36]. As mentioned before, the two main applications of any railway system are train traffic management and operational monitoring. The research work in the case of train traffic management includes speed and position precision improvement along with external applications such as security and other activity monitoring alongside railway tracks, whereas the operational monitoring applications involve continuous monitoring of both the train and tracks for their faults. **Table 3** includes the research work by exploiting DAS and ϕ-OTDR in traffic management of the railway system. Another important set of railway applications involves the operational monitoring of both the trains and tracks. **Table 4** shows the literature work regarding the use of the DAS system for operation monitoring of the railway system.

2.3 Other OFS technologies

The probable future technologies in the railway sector related to OFS are FBGs and DAS, as discussed. Other technologies linked with Brillouin scattering and Raman scattering have been investigated quite often in the railway sector. The reason is that

S. no	References	Year	Work purpose
1	[36]	2014	Traffic management along with the safety monitoring of the trains in the case of two parallel railway tracks.
2	[52]	2014	Traffic management along with the safety monitoring of the trains in the case of two parallel railway tracks.
3	[53]	2020	Speed precision improvement for a speed of up to 160 km/hr. train.
4	[54]	2015	Location, mass, and speed precision improvement
5	[55]	2015	Location precision monitoring
6	[56]	2016	Machine learning methods for alleviating false Alarms in train detection system using long-time recording.
7	[57]	2017	Intrusion detection system implementation for train and track security
8	[58]	2017	Construction monitoring alongside the railway infrastructure
9	[59]	2018	Segregating trains with the help of signatures
10	[60]	2019	Noise removal for increasing processing time for real-time train traffic management applications

Table 3.
Train traffic management applications of DAS in refs to literature.

S. no	References	Year	Work purpose
1	[51]	2013	Study of ballastless track structure monitoring by distributed optical fiber sensors on a real-scale mockup in a laboratory.
2	[61]	2016	Distributed acoustic monitoring to secure transport infrastructure against natural hazards—Requirements and new developments.
3	[62]	2018	Measurement of distributed dynamic rail strains using a Rayleigh backscattered based fiber optic sensor: Lab and field evaluation.
4	[63]	2020	Operational Monitoring of Railway Infrastructure

Table 4.
Train traffic management applications of DAS in references to previous work.

the technologies related to Brillouin-based scattering cannot differentiate between static sensings such as stress and temperature. An abrupt change in the temperature occurs after the train traverses the railway track, which provides confusion between the real measured variable (stress) and temperature. Though some efforts are made in this regard to adopt the Brillouin-based sensing technologies in railways such as traffic monitoring [64], railway infrastructure [65], track deformations monitoring [66], and operation monitoring of the railway infrastructure [67]. However, the mentioned research work is insufficient for these technologies to be adopted in the railway sector anytime soon. The reason the OFS technologies related to Raman scattering are not used in the railway sector is because of their use in temperature sensing only. A long-distance infrastructure monitoring with a temperature sensing-only application is not the cost-effective solution, and hence this sensing technology has not been investigated so far in this sector. Beside, the interferometric-based OFS cannot be used in the railway sector due to the nonlinear nature of this sensing, as well as the non-distributed nature. Though, [68, 69] have presented their work in interferometric-based sensing, but these techniques were not verified with a distributed or quasi-distributed sensing techniques over a very long range of railway track or vehicle. OFDR is another type of interferometric-based sensing technology with distributed nature of its sensing. However, due to its short-range applications, this type of interferometric-based sensing cannot be used in railway applications.

3. Summary

This chapter has presented a detailed description of the use of the OFS systems and their advantages as compared to the conventional sensing system for use in the railway sector. In a broad sense, the two categories of applications in any railway sensing system are comprised of operational monitoring and traffic management. Normally, dedicated sensing systems are utilized to implement these two types of applications using conventional methods. Postaccidental investigation can be termed as a third category of applications that is related to acquiring data from spatial locations with a minute gap among these locations, which was otherwise impossible in the case of a conventional sensing system. With OFS all three categories of applications are possible with a single sensing system due to its features such as the best reliability and cost-effectiveness to employ all these solutions in a long-range railway system. Moreover, there are additional benefits of OFS in the railway sector, that outweigh this sensing system as compared to conventional sensing systems. There are many sensing systems in OFS; however, in the railway industry, the popular quasi-distributed sensing system is FBG and one of the popular distributed sensing systems is DAS. The two sensing systems have been explained.

Author details

Muhammad Adeel[1*], Aadil Raza[2] and Muhammad Muaz[3]

1 The Hong Kong Polytechnic University, Hong Kong

2 Department of Physics, COMSATS University Islamabad (CUI), Islamabad, Pakistan

3 Hong Kong Science Park, Hong Kong Industrial Artificial Intelligence and Robotics Centre Limited, Hong Kong

*Address all correspondence to: m.adeel@connect.polyu.hk

References

[1] Rogers AJ. Distributed Optical-Fibre Sensors. Dordrecht: Springer Netherlands; 1987. pp. 143-163

[2] Lee K, Lee K, Ho SL. Exploration of using FBG sensor for derailment detector, 3. 2004

[3] Lee K, Lee K, Ho SL.. Exploration of using FBG sensor for axle counter in railway engineering, 3. 2004

[4] Filograno ML, Corredera P, Rodrıguez-Plaza M, et al. Wheel flat detection in high-speed railway systems using fiber Bragg gratings. IEEE Sensors Journal. 2013;**13**(12):4808-4816

[5] Roveri N, Carcaterra A, Sestieri A. Real-time monitoring of railway infrastructures using fibre bragg grating sensors. Mechanical Systems and Signal Processing. 2015;**60**

[6] Kouroussis G, Kinet D, Moeyaert V, Dupuy J, Caucheteur C. Railway structure monitoring solutions using fibre Bragg grating sensors. Journal of Rail Transportation. 2016;**4**(3):135-150

[7] Mennella F, Laudati A, Esposito M, Cusano A, Cutolo A, Giordano M, et al. Railway monitoring and train tracking by fiber Bragg grating sensors. In: Cutolo A, Culshaw B, Higuera ML, editors. Third European Workshop on Optical Fibre Sensors. Vol. 6619. International Society for Optics and Photonics, SPIE; 2007. p. 66193H

[8] Kinet D, Kouroussis G, Dupuy J, Moeyaert V, Verlinden O, Caucheteur C. Cost-effective fbg interrogation combined with cepstral-based signal processing for railway traffic monitoring. 2016; 989915

[9] Qiushi M, Gao X, Zhu H, Zeyong W, Quanke Z. Composite railway health monitoring system based on fiber optic Bragg grating sensing array. 2014; 259–264

[10] Wang P, Xie K, Shao L-Y, Yan L, Xu J, Chen R. Longitudinal force measurement in continuous welded rail with bi-directional fbg strain sensors. Smart Materials and Structures. 2016;**25**: 015019

[11] Kang D, Kim D-H, Jang S. Design and development of structural health monitoring system for smart railroad-gauge-facility using fbg sensors. Experimental Techniques. 2012;**38**:6

[12] Zhang Y, Liu F, Jing Y, Li W. Application of FBG sensing technique for monitoring and early warning system of high-speed railway track conditions. In: 25th International Conference on Optical Fiber Sensors. 2017. p. 10323

[13] Hussaini S, Indraratna B, Vinod JS. Application of optical-fiber bragg grating sensors in monitoring the rail track deformations. Geotechnical Testing Journal. 2015;**38**:20140123

[14] Lai C, David A, Liu S-Y, Ho SL, Tam H. Development of level sensors based on fiber bragg grating for railway track differential settlement measurement. IEEE Sensors Journal. 2016;**16**:1

[15] Buggy S, James S, Staines S, Carroll R, Kitson P, Farrington D, Drewett L, Jaiswal J, Tatam R. Railway track component condition monitoring using optical fibre bragg grating sensors, 27. 2016

[16] Wei YJ, Zhang JT, Zhang YL, Xi XC, Li K, Liu SC. Research on evaluation method of the bridge strengthening

effect based on fiber optic sensor. Advanced Materials Research. 2013;**79**: 1901-1904

[17] Bocciolone M, Bucca G, Collina A, Comolli L. Pantograph–catenary monitoring by means of fibre bragg grating sensors: Results from tests in an underground line. Mechanical Systems and Signal Processing. 2013;**41**:226-238

[18] Martinek R, Nedoma J, Fajkus M, Kahankova R. Fiber-optic bragg sensors for the rail applications. International Journal of Mechanical Engineering and Robotics Research. 2018;7:292-295

[19] Van Esbeen B, Finet C, Vandebrouck R, Kinet D, Boelen K, Guyot C, et al. Smart railway traffic monitoring using fiber bragg grating strain gauges. Sensors. 2022;**22**(9):3429

[20] Shao L-Y, Zhang M, Xie K, Zhang X-P, Wang P, Yan L. The longitudinal force measurement of cwr tracks with hetero-cladding fbg sensors: A proof of concept. Sensors. 2016;**16**:2184

[21] Yan L, Zhang Z, Wang P, Pan W, Guo L, Luo B, et al. Fiber Sensors for Strain Measurements and Axle Counting in High-Speed Railway Applications. IEEE Sensors Journal. 2010;**11**(7): 1587-1594

[22] Zhou Y, Yan L, Wang P, Chen R, Li Z, Ye J, et al. Discriminative monitoring of seamless rail force by a high-birefringence effect-based fiber optic sensing method. Frontiers in Physics. 2022;**10**(May):1-7

[23] Kouroussis G, Kinet D, Mendoza E, Dupuy J, Moeyaert V, Caucheteur C. Edge-filter technique and dominant frequency analysis for high-speed railway monitoring with fiber bragg gratings. Smart Materials and Structures. 2016;**25**:075029

[24] Filograno M, Corredera P, Rodriguez-Barrios A, Martin-Lopez S, Rodriguez-Plaza M, Andres-Alguacil A, et al. Real-time monitoring of railway traffic using fiber bragg grating sensors. Sensors Journal, IEEE. 2012;**12**:85-92

[25] Yuksel K, Kinet D, Moeyaert V, Kouroussis G, Caucheteur C. Railway monitoring system using optical fiber grating accelerometers. Smart Materials and Structures. 2018;**27**(10):105033

[26] Tam H, Liu S-Y, Guan B-O, Chung W-H, Chan T, Cheng L. Fiber bragg grating sensors for structural and railway applications. In: Proceedings of SPIE - The International Society for Optical Engineering. 2005

[27] Hwa T, Lee T, Ho SL, Haber T, Graver T, Mendez A. Utilization of fiber optic bragg grating sensing systems for health monitoring in railway applications, 2. 2007

[28] Tosi D, Olivero M, Perrone G. Performance analysis of peak tracking techniques for fiber bragg grating interrogation systems. Journal of Microwaves, Optoelectronics and Electromagnetic Applications. 2012;**11**: 252-262

[29] Lai C, Kam J, Leung D, Lee T, Aiken T, Ho SL, et al. Development of a fiber-optic sensing system for train vibration and train weight measurements in Hong Kong. Journal of Sensors. 2012;**2012**

[30] Wagner R, Maicz D, Viel W, Saliger F, Saliger C, Horak R, Noack T. A fibre optic sensor instrumented pantograph as part of a continuous structural health monitoring system for railway overhead lines. 2014

[31] Adeel M, Shang C, Hu D, Wu H, Zhu K, Raza A, et al. Impact-based

feature extraction utilizing differential signals of phase-sensitive OTDR. Journal of Lightwave Technology. 2020;**38**: 2539-2546

[32] Adeel M, Shang C, Zhu K, Lu C. Nuisance alarm reduction: Using a correlation based algorithm above differential signals in direct detected phase-OTDR systems. Optics Express. 2019;**27**:7685

[33] Adeel M, Tejedor J, Macias-Guarasa J, Lu C. Improved perturbation detection in direct detected -OTDR systems using matched filtering. IEEE Photonics Technology Letters. 2019;**31**:1689-1692

[34] Mahmoud SS, Visagathilagar Y, Katsifolis J. Real-time distributed fiber optic sensor for security systems: Performance, event classification and nuisance mitigation. Photonic Sensors. 2012;**2**(3):225-236

[35] Peng F, Wu H, Jia X-H, Rao Y-J, Wang Z-N, Peng Z-P. Ultra-long high-sensitivity ϕ-OTDR for high spatial resolution intrusion detection of pipelines. Optical Express. 2014;**22**(11): 13804-13810

[36] Peng F, Duan N, Rao Y, Li J. Real-time position and speed monitoring of trains using phase-sensitive OTDR. IEEE Photonics Technology Letters. 2014; **26**(20):2055-2057

[37] Adeel M, Tejedor J, Macias-Guarasa J, Shang C, Chao L. Segregating the true perturbation position from ghost energy points region in OTDR systems. Optics Express. 2020;**28**(3):2699

[38] Adeel M, Tejedor J, Macias-guarasa J, Zhu K. Undiscovered issues and solutions for direct detected ϕ -OTDR systems. Optical Fiber Technology. 2020;**60**

[39] Qin Z, Liang C, Bao X. Continuous wavelet transform for non-stationary vibration detection with phase-OTDR. Optics Express. 2012;**20**(18): 20459-20465

[40] Martins HF, Martin-Lopez S, Corredera P, Salgado P, Frazao O, Miguel GH. Modulation instability-induced fading in phase-sensitive optical time-domain reflectometry. Optical Letters. 2013;**38**(6):872-874

[41] Tejedor J, Macias-Guarasa J, Martins HF, Pastor-Graells J, Corredera P, Martin-Lopez S. Machine learning methods for pipeline surveillance systems based on distributed acoustic sensing: A review. Applied Sciences. 2017;**7**(8):841

[42] Dong Y, Chen X, Liu E, Cheng F, Zhang H, Zhiwei L. Quantitative measurement of dynamic nanostrain based on a phase-sensitive optical time domain reflectometer. Applied Optics. 2016;**55**(28):7810-7815

[43] Zhou Y, Wang Z, Zhang L, Li J, Wu H, Li Y, et al. Phase-sensitive optical time-domain reflectometry assisted by gated Raman amplification. In: Asia Communications and Photonics Conference 2014. Optical Society of America; 2014. p. 192

[44] Zhong X, Zhang C, Li L, Liang S, Li Q, Lu Q, et al. Influences of laser source on phase-sensitivity optical time-domain reflectometer-based distributed intrusion sensor. Applied Optics. 2014; **53**(21):4645-4650

[45] Izumita H, Koyamada Y, Furukawa S, Sankawa I. The performance limit of coherent OTDR enhanced with optical fiber amplifiers due to optical nonlinear phenomena. Journal of Lightwave Technology. 1994; **12**(7):1230-1238

[46] Qin Z, Zhu T, Chen L, Bao X. High sensitivity distributed vibration sensor based on polarization-maintaining configurations of phase-OTDR. IEEE Photonics Technology Letters. 2011; **23**(15):1091-1093

[47] Ren M, Zhou D-P, Liang C, Bao X. Influence of finite extinction ratio on performance of phase-sensitive optical time-domain reflectometry. Optical Express. 2016;**24**(12):13325-13333

[48] Ran ZL, Yue JF, Luo XD, Zhou Z, Rao YJ, Luo J. Long-distance fiber-optic phase-OTDR intrusion sensing system. 2009

[49] Wang ZN, Zeng JJ, Li J, Fan MQ, Wu H, Peng F, et al. Ultra-long phase-sensitive OTDR with hybrid distributed amplification. Optical Letters. 2014; **39**(20):5866-5869

[50] Adeel M, Tejedor J, Iqbal S, Muaz M, Raza A, Macias-Guarasa J. Differentiating trace-to-trace noise effects using novel signal characteristics in phase-sensitive OTDR systems. Optical and Quantum Electronics. 2023; **55**(1):49

[51] Chapeleau X, Sedran T, Cottineau L-M, Cailliau J, Taillade F, Gueguen I, et al. Study of ballastless track structure monitoring by distributed optical fiber sensors on a real-scale mockup in laboratory. Engineering Structures. 2013; **56**:1751-1757

[52] Duan N, Peng F, Rao Y-J, Jiang D, Lin Y. Field test for real-time position and speed monitoring of trains using phase-sensitive optical time domain reflectometry (-OTDR). In: Lopez JM, Jones JDC, Loopez-Amo M, Santos JL, editors. 23rd International Conference on Optical Fibre Sensors. Vol. 9157. International Society for Optics and Photonics, SPIE; 2014. p. 91577A

[53] Kowarik S, Hussels M-T, Chruscicki S, Muunzenberger S, Laammerhirt A, Pohl P, et al. Fiber optic train monitoring with distributed acoustic sensing: Conventional and neural network data analysis. Sensors. 2020;**20**(2)

[54] Timofeev A. The rail traffic management with usage of C-OTDR monitoring systems. 2015

[55] Timofeev A. Monitoring the railways by means of C-OTDR technology. 2015

[56] Papp A, Wiesmeyr C, Litzenberger M, Garn H, Kropatsch W. Train detection and tracking in optical time domain reflectometry (OTDR) signals. 2016;**9796**:320-331

[57] Catalano A, Bruno F, Galliano C, Pisco M, Persiano GV, Cutolo A, et al. An optical fiber intrusion detection system for railway security. Sensors and Actuators A: Physical. 2016;**253**:14

[58] Wang Z, Lu B, Zheng H, Ye Q, Pan Z, Cai H, et al. Novel railway-subgrade vibration monitoring technology using phase-sensitive OTDR. 25th International Conference on Optical Fiber Sensors. 2017;**10**:103237G

[59] Cedilnik G, Hunt R, Lees G. Advances in train and rail monitoring with das. 2018; ThE35

[60] He M, Fan J. A method for real-time monitoring of running trains using -OTDR and the improved canny. Optik. 2019;**2019**:184

[61] Kogelnig A, Koenig U, Neunteufel G, Schilcher H. Distributed acoustic monitoring to secure transport infrastructure against natural hazards-requirements and new developments. 2016

[62] Lisa NW, Pannese E, Hoult NA, Take WA, Le H. Measurement of distributed dynamic rail strains using a Rayleigh backscatter based fiber optic sensor: Lab and field evaluation. Transportation Geotechnics. 2018;**14**: 70-80

[63] Vidovic I, Marschnig S. Optical fibers for condition monitoring of railway infrastructure—encouraging data source or errant effort? Applied Sciences. 2020;**10**(17)

[64] Minardo A, Porcaro G, Giannetta D, Bernini R, Zeni L. Real-time monitoring of railway traffic using slope-assisted brillouin distributed sensors. Applied Optics. 2013;**52**:3770-3776

[65] Yoon HJ, Song KY, Choi C, Na HS, Kim JS. Real-time distributed strain monitoring of a railway bridge during train passage by using a distributed optical fiber sensor based on brillouin optical correlation domain analysis. Journal of Sensors. 2016;**2016**

[66] Klug F, Lackner S, Lienhart W. Monitoring of railway deformations using distributed fiber optic sensors. 2016

[67] Bao Y, Chen G, Meng W, Tang F, Chen Y. Kilometer long optical fiber sensor for real-time railroad infrastructure monitoring to ensure safe train operation. In: 2015 Joint Rail Conference, JRC. 2015

[68] Nedoma J, Stolarik M, Fajkus M, Pinka M, Hejduk S. Use of fiber-optic sensors for the detection of the rail vehicles and monitoring of the rock mass dynamic response due to railway rolling stock for the civil engineering needs. Applied Sciences. 2019;**9**:134

[69] Kepak S, Cubik J, Zavodny P, Siska P, Davidson A, Glesk I, et al. Fiber optic track vibration monitoring system. Optical and Quantum Electronics. 2016; **48**(7):1-8